Concise Color
Encyclopedia of
NATURE

Left: The giraffe's long
neck enables it to browse
on the juicy upper leaves
of trees.

Right: Pangolins are able
to roll themselves up into a
ball when danger threatens.
The horny plates then
provide them with an
effective suit of armor.

Michael W. Dempsey B.A. **Editor-in-Chief**
Angela Sheehan B.A. **Executive Editor**
Peter Sackett **Production Manager**
Marion Gain **Picture Researcher**
Sarah Tyzack **Picture Editor**

Concise Color
Encyclopedia of
NATURE

Michael Chinery B.A.

Thomas Y. Crowell Company

New York • Established 1834

Contents

PART 1

THE LIVING ORGANISM
Living Things   9
The Living Cell   10
The Origins of Life   11
Plants and Animals   12
Plant and Animal Species   13

PART 2

PLANT PHYSIOLOGY
The Soil   14
Roots and Stems   15
How Plants Make Food   17
Food Storage in Plants   20
Plant Growth and Movement   22
Vegetative Reproduction   24

THE PLANT KINGDOM
Plant Classification   25
The Algae   28
Fungi   30
Lichens   32
Mosses and Liverworts   33
Ferns and Horsetails   34
Conifers   36
Flowering Plants   38
Flowering Trees   42
Grasses   44

PLANT ADAPTATIONS
Insect-Eating Plants   46
Parasitic Plants   47
Climbing Plants   49
Protection in Plants   50

DICTIONARY OF FLOWER FAMILIES   52

PART 3

ANIMAL PHYSIOLOGY
Skeletons and Muscles   72
The Senses of Animals   74
Feeding and Digestion   76
How Animals Breathe   78
The Blood System   80
Reproduction   82

THE ANIMAL KINGDOM
Animal Classification   84
The Microscopic World   86
The Reef Builders   88
Worms and Leeches   90
Crabs and Their Relatives   92
The Insect World   94
Butterflies and Moths   98
Insect Pests   100
The Honey Bee   102

Ants and Termites 104
Spiders and Scorpions 106
Slugs, Snails and Bivalves 108
Squids and Octopuses 110
Spiny-Skinned Animals 112
Fishes 114
Amphibians 117
Reptiles 119
Birds 122
Mammals 125

ANIMAL BEHAVIOR

Instinct and Learning 129
Animal Parasites 131
Animal Partnerships 133
Survival 136
Migration 139
Hibernation 141
Nocturnal Life 142
Animal Electricity 144

DICTIONARY OF ANIMALS 145

PART 4

ECOLOGY

Homes and Habitats 193
Life on the Grasslands 194
Life in the Rain Forests 196
Desert Life 198

Life in Northern Forests 200
Mountain Life 202
Polar Life 204
Life in the Sea 206
Life on the Seashore 208
Life in Fresh Water 211

PREHISTORIC LIFE

Fossils 214
The First Animals 218
The Age of Fishes 220
Live Moves on Land 222
The Age of Reptiles 224
The Age of Mammals 226
The Origins of Man 228
Prehistoric Plants 230

EVOLUTION

The Theory of Evolution 232
Living Fossils 236

MAN AND NATURE

The Balance of Nature 237
Conservation 239
Domestic Animals 243

INDEX 244

PART 1

Living Things

Nature or natural history is sometimes taken to mean the study of all natural things—things that have not been made by man. More commonly, however, it means the study of living things. Scientists often use the word *organism* for anything that is alive or that has been alive at some time. Trees and flowers, butterflies and tortoises, fishes and elephants are all living things or organisms. Rocks, metals, and water are not organisms.

The first thing we must do if we are going to study living things is to find out what it is that makes living things different from non-living objects. We do not know the vital ingredient necessary to make something alive, although some scientists think they are near to discovering it, but we can list seven features possessed by all living things. These features are not shown by any non-living object, and we can call them the *characteristics of living things*.

Characteristics of Life

One of the most obvious features of living things is that they can *move* by themselves, in search of food perhaps or to escape from their enemies. *Feeding* (including drinking) is another feature of all living things. They have to take in food to provide themselves with energy and with body-building

materials. Energy is released from food through the continual chemical process of *metabolism*. Food is burned up by being combined with oxygen through breathing or *respiration*. Body-building materials in the food are taken to all parts of the organism's body and added to it, so that it *grows*. Stalactites and stalagmites in a cave get longer and thicker with time, but only because extra material is added to the outside. There is no production of new material inside the stalactites, and their growth is not like that of a living organism. Breathing and the other processes that go on inside the living organism all result in the production of waste materials. These must be removed, and so all organisms have some way of *excreting,* or getting rid of these waste materials. Living things are also *sensitive* to changes in their surroundings. A fly will soon be off if it senses the approaching fly swatter, and a potted plant will soon turn upwards again if you lay the pot on its side. Finally, living things can all *reproduce* themselves. Organisms cannot live for ever because their bodies gradually wear out, but they ensure that more of the same kind will take their place by producing seeds or eggs or babies.

The seven characteristics of living things are therefore: movement, feeding, breathing, growing, excretion, sensitivity, and reproduction. These seven features are exhibited by all living things, but no non-living object shows any of them.

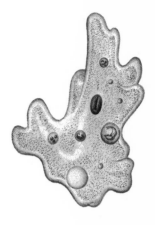

Amoeba, a tiny speck of jelly, is an animal with all the characteristics of living things.

This serval cat is showing his anger. He is obviously alert to what is going on, an important feature of living things.

The white-handed gibbon (left) spends the major part of its active time foraging for food in the tree tops of its jungle home. Plants are just as much alive, although many plants produce their own food.

The Living Cell

Just as a house is built of bricks or building blocks, so animals and plants are made of little units called cells. Cells are generally extremely small and they cannot usually be seen without a microscope. Some animals and plants—known as the Protozoa or Protista—consist of only one tiny cell, but larger creatures consist of many, many cells. There are many different kinds of cells and each kind has a particular job to do in the body. In our bodies, for example, there are skin cells covering the outside and lining the various internal organs; nerve cells carrying messages from place to place; bone cells forming our skeletons: and muscle cells enabling us to move. Plants have a variety of cells too. There are food-making cells in the leaves, tube-like water-carrying cells in the roots and stems, and food-storing cells in various places.

Plants and Animals

Plant cells differ from animal cells mainly in having a stiff cellulose wall around them. This gives them a fixed, regular shape. Animal cells do not have cell walls and they are less rigid than plant cells. Plant cells also normally contain a large sap-filled space called a vacuole. Animal cells may have small fluid-filled spaces, but they never have large vacuoles. Plant cells often contain small green objects called chloroplasts. These give them their green color and they also help the plants to make food. Animal cells never contain chloroplasts, although some animals may have tiny green plants called algae growing inside them.

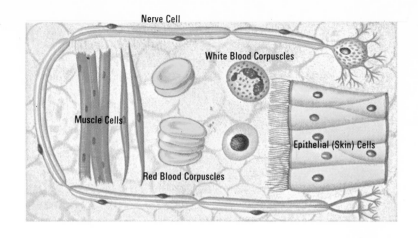

Animals and plants are built of cells just as a house is built of bricks. The cells are not all the same shape. Different shaped cells have different jobs to do. The diagram above shows a selection of human body cells.

Although cells are so tiny, they have complicated structures inside them. Fine channels carry chemical materials from the nucleus to all parts of the cell.

Inside the Cell

All cells are bounded by a very thin cell membrane. In plants it is just inside the cell wall. Inside the membrane is the protoplasm. This is a watery material in which there are very complicated mixtures of proteins and other materials. It is here that all the processes of life take place. Protoplasm is a sort of 'living jelly'.

In the center of most living cells there is a darker, dense region called the nucleus. This is made of a special type of protoplasm and it is the 'brain' of the cell. Chemical materials made in the nucleus spread out into the protoplasm through minute channels and they 'tell' the protoplasm what to do. This ensures that the cells make the right sort of materials.

Inside the nucleus are the chromosomes. These are minute thread-like structures which carry the genes. Genes are very complicated chemical materials and each one carries the 'plans' for a part of the body. The nucleus instructs its cell to make things according to these 'plans', and so the whole body is made up in the correct way. The genes and chromosomes in an apple seed will ensure that it grows into an apple tree and not into an oak tree.

Cell Division

Living things normally start out as single cells called egg cells. The single cell then begins to divide into two. The two cells then divide into four, and so it goes on. Cell division of this sort takes place even in the fully grown plant or animal because new cells have to be produced to replace those that have worn out. The chromosomes play a very important part in cell division because it is essential that the new cells have the same 'instructions' as the old cells. Each chromosome produces an exact copy of itself and its genes and then, when the cell divides into two, one set of chromosomes goes to each new cell.

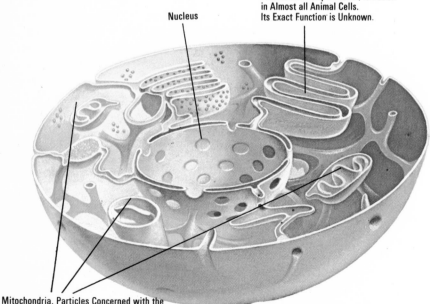

Nucleus

The Golgi Apparatus, a Collection of Protein and Fatty Material, is Found in Almost all Animal Cells. Its Exact Function is Unknown.

Mitochondria, Particles Concerned with the Respiratory Mechanisms of the Cell

The Origins of Life

When the Earth first came into existence some 4,500 million years ago there was no life on the planet. Starting probably as a dense ball of dust, the Earth passed through a very hot and partly liquid stage. Volcanoes belched out carbon dioxide and the atmosphere also contained large amounts of methane (marsh gas) and ammonia. The whole of the Earth was surrounded by a thick blanket of steam. Things gradually cooled down after this and much of the steam blanket turned into water and formed the ocean. The Earth was ready to support its first life.

Virus particles come in many different shapes when they are crystallised, as shown by the two types of 'flu' virus (top) and the smallpox virus.

Microscopic algae may have been the earliest life forms.

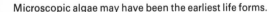

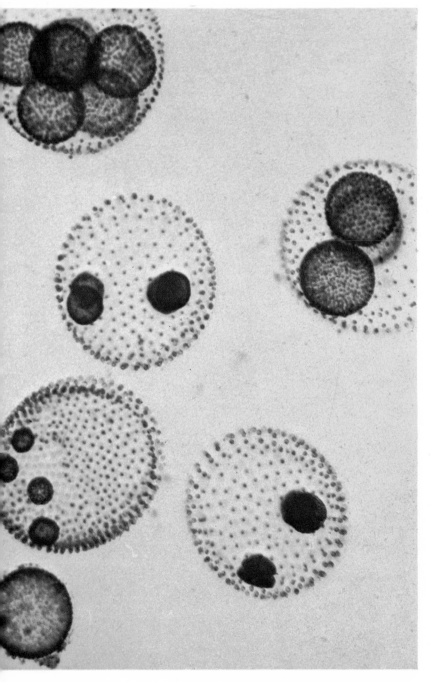

Viruses

Viruses are minute germs that can be seen only with the powerful electron microscope. Some of them are not much more than one millionth of a millimeter across. They appear to be on the borderline between living and non-living matter. They can be crystallised, dried, and then dissolved again like ordinary chemicals, and yet they can 'come to life' and multiply themselves if given the right conditions. Viruses can grow and reproduce only inside living cells. Some use plant cells and some use animal cells. The virus particles contain 'instructions' rather like those of the ordinary cell nucleus, and the virus then 'instructs' the cell to make more virus material instead of normal cell material. This interferes with the normal working of the cell and produces illness or disease in the animal or plant. The damaged cells eventually break down and release the viruses which move on to other cells in the body. Foot and mouth disease, fowl pest, and myxomatosis are among the many animal diseases caused by viruses. Plant diseases caused by viruses include sugar beet yellows, tobacco mosaic, and many others. The viruses are often carried from one host to another by insects.

The Primeval Soup

How did life start on the Earth? We shall never know exactly how it happened, but we can make some intelligent guesses as to how it could have happened. The atmosphere of our planet in its younger days consisted largely of methane, carbon dioxide, ammonia and water. These are simple substances but they contain all the major elements of life—carbon, oxygen, hydrogen and nitrogen. The sun's ultra-violet light was able to get through this atmosphere and, under the influence of this radiation, some of the simple materials probably linked up to form sugars and amino-acids. These substances are the building bricks of living material.

There is some experimental evidence to support this idea, because bombardment of methane/ammonia mixtures with ultra-violet rays and artificial lightning in the laboratory has produced sugar and amino-acid. If these reactions went on for a long time on the young Earth, we can imagine the sugars and amino-acids being concentrated around the sea-shores or in the surface layers of the oceans. They would have formed what is known as the 'primeval soup'—a rich mixture of life's building bricks.

We can only guess at the next stage in the development of life, but presumably some of the sugars and amino-acids linked up and produced proteins and other complicated materials that could actually duplicate themselves. If these materials then formed some sort of membrane around themselves they would have produced the first cell. It would have taken a long time for the right combination of materials to appear, but it could have happened. In fact, recent reports suggest that scientists have succeeded in producing a living cell in the laboratory.

Plants and Animals

We do not know what the earliest living things were like, and we probably never will. They might have been something like today's bacteria, little more than specks of protoplasm obtaining energy by carrying out chemical processes. Whatever they were like, we can be fairly certain that they have given rise to all of today's living things with the possible exception of the viruses.

Apart from the viruses, all of today's living things fall into one of two large groups—plants and animals. Examples of plants include mosses, seaweeds, trees, grasses, and cabbages. Examples of animals include jellyfishes, spiders, snakes, birds, and dogs. Man is also an animal.

It is easy to tell the difference between a cabbage and a dog, but it is not so easy to tell the difference between all plants and animals. You might think that animals move about and plants stay still, but this is not always so. There are many animals, such as the sea anemone, that stay in one place and there are many tiny plants that swim about in the water. Then there is the insect-eating sundew—a plant whose sticky 'fingers' can

It is easy to distinguish a typical plant, such as a buttercup, from a typical animal . . . .

. . . . but less easy to say which of these is a plant and which is an animal. The sea anemone (left) is an animal, and the other organism is a microscopic plant that swims about in the water.

move quickly to trap a small fly. You might say that animals have eyes and ears, while plants do not. Plants certainly do not have such sense organs, but there are also many animals without them, for instance the earthworm and the jellyfish.

It is true that most animals grow to a definite size and shape and then stop growing, whereas plants generally go on growing indefinitely, but the most important distinction between the two groups concerns the way in which they get their food. Plants are nearly all able to make their own food, and those that cannot do this are obviously plants from their other features. Animals cannot make food at all and they always have to feed on plants or upon other animals that have themselves eaten plants.

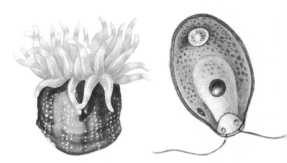

The Grant's gazelle, one of many herbivores, feeds entirely on plants.

Plant and Animal Species

Each of the many different kinds of animals and plants is known as a species. There are probably well over a million different kinds of animal species, and something like half a million plant species.

Each species has its own appearance and, although there might be minor variations, all the individuals in a species look and behave alike. All polar bears, for example, look like polar bears and act like polar bears. A polar bear will never look like a dog.

We know today that the appearance and behavior of an organism depend upon and are controlled by the genes present in its cells. All members of the species have a similar arrangement of genes and chromosomes in their cells. When the animals mate or the plants are fertilized this same pattern is passed on to the next generation, and so they will also conform to the general appearance and behavior of the species. The structure and behavior of different animals normally prevents them from mating other than with members of the same species. Even if animals of two species could mate together, the chances are that the arrangement of the genes in the two animals would be so different that they could not join up to give a new animal. Each species is thus maintained as a separate kind of organism. A species, therefore, is a group of plants or animals which all have the same appearance and behavior and which can breed among themselves.

The Species and Evolution.

Although all the members of a species look basically alike, there are always some minor variations. You have only to look at people to see this variation: they all belong to the same species, but none of them look *exactly* alike. The same thing occurs among other animals and plants. When all the members of the species are free to breed among themselves these small differences are evened out and the species stays more or less the same. But suppose that the members of a species become separated into two regions—by a chain of mountains for example. Very small changes may occur in one population and not in the other. Over a very long time these changes might spread throughout one population and it would be slightly different from the other population. Perhaps the difference might simply be that one population had a slightly different color. At this stage, the members of the two populations would probably be able to

Closely related species have many features in common because they have evolved fairly recently from a common ancestor. The picture shows a great tit (left), a varied tit (top right) and a blue tit (bottom right)

Species and Genera
Linnaeus, the distinguished Swedish naturalist of the seventeenth century, was one of the first people to try to classify things. He recognised that some species were very alike (these were the ones that had most recently separated from each other) and he concluded that they were related to each other. He grouped the similar species into larger units called genera (singular genus). He then gave a scientific name to every plant and animal that he knew. The scientific name was made up of the name of the genus and the name of the individual species. We still use Linnaeus' method today. The scientific name of the large white butterfly is *Pieris brassicae*, while that of the small white butterfly is *Pieris rapae*. Because they both have the same generic name, a biologist in any country knows at once that the two butterflies are related to each other.

Although one species does not normally mate with another, closely related species do sometimes mate and produce offspring. These offspring are called *hybrids*. One of the best known examples is the mule, the offspring of a male donkey and a female horse. The mule is a strong animal, but, like most hybrids, it is sterile and unable to reproduce.

interbreed. They would still belong to the same species, and would probably be known as geographical races or varieties of that species. If the two populations continued to be separated, the differences between them might well increase until there came a time when they were no longer able to interbreed. The two populations would then be separate species. This sort of process has obviously taken place many times during the history of animals and plants and has given rise to all the many species we know today. Species which are very similar to each other, for example the horse and the ass, are those that have most recently been separated from each other.

13

PART 2

The Soil

The soil is a very complex material that is formed mainly from the underlying rocks. You can see the soil-forming processes at work in old quarries and in road or railway cuttings. Very few plants can get a hold on the bare rock, although some lichens and mosses manage it. Wind and rain attack the bare surfaces and small pieces begin to crumble away. If the slope is very steep the particles may roll to the bottom, but otherwise they may stay where they are. They are often trapped by the mosses and lichens. These little rock particles form what is known as the soil skeleton. Small grasses and other plants can take root among them and the roots find their way into tiny cracks in the rock, thus helping to break it up further. When the plants die their remains decay and add organic matter called humus to the soil skeleton. A true soil is now beginning to form and larger plants can take root in it. The underlying rock is broken down further and the soil gets deeper. In time it becomes a mature soil, with plenty of humus. As well as rock particles and humus, a soil contains air spaces, water, and hordes of tiny organisms which help to break down dead plants and animals.

Because the soil is composed mainly of rock particles, different types of rocks produce different kinds of soils. Soft clay rocks provide very tiny mineral particles which hold a lot of water and which stick together very readily. Such particles produce wet sticky soils. Sandy rocks, on the other hand, produce large particles and light, well-drained soils.

These different types of soil support very different types of vegetation. Sticky clays, for example, support oak woods. Sandy soils support pine woods or grassland. Chalk and limestone rocks are composed almost entirely of limestone, and their soils therefore contain a great deal of lime. Only lime-loving plants will be able to grow on such soils. The chalk lands of Europe originally carried forests but, through the activity of man and his grazing animals, the forests have now been converted into grasslands.

Climate and Soil

Although the nature of the underlying rock has a very great effect on the nature of the soil, the climate has the final say as long as the ground remains undisturbed. In the cold temperate regions, the rainfall exceeds the evaporation of water from the soil. Humus and minerals such as iron and calcium are therefore continually washed down from the surface layers. This action, known as leaching, is especially obvious in sandy soils and the surface layers become quite pale. Leached soils are called *podsols.*

In the warm temperate regions the evaporation of water more or less equals the rainfall and so there is not much downward movement of humus and minerals. The surface layers are brown and the soils are called *brown earths. Black earths* are found under the grass of the steppes and prairies. These regions are fairly dry and water is nearly always moving up to the surface. Minerals and humus remain at the surface and the surface layers are black.

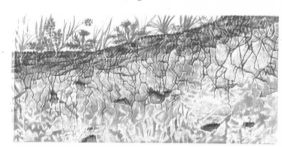

Chalk and limestone rocks do not provide much material for the soil skeleton. The soils are therefore rather thin, but they carry a wealth of attractive lime-loving plants.

The chalk downs of England are famous for their orchids, although there are not nearly so many now as there were 100 years ago. The fragrant and pyramidal orchids, seen here with yellow bird's foot trefoil, are among the commonest species.

Roots and Stems

If you have ever tried to pull up a tree seedling, or even a dandelion plant, you will know that it is hard work. The root holds the plant very firmly in the ground. This is actually one of the main jobs of the root. It anchors the plant so that the wind cannot pull it out of the ground. The other job of the root is to absorb water for the plant.

Dandelions, thistles, and many other plants have *tap roots*. Each plant has just one main root which grows more or less straight down into the ground. There may be a few branches, but they are never as large as the main root. Not all plants have tap roots. If you pull up a grass or a daisy plant you will see that it has lots of roots, all more or less the same size. Roots of this type are called *fibrous roots*.

Whether a root is a fibrous root or a tap root, it anchors the plant by growing down between the soil particles and filling the spaces very tightly. Both types of roots also absorb water from the soil. The finer regions of the roots, a little way behind the root tip, bear thousands of *root hairs*. These are very tiny and they are usually broken off when a plant is uprooted, but you can see them very well on seedlings if you grow them on blotting paper. The root hairs grow between the soil particles and, because they have such thin walls, water can seep into them from the soil. This water, containing dissolved minerals, passes to the center of the root and enters the long conducting tubes. It then begins its journey up to the leaves. The conducting tubes have very tough walls and they give the root its strength. You can see them in the center of the root if you slice through a carrot or a parsnip.

Roots hairs grow out from the outer cells of the root and greatly increase its surface area. This increases the amount of water that the root can absorb.

The conducting tubes in a young stem are collected into little bundles and arranged in a ring around the stem. The outer part of each bundle consists of numerous tough fibers which support the stem.

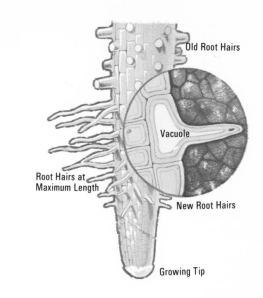

Old Root Hairs

Vacuole

Root Hairs at Maximum Length

New Root Hairs

Growing Tip

Each Region Between Two Vascular Bundles is Called a Ray

Pith

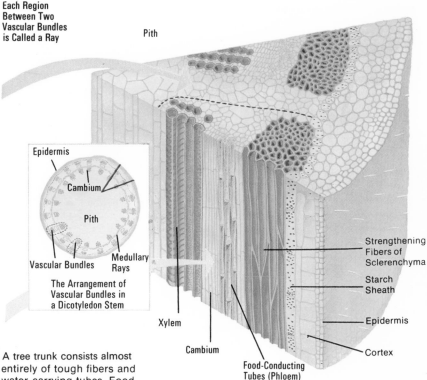

Epidermis

Cambium

Pith

Vascular Bundles

Medullary Rays

The Arrangement of Vascular Bundles in a Dicotyledon Stem

Xylem

Cambium

Strengthening Fibers of Sclerenchyma

Starch Sheath

Epidermis

Cortex

Food-Conducting Tubes (Phloem)

A tree trunk consists almost entirely of tough fibers and water-carrying tubes. Food is carried only by tubes in the inner bark. The dark region in the center of this section is heartwood, which no longer carries water.

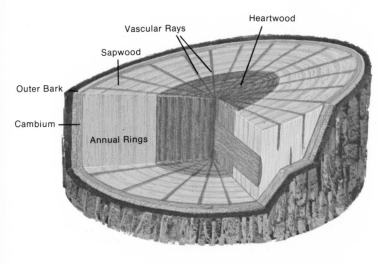

Vascular Rays

Heartwood

Sapwood

Outer Bark

Cambium

Annual Rings

The Stem

The stem of a plant has to carry the water from the roots to the leaves. It also has to hold up the leaves (and the flowers if there are any) so that they get plenty of light. Stems are usually upright and they may or may not have branches. Some stems are quite soft and juicy. They are called *herbaceous* stems (above). Other stems are woody (left).

The water is carried in conducting tubes which are continuous with those in the roots, although the arrangement is different in the stem. The tubes of a young stem are arranged in little clusters called *vascular bundles*. They form a ring in the outer half of the stem. The tough walls of the tubes, to-

15

A tap root system, shown by the thistle, has one main root with several smaller branches. A fibrous root system consists of many similar sized roots, none of which can be called a main root. Grasses all have fibrous roots.

gether with the tough fibers that surround them, give great strength to the stem yet they still allow it to bend in the wind. As well as the water-carrying tubes, the vascular bundles contain food-carrying tubes. These carry food from the leaves to other parts of the plant.

As a stem gets older its vascular bundles join up to form a complete ring of tubes and fibres. Woody plants form more tubes and fibres each year and the trunk gets thicker and thicker. It consists almost entirely of tubes and fibers. The older ones get squashed in the middle of the trunk and they form the *heartwood*. They no longer carry water. The outer regions, which do carry water, form the *sapwood*. Sapwood is much less durable than heartwood and decays much more easily when it is dead.

Annual Rings

In those parts of the world where there are definite seasons, the trees slacken their growth in the autumn. Only narrow conduct-

ing tubes are formed at this time. No new tubes are formed in the winter, but large ones are formed when the trees 'wake up' in the spring. There is a distinct boundary between the autumn wood and the spring wood and so, when a tree is cut down, we can see the *annual rings*. It is sometimes possible to count the rings and find out exactly how old the tree is.

Why a Plant Needs Water

All living things need water because protoplasm, the 'living jelly' of which all plants and animals are formed, is nearly all

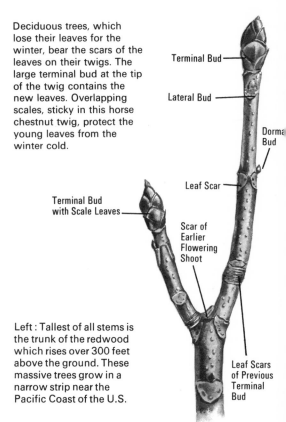

Deciduous trees, which lose their leaves for the winter, bear the scars of the leaves on their twigs. The large terminal bud at the tip of the twig contains the new leaves. Overlapping scales, sticky in this horse chestnut twig, protect the young leaves from the winter cold.

Terminal Bud

Lateral Bud

Dorma Bud

Leaf Scar

Terminal Bud with Scale Leaves

Scar of Earlier Flowering Shoot

Leaf Scars of Previous Terminal Bud

Left : Tallest of all stems is the trunk of the redwood which rises over 300 feet above the ground. These massive trees grow in a narrow strip near the Pacific Coast of the U.S.

water. Water accounts for a large proportion of the weight of any plant or animal. Water is necessary to dissolve food materials so that they can be absorbed from the soil and carried about in the plant. Water is also one of the basic raw materials from which plants make sugar and starch (see page 17). If you forget to water a plant growing in a pot it will soon begin to droop or wilt. This shows us yet another way in which plants use water: they use it for support. This is very important in herbaceous plants which do not have many woody fibers to support them. The water blows out the cells and makes them firm, or turgid. When the cells of a plant are turgid the whole plant stands up, but if the water is lost the cells collapse and the plant droops.

16

How Plants Make Food

The sun provides almost all of the energy on the earth, including all the energy that we use in our bodies to move about and to keep warm. But our bodies cannot harness the sun's energy directly, and nor can any other animal. This can be done only by green plants. They are the primary converters of energy. They make use of the sun's light energy and they convert some of it into chemical energy which they store in their bodies. Animals obtain some of this energy by eating the plants. If it were not for the plants, animals would be unable to survive, because only the plants can make energy-containing foods from the simple materials in the air and the soil. The prophet Isaiah did not know much about energy conversion, but he was quite right when he said 'all flesh is grass': all animal flesh does come from plant material. This holds true in the sea as well as on the land. The fishes and the whales all depend upon the millions of tiny floating plants for their food.

Photosynthesis

The basic source of all food is the remarkable process called photosynthesis, which takes place in all green plants. Photosynthesis means 'building with light' and it is the process by which green plants make sugars from carbon dioxide and water. It takes place only in the presence of sunlight and in the presence of the green coloring matter called chlorophyll.

The chlorophyll absorbs some of the light energy falling on it and it uses it to power a complicated chain of reactions involving water and carbon dioxide. Energy is passed from stage to stage of the process and finally ends up in the sugar that is made. Oxygen is given off in the process, but the chlorophyll is unchanged and it is able to go on absorbing more light and passing it on to more water and carbon dioxide.

Leaves

Most of the photosynthesis that takes place in an ordinary plant takes place in the leaves. The leaves are the 'food factories' of the plant and they are specially constructed for this work. Most leaves sit horizontally on the plant and the stems ensure that the leaves are arranged so as to catch the greatest possible amount of light. You rarely see one leaf sitting directly on top of another and, if you look up from the base of a tree, you will see how efficiently the leaves trap the light.

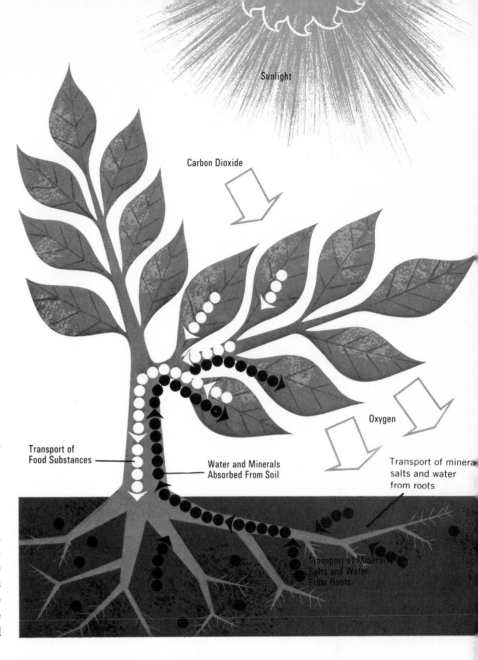

Sunlight

Carbon Dioxide

Oxygen

Transport of Food Substances

Water and Minerals Absorbed From Soil

Transport of mineral salts and water from roots

Transport of Mineral Salts and Water From Roots

Carbon dioxide enters the plant through the leaves. Water enters through the roots. They combine in the leaves to form sugar, which is then carried away to all other parts of the plant. Oxygen is given out during the process and it escapes through the leaves. The whole process is powered by sunlight and therefore takes place only during the daytime.

The leaf is covered with a thin, transparent skin or cuticle and the inner region consists of numerous rather loosely packed cells. Most of the chlorophyll is concentrated in the upper layers where there is the most light. It is not spread throughout the cells, but packed into little 'envelopes' called chloroplasts. The water needed for photosynthesis comes from the roots and enters the leaves through the veins. The carbon dioxide comes from the air and enters the leaves through tiny breathing pores called stomata. Most of these are on the lower surface of the leaves. As well as letting carbon dioxide into the leaf, they let the unwanted oxygen out. They also let the excess water vapor out, but if water is in short supply they will close up and prevent the plant from losing water.

The cells inside the leaf are always surrounded by a film of water and the carbon dioxide dissolves in this film. It can then pass through the cell walls and into the protoplasm where the chloroplasts are. The pro-

17

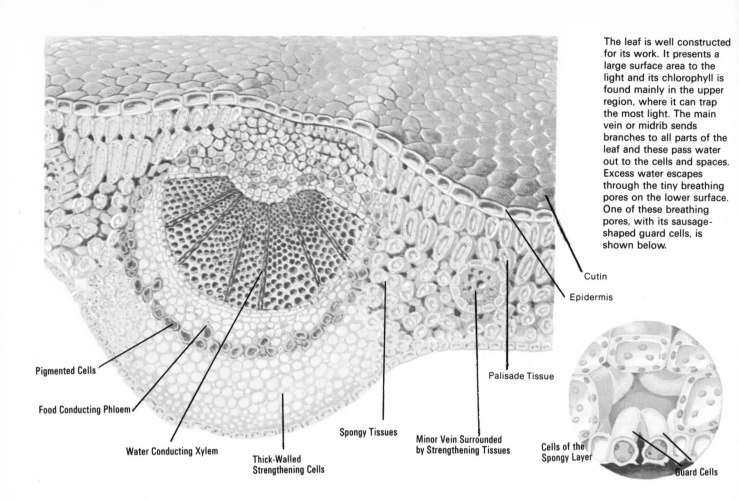

Cutin

Epidermis

Pigmented Cells

Food Conducting Phloem

Water Conducting Xylem

Thick-Walled Strengthening Cells

Spongy Tissues

Minor Vein Surrounded by Strengthening Tissues

Palisade Tissue

Cells of the Spongy Layer

Guard Cells

toplasm consists largely of water and so, as long as the sun is shining, the plant can carry out photosynthesis. The speed at which photosynthesis takes place depends upon several things, including the amount of carbon dioxide entering the plant, the temperature, the brightness of the light, and the amount of chlorophyll in the leaves. Plants which grow in shady places often make up for the low light intensity by having more chlorophyll in their leaves. This makes them somewhat darker than plants growing in full light.

After Photosynthesis

The process of photosynthesis provides the plants with sugars. These can be carried to other parts of the plant and used to provide energy. They can also be converted into starch and stored up until they are needed. Some of the sugars are converted into cellulose and used to build up more plant material. But the plant cannot exist just on the sugars and starches. It must make more protoplasm if it is to make more cells and grow. Protoplasm contains proteins, and proteins contain nitrogen and various other materials. The plant must obtain these materials from the soil. They are called minerals and they are combined with the sugars to form the proteins that the plant needs.

Mineral Nutrition

Every gardener and farmer knows that his plants will not grow very well and will not give good yields unless they are supplied with manure or fertilizer. The plant body consists mainly of the four elements hydrogen, carbon, oxygen, and nitrogen. The first three elements are obtained from water and carbon dioxide and they are always freely available. There is plenty of nitrogen in the air as well, but plants cannot use this atmospheric nitrogen. They have to absorb nitrates or other nitrogen-containing compounds from the soil.

Nitrogen is essential for the growth of plants because nitrogen is a constituent of all proteins, and proteins are the basis of all living matter. If a plant is not able to get enough nitrogen it will not be able to grow. It will become weak and stunted.

If a plant is analyzed chemically it will be found to contain much more than the four elements mentioned above. Among the other elements will be iron, magnesium, calcium, potassium, sodium, silicon, phosphorus, sulphur, and chlorine. These substances are generally called minerals. The amounts vary

Right: The effect of nitrogen on plant growth. A. Nitrogen encourages rapid leaf growth. B. Insufficient nitrogen means poor weak growth. C. Excessive nitrogen can cause soft rank growth.

Below : Spraying trace elements on pasture land in New Zealand. Large areas have been made more fertile by the addition of small, but vital amounts of 'trace' elements such as manganese, molybdenum, zinc, copper and iron.

a great deal, and some of the minerals may not be essential—they are there in the plant simply because the plant had absorbed them in water taken up from the soil. Scientists have found out which elements are essential and which are not by carrying out water culture experiments. In these experiments plants are grown in bottles containing solutions of various elements. One bottle has all the elements in it, and the plant in this bottle grows well. The other bottles each have one element missing. If a plant does not grow well in any of these bottles it is because it needs the missing element. Apart from hydrogen, carbon, oxygen, and nitrogen, the main elements or minerals needed by plants are: magnesium, iron, calcium, potassium, phosphorus, and sulphur. A lack of any of these in the soil will interfere with the proper growth and development of the plant.

In nature, plants grow and die and their remains decay on the ground. The minerals are then returned to the soil for a new generation of plants. Animals also leave droppings which decay and release minerals. The farmer and the gardener, however, do not leave their crops to rot on the land. They take them away to sell or to eat. Minerals are not returned to the soil and so the farmer and the gardener have to add extra minerals every now and then in the form of manure or fertilizer. Most soils contain plenty of calcium and they also have enough iron and sulphur for the plants' needs. The only minerals that normally have to be added are nitrogen, phosphorus, and potassium. Most fertilizers provide one or more of these elements.

Food Storage in Plants

Some of the food made by the plant is used for building new parts and some is used to provide the plant with energy. This food is burned in the plant tissues just as food is burned in our own bodies and the reaction releases energy. Plants do not need a lot of energy, however, and much of the food they make is stored for later use. It may be stored up to tide the plant over the winter and to get it going again in the spring, or it may be stored up in readiness for flowering and the production of fruits and seeds.

During the process of photosynthesis the plant makes sugar, but relatively few plants actually store sugar. Most of them convert the sugar to starch before they store it. Proteins and fats or oils are stored by some plants, but these materials are most frequently stored in seeds.

It is very lucky for mankind that plants do accumulate stores of food materials because we can then step in and take the food for ourselves. Most of the plants store food in one certain region and we normally use only one part of a plant as food. Our vegetable crops have all come from wild plants and most of them have been cultivated for a very long time. During that time, the size of the various food-storing regions has been greatly increased. The cultivated plants therefore store more food and give a much greater yield than their wild cousins.

Storage in Roots

Many plants store their food in swollen roots. Two well known examples are the carrot and the sugar beet. Each plant has one main root, known as a tap root, and a

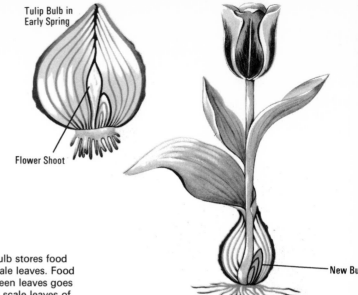

Tulip Bulb in Early Spring

Flower Shoot

New Bud

The tulip bulb stores food in fleshy scale leaves. Food from the green leaves goes back to the scale leaves of the new bud which makes the next year's bulb. The old scale leaves remain as thin brown coating scales.

few smaller branches. These plants, and most of the others like them, are biennials. This means that they live for two seasons. During the spring and summer of the first year the plants make food and store it in their swollen tap roots. The plants die down in the autumn and the food reserve in the roots is used to produce new leaves, flowers, and seeds in the spring. The plants then die.

Other plants storing food in their roots include the dahlia and the lesser celandine, but these do not have tap roots. The food is stored in a number of swollen roots called root tubers. It is used when the plant shoots up in the spring, but the plant does not die. It lives from year to year and produces new tubers each summer.

Storage in Stems

Many plants store food in their stems without any special modifications. Others however, develop specially swollen stems. The kohlrabi, a member of the cabbage family, stores food in a tennis ball sized swelling just above the ground. Irises and several other plants possess horizontal underground stems called rhizomes. These are usually full of food. Potato tubers are also underground stems used for storing food. They develop at the tips of finer stems growing horizontally through the soil.

Corms are underground stems as well. Although often called bulbs, they are really quite different. The crocus corm is a short, rounded stem with one or more buds on the top. The stem is full of food and this is used up when the plant throws up its leaves and flowers in spring. After flowering the plant makes more food and a new corm develops at the base of the leaves.

Leaves and Flowers

Cabbages are some of the best known plants that store food in their leaves. The

A corm stores food in the swollen stem. Buds develop on the corm which forms each year at the base of the flower shoot.

In Spring Food Passes into Leaves.

After Flowering Food is Passed into the New Corm from the Withering Leaves

Bud

Winter Corm of Crocus

Next Year's Bud

New Corm Forms Here

Next Year's Bud

New Corm

Old Corm

ordinary cabbage consists of a tightly packed bunch of leaves, all full of food. This food is passed to the flowers when they develop and the leaves become thinner. Bulbs also store food in leaves, although these are rather special leaves. A bulb consists almost entirely of swollen, juicy leaves called scales. These surround one or more buds. When the bulb starts to grow the food is passed from the scales to the buds and it is used to feed the leaves and flowers as they develop. The scales then wither, but a new set of scales is formed inside them. This new set will be the next year's bulb. The papery coverings on it are the shrivelled old scales.

Some plants even store food in their flowers. Best known of these is the cauliflower. The large creamy heads that we eat are the young flowers, tightly packed together because their stems have not grown much.

Fruits and Seeds

All seeds must have some sort of food reserve which will enable the young seedlings to grow to the stage at which they can start making their own food. Many of the larger seeds—peas, beans, cereal grains, nuts, and so on—are used for human food. Many fruits also contain food materials, and the plants really do provide this food for animals. Sweet, juicy fruits attract animals and are readily eaten. But, while eating the fruits, the animals scatter the seeds inside them. This is of great advantage to the plants and it is therefore worth their while to provide food for the animals.

Above right : Collecting sugar beet for processing. The swollen tap roots contain the plants' food reserves, designed to produce new leaves, flowers, and seeds in the spring.

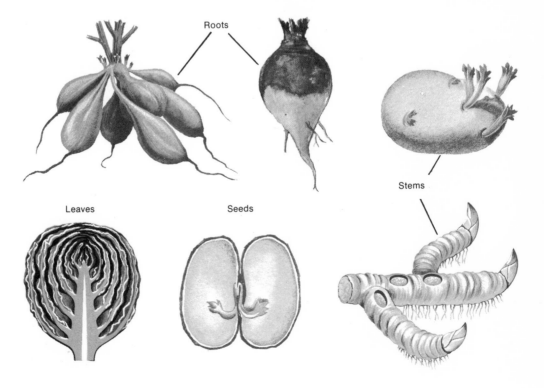

Right: A selection of plant food stores including roots (dahlia tubers and sugar beet), stems (potato tuber and rhizome), seeds (bean seed) and leaves (cabbage).

Roots

Stems

Leaves

Seeds

Plant Growth and Movement

One of the main differences between the higher plants and the higher animals is that the plants do not have a fixed shape or size. A lion always has four legs, two eyes, one tail, and so on. But we cannot say how many branches an oak tree will have because the growth pattern of a plant is so flexible. A tree growing in the middle of a wood has a very different shape from the same kind of tree growing by itself in the middle of a field. Plants also differ from animals in that they go on growing throughout their lives.

Stems and Roots

You can learn quite a lot about plant growth if you watch buds, such as those of the horse chestnut, opening in the spring. The buds are actually very short stems surrounded by young leaves. The leaves on the outside protect the bud during the winter. The bases of all the leaves are very close together. In other words, the *internodes* of the stem – the regions between the leaf bases – are very short. When spring comes the leaf scales fall from the outside of the bud and the

Many climbing plants have tendrils which grow rapidly when touched and coil tightly around a support.

Many flowers close up at night or when the weather is cold. The daisy closes its flowers when the light begins to fade in the evening. These movements are called nastic movements.

By marking a root or a shoot with equally spaced lines you can find out the region in which most of the growth takes place.

internodes begin to lengthen. The new leaves become separated and they begin to unfold. If the bud contains a flower, as most of the terminal buds do, its growth soon stops. The growth of the stem continues through one or more of the subsidiary buds. The stem tips of these buds continue to grow and produce more cells. The cells get longer and so the stem itself grows. The actual tip of the stem is never seen because it is always sheathed

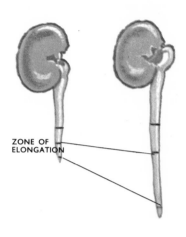

ZONE OF ELONGATION

by developing leaves. Later in the summer more scale leaves are produced around the stem tip and growth slows down. The scales protect the stem tip and its leaves during the next winter.

Many of the subsidiary buds on a stem never open. They are reserve buds, which develop only if the main stem is damaged. They may grow if the plant is damaged by insects, or if branches are broken by the wind or by man. Some trees produce a great many new shoots when the branches are cut. The European linden is a good example. It is often grown along drives and pathways and it is regularly cut back to the main branches. The cut ends produce dense clusters of twigs. Treating trees in this way is called *pollarding*.

Roots grow in a similar way to stems, although they are simpler because there are no leaves. The tip of the root continuously produces new cells which then begin to lengthen. As the top of the root is fixed, the tip is driven down into the soil. The tip is protected by a special layer of cells called the *root cap*. These cells are always being worn away, but new ones replace them.

The Mechanism of Growth

A plant grows basically by producing new cells at the tips of its stems and roots. Most of the actual growth, however, takes place a little way behind the tip. It is possible to show this by marking a young root with dye as shown in the diagram. The marks are equally spaced to start with, but after a while those that were just behind the tip become widely separated. Although growth actually takes place just behind the tip, it is controlled by the tip. The tip of the root or shoot produces a material called *auxin*. The auxin causes the cells to grow longer. If the shoot tip is removed, the auxin supply is lost and the cells do not grow.

Growth Movements

If you grow a plant in a room with only one window, you will soon notice that the plant bends towards the window. This bending movement is very useful for the plant because it ensures that it gets as much light as possible. It is produced by the auxin. When a plant receives light from one side only the light somehow causes the auxin to move across to the shady side. The cells on that side therefore grow longer than those on the lighter side and so the plant bends over towards the light. Bending movements of this kind, in which the movement depends on the direction of the stimulus, are called *tropisms*. Bending movements caused by light are called *phototropisms*.

If you grow some cress seedlings and then lay some of them down flat in the dark, you will find that the stems start to grow up again and the roots try to grow down again. These are some more bending movements, but this time the plants are responding to gravity. The movements are called *geotropisms*. We do not really know what makes the stem grow upwards and the root grow downwards, but one of the most likely explanations is that tiny starch grains in the cells are concerned. If the plant were laid down, the starch grains would accumulate on the lower sides of the cells and they could stimulate growth in some way.

Other bending movements include the bending of roots towards water, and the curling of tendrils around sticks and other supports when they are touched. All of these movements take place in the growing region, just behind the tip of the root or shoot. They are therefore called *growth movements* as well as tropisms.

Another kind of bending movement is shown by the leaves of several plants. The garden nasturtium, for example, bends its leaves round during the day and keeps them facing the sun. They get the maximum amount of light in this way. The movements are controlled by changes in the pressure of water in the cells of the leaf stalks.

Right: A number of plants bend round during the day and thus remain facing the sun. This is particularly noticeable in a field of sunflowers where nearly every flower faces in the same direction.

Vegetative Reproduction

Many of the simplest plants, including the bacteria, can reproduce themselves simply by splitting into two similar halves. This is very similar to the reproductive methods of some of the simplest animals—the Protozoa. The more advanced plants, however, generally have special reproductive organs and their reproductive processes involve the joining together of two cells. For example, among the flowering plants, pollen cells have to join with egg cells in the flowers before seeds can be formed (page 40). This method is known as sexual reproduction, but the process is not exactly the same in all plants. Some of the variations are described in the following pages.

In addition to the sexual process, many plants have a second method of reproduction. It is known as vegetative reproduction. Parts of the plant, usually the stem, become separated and grow into new plants.

The strawberry provides one of the best examples of vegetative reproduction. The plant sends out slender stems. called runners, that grow over the surface of the ground and then develop roots at their tips. Leaves develop there as well and the runners decay, leaving a number of new plants all around the parent. Many plants spread in this way, but with underground stems instead of runners. These underground stems are called rhizomes. If you have mint in your garden you will know how efficiently the rhizomes

Bracken fern is a serious weed when it gets on to farm land because its underground rhizomes creep through the soil and soon cover a very large area. One original plant can become several plants if part of the connecting rhizome dies away.

Several plants, such as bryophyllum (near right) and coral root (far right), produce detachable buds or plantlets which grow into new plants.

spread through the soil. Many rhizomes store food as well. Some of the other common methods of vegetative reproduction are pictured on this page.

Vegetative reproduction is an efficient way of spreading a plant over a small area, as we can see by the way in which couch grass or mint soon forms a large clump. But it does not carry the plants into new regions. A more important point is that the new plants are all the same, because they are all really just pieces of the parent plants. There is no variation of the sort that we get when cells join together in sexual reproduction. And without variation there can be no improvement. Vegetative reproduction has not, therefore, been very important in the history of plants.

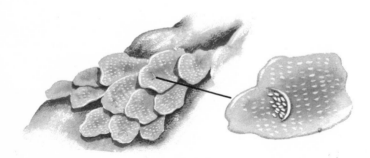

Some liverworts (page 33) produce detachable buds called gemmae. They are formed in little cups and they are washed away by the rain.

Strawberry runners grow very rapidly and soon cover the ground with young plants.

24

Plant Classification

There are more than 350,000 different kinds of plants in the world. Each kind is called a species. Unlike most animals, plants do not have a definite shape and no two plants will ever look exactly alike. But the members of one species are all quite similar. Two bluebell plants have more in common than a bluebell and a daffodil. There is a closer connection between a bluebell and a daffodil, however, than between a bluebell and a primrose. And there is a closer connection between a bluebell and a primrose than between a bluebell and a fern. These similarities and differences are used to classify plants, or to divide them into groups.

Most people who study plants divide the plant kingdom into four major groups. They are called thallophytes, bryophytes, pteridophytes and spermatophytes. Each group is further divided into orders and families as shown on the chart or 'family tree' on the next page.

Thallophytes

The thallophytes are the simplest of the plants alive today, and the earliest plants must have been of this type. There are no flowers, and the plant body is not even divided into root, stem and leaf. There are two main groups: the Algae and the Fungi. The Algae nearly all live in water. They all have green chlorophyll, although this may be masked by other colors. Seaweeds are algae, and so are the slimy green threads that choke fish ponds in the summer and make the water green. Many algae are microscopic organisms that swim freely in the water. The Fungi have no chlorophyll and they cannot carry out photosynthesis to make their own food (see page 17). Fungi include moulds and toadstools.

Bacteria are usually classified as thallophytes as well, although they are rather different from the other members of the group.

Bryophytes

The bryophytes are green plants which usually have distinct stems and leaves. There are no true roots, however, and no flowers. Bryophytes are divided into two main sections: the mosses and the liverworts.

Pteridophytes

This group includes the ferns, horsetails and club mosses. They are green plants and they are generally much larger than the mosses and liverworts. They have stems and

Parallel Veins

Branched Veins

The leaves of a monocotyledon such as an iris normally have narrow leaves with parallel veins. The leaves of a dicotyledon such as a rose are normally broader and have a network of veins.

leaves as a rule, and they also have proper roots. Associated with their larger size, they also have special tubes which carry water and food around the plant. There are no flowers.

Spermatophytes

These are the seed-bearing plants. Roots, stems and leaves are normally well developed. Apart from a few parasites (see page 47), they all contain chlorophyll and make their own food. There is a tremendous variation in size in this group, from minute floating duckweeds to huge redwood trees nearly 400 feet tall.

The two major sections of the seed-bearing plants are the gymnosperms and the angiosperms. Most of the gymnosperms carry their seeds in cones and are therefore called conifers. Other gymnosperms include the cycads, the yews, and the ginkgo or maidenhair tree.

The angiosperms are the flowering plants. They account for something like 250,000 of the known plant species. There are two main sections: the monocotyledons and the dicotyledons. The basic difference between the two sections is found in the seed. The seed of a monocotyledon has only one seed leaf (cotyledon) in it, whereas the seed of a dicotyledon has two seed leaves. Monocotyledons include plants such as grasses, irises, daffodils, tulips and orchids. Their leaves are normally narrow, and the veins run parallel to each other. The flowers frequently have either three or six petals. The dicotyledons normally have broader leaves, with a network of veins. Their flowers usually have either four or five petals. All the flowering trees are dicotyledons, and so are most of the flowers and vegetables that we grow in our gardens.

Within the monocotyledons and dicotyledons, the plants are arranged in many families. The arrangement depends very much on the structure of the flower. The leaves and stems are not of much use in classifying plants because they vary so much according to the conditions under which the plants live. Unrelated plants growing in the same habitat often look alike because they are adapted to the same conditions. Conversely, related plants growing in different habitats may look very different. This can be so even within one species. For example, a dandelion living in a damp meadow may have large upright leaves and long flowering stems. If transplanted to a windswept mountainside, this same plant would start to produce much shorter leaves and they would lie flat and form a rosette on the ground. The flowering stems would also be shorter, although the flowers themselves would remain the same.

25

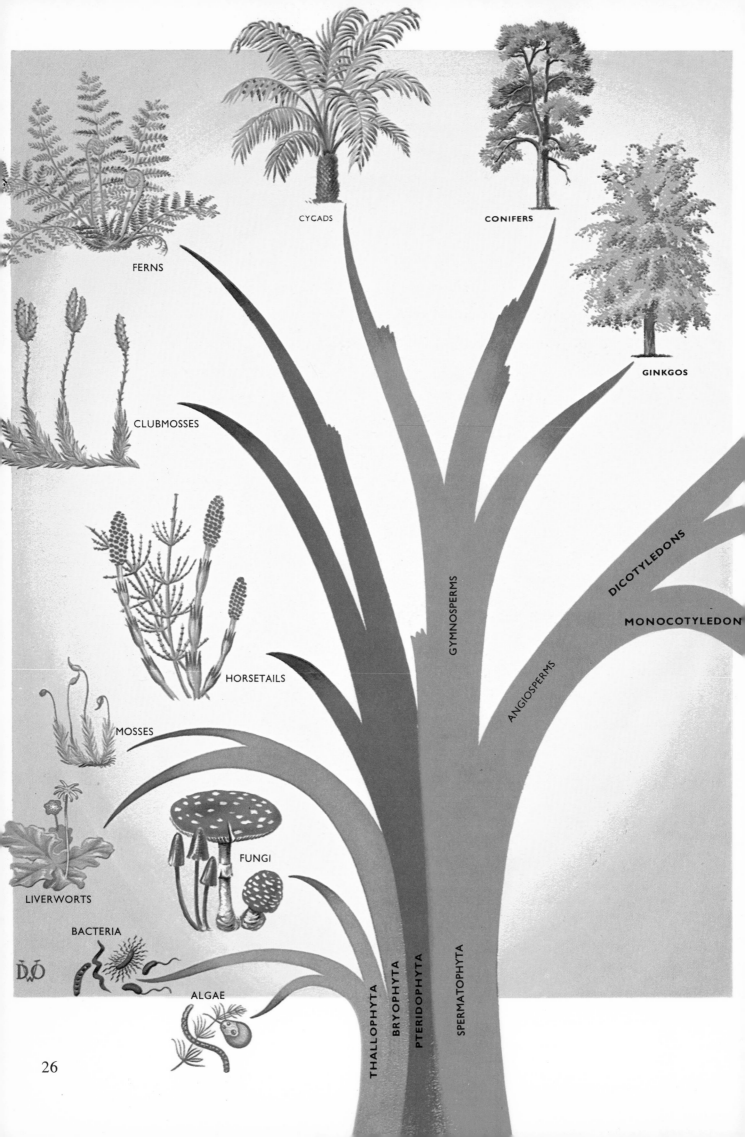

FERNS

CYCADS

CONIFERS

GINKGOS

CLUBMOSSES

DICOTYLEDONS

MONOCOTYLEDON

GYMNOSPERMS

HORSETAILS

ANGIOSPERMS

MOSSES

FUNGI

LIVERWORTS

BACTERIA

ALGAE

THALLOPHYTA

BRYOPHYTA

PTERIDOPHYTA

SPERMATOPHYTA

26

LEGUMINOSAE
Vetch

RANUNCULACEAE
Buttercup

UMBELLIFERAE
Hogweed

ROSACEAE
Rose

PRIMULACEAE
Primrose

COMPOSITAE
Dandelion

SOLANACEAE
Woody Nightshade

LILIACEAE
Bluebell

ORCHIDACEAE
Orchid

Bacteria

Bacteria are minute organisms, each one consisting of just one cell. They are usually regarded as very simple plants, although they do not contain chlorophyll. The largest bacteria are only about 1/2000th of an inch across, and they can be seen only with a good microscope. There are three main groups of bacteria, known as bacilli, cocci, and spirilli. Bacilli are rod-shaped, cocci are spherical, and spirilli are like minute curved or twisted sausages. Many bacilli and most of the spirilli possess flagella, which enable them to swim about. The bacteria are so small that they can swim in even the thinnest film of water.

We often think of bacteria as disease-causing organisms or germs. Many certainly do cause diseases, such as tuberculosis and tetanus, but the majority are very useful. Whereas green plants get their food and energy from sunlight and the process of photosynthesis, the bacteria get their energy from various chemical reactions. The soil teems with bacteria and they get their energy by breaking down dead plants and animals. In doing so, they release minerals which growing plants can take up again. Bacteria therefore play a vital role in the economy of nature. Life would soon come to a stop without them.

The Algae

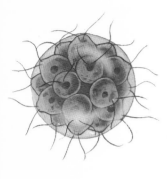

A diatom, one of the algae with glassy shells.

The algae are simple plants belonging to the large group known as the thallophytes. This group also includes the fungi and the bacteria, although the latter are not closely related to any other plants. The plant body is called a *thallus*. It is not divided into roots, stems, and leaves, although some of the seaweeds have root-like holdfasts which fix them to the rocks. There are no special cells for carrying water or food around the plants. And there are, of course, no flowers. All the algae possess chlorophyll, although it is often masked by brown or red pigments, and they make their own food just like any other green plants.

Plants That Move About

The simplest algae are single-celled organisms and most of them are extremely small. Some of them live in damp soil and some of them clothe tree trunks in damp places, but most of them live in water. They possess little whip-like hairs called flagella, and they swim about by lashing these hairs.

One of the commonest algae in fresh water is a little thing called *Chlamydomonas*. Its single oval cell is so tiny that fifty of them would only just about stretch across a pin's head. But this minute creature is a complete plant. It has a cellulose cell wall, a nucleus to control its activity, and a chloroplast in which it can make food. The little plant also

Above: Pandorina, a colonial alga consisting of 16 individual plants embedded in a sphere of jelly.

has a red 'eye spot' which is sensitive to light. This eye spot controls the movement of the plant, directing it towards areas of moderately bright light. This ensures that the plant remains in a position where it can carry out photosynthesis efficiently. When the water is warm Chlamydomonas grows rapidly and it splits into two every day or so. Its numbers build up so rapidly that, although each individual is so small, it soon turns the water into a thick green 'soup'.

Chlamydomonas is a solitary plant, with each individual leading a separate existence. There are, however, several common pond-dwelling algae which live in little clusters or colonies. The colonies consist of eight or more cells, all more or less like Chlamydomonas. Although the cells are all loosely connected in a mass of jelly, they each act as separate individuals.

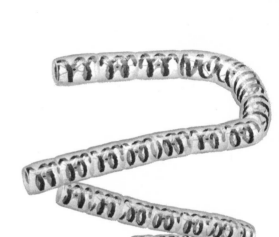

Above: Spirogyra, showing the spiral chloroplast which gives the plant its name.

Spirogyra

During the summer many ponds become choked with tangled green strands known as blanketweed. Each strand is an alga, and one of the commonest kinds is called *Spirogyra*. The strands are unbranched and each is composed of a number of cylindrical cells. These cells are all more or less alike and each acts independently, making food in its spiral chloroplast. It is this spiral chloroplast that gives the plant its name. The cells divide into two every now and then, but the two new cells do not separate: they remain in the strand and continue to grow. The strand therefore increases in length. Spirogyra can also reproduce in another way. Neighboring strands send side branches into each other, forming a sort of 'ladder'. The contents of one cell moves across into the neighboring cell and forms a tough walled 'egg'. The parent strands then decay and the 'eggs' each develop into new strands.

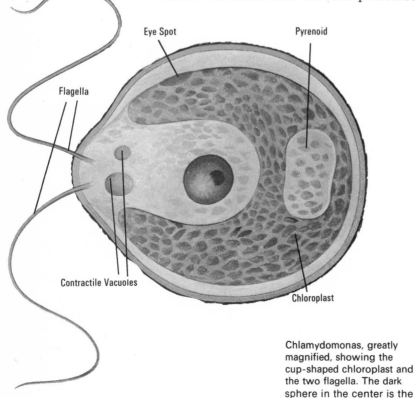

Eye Spot

Pyrenoid

Flagella

Contractile Vacuoles

Chloroplast

Chlamydomonas, greatly magnified, showing the cup-shaped chloroplast and the two flagella. The dark sphere in the center is the nucleus

Above : The algae are prominent on rocky sea-shores where seaweeds grow in profusion. They provide a retreat for many small animals when the tide goes out.

Below : A few of the many varieties of seaweed.

there is usually a zone of brown seaweeds. These include the largest of the algae, some of the oarweeds or kelps being many yards long. One of the commonest of the brown seaweeds is the bladder wrack, so called because its fronds bear numerous air bladders which pop if squeezed. The bladders help to buoy the plants up when the tide comes in. The fronds of the brown seaweeds are covered here and there, especially at the tips, with pale spots. These are tiny pits in which the reproductive bodies develop. Tiny free-swimming spores are released from the pits. Some of them have to pair up before growing into new seaweeds, but others can grow directly into new plants.

Red seaweeds are generally much smaller than the brown ones and many of them are constructed in a different way. Instead of having broad blades, many of them have slender branching filaments made up of chains of cells. Some secrete chalky skeletons around themselves and these seaweeds play an important part in building coral reefs in the warmer seas. Red seaweeds normally grow further down the shore than the brown ones and, unless there are rock pools, they do not very often reach above low tide level. They cannot withstand exposure to the air as well as the other seaweeds. They can extend into deeper and darker water, however, because the red pigment helps the chlorophyll to absorb light more efficiently.

Seaweeds

Seaweeds include the largest and the most complicated algae. Some are red, some are brown, and some are green, but they all contain chlorophyll and they all make their own food with the aid of sunlight. The need for sunlight means that the seaweeds can grow only in shallow water, where light can reach the bottom. Most of them are to be found on rocky shores, and many are good to eat.

Green seaweeds, such as the common sea lettuce, normally grow on the upper parts of the sea shore and they make the rocks very slippery. A little lower down on the shore

The bladder wrack is one of the commonest brown seaweeds and it coats many of the rocks of the middle shore. The bladders contain air which helps to buoy up the plant. The tips of the thallus bear pimply swellings. Each pimple is the opening of a reproductive organ.

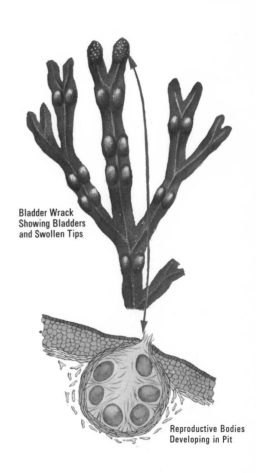

Bladder Wrack Showing Bladders and Swollen Tips

Reproductive Bodies Developing in Pit

29

Fungi

Above: The mushroom begins as a small, round cap and grows into the familiar parasol shape.

Moulds and toadstools belong to the large group of plants called Fungi. Some of them are clearly similar to the algae, but none of the fungi ever has any chlorophyll. Fungi cannot, therefore, make their own food. They have to get 'readymade' food just like animals. Most of them get it from dead and decaying matter, such as dead leaves and rotting wood. Fungi getting food in this way are called *saprophytes*. Some fungi, however, are *parasites* and they obtain food from living plants or animals. Many serious crop diseases are caused by parasitic fungi. Potato blight is a well known one, and the rust disease of wheat is another. Fungi also produce various skin diseases, such as ringworm, in animals.

Because the fungi do not make their own food they do not need sunlight. This is why we can grow mushrooms in dark damp cellars. Fungi always need damp conditions because their soft bodies rapidly shrivel in dry air. Like the algae, the fungi belong to the division of plants called the thallophytes. They have no roots, stems, or leaves. Their bodies are composed basically of slender threads called hyphae. These creep over or through the food materials and secrete digestive juices which dissolve the food. The resulting solution is then absorbed. Fungi are a nuisance when they start to grow on our jam or our floorboards, but they are of great importance in nature. They play a major part in breaking down dead leaves and returning the goodness to the soil.

Moulds

Moulds are relatively simple fungi. They consist of a mass of fluffy threads that cover the material they are growing on. One of the common moulds is the pin mould, scienti-

Above : The beautiful but poisonous fly agaric.

Few organic materials are safe from the attacks of fungi. Mucor, the pin mould, grows on stale bread, cheese, leather, animal dung, and many other substances. The left pictures show the little black capsules which contain the spores greatly enlarged.

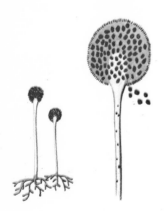

Penicillin

One of the biggest advances in medical science was the discovery of penicillin. It was discovered quite by accident in 1928, when Alexander Fleming found that a little green mould was growing among some bacteria (germs) he was studying at a London hospital. Fleming then noticed that the germs near the mould were dying. In fact, the mould was producing a germ-killing substance. It was a long time before scientists managed to isolate this substance, but they succeeded in the end and found that it could be used to fight a wide variety of diseases. The mould was called *Penicillium notatum*, and the substance was therefore called penicillin. Large amounts of penicillin are now produced by growing the mould in huge tanks.

Yeasts

Yeasts are rather special kinds of fungi which consist of tiny egg-shaped cells instead of threads. The cells often join up to form chains. However, there are many different kinds and they produce digestive juices which turn sugar into alcohol. Yeasts are therefore used in making wine and beer.

fically known as *Mucor*. It grows on a variety of materials and can be grown very easily on a piece of damp bread on a saucer. Like nearly all fungus threads, the hyphae are white, but upright branches form after a while and they bear little black capsules at their tips. These look like tiny pinheads and give the mould its common name. Each capsule contains hundreds of tiny spores and, when the capsule splits, the spores are scattered like dust into the wind. If they land in a suitable place they will grow into new threads. This is a form of vegetative or asexual reproduction. Sexual reproduction takes place when side branches join together and form tough-walled 'eggs'. These can survive cold and drought and, when good conditions return, they start to grow. A spore capsule develops and scatters lots of spores which can then grow into new threads.

Toadstools

Toadstools are more complicated fungi than the moulds, but the familiar umbrella-shaped toadstool is only a part of the fungus. For most of the year the fungi exist as slender threads just like those of the moulds. They are usually under the ground and many of them are associated with tree roots. They grow and branch and then, usually in the autumn, they become tightly packed together and they start to form the toadstool. At first it is a little button-shaped object in the ground, but it grows rapidly and breaks through the surface. The cap then opens out to reveal either the familiar gills radiating out from the stalk or a mass of tiny pores. Spores are formed on the gills or in the pores and when they fall out they are blown away.

The mushroom is simply one particular kind of toadstool that happens to be very good to eat. There are many other toadstools that are good to eat, but some of them are very poisonous. You should never eat fungi unless you are absolutely sure that they are edible. Two of the most deadly kinds are the death cap and the red and white fly agaric. Both have little 'cups' at the base of the stalk and both have rings higher up on the stalk.

Two other groups of fungi that are commonly found are the bracket fungi and the puffballs. Bracket fungi grow on living and dead trees and, instead of forming umbrella-shaped spore-bearing bodies, they form shelf-like structures on the trunks. Puffballs are ball-shaped or club-shaped fungi that grow on the ground. Some reach the size of footballs or even more. When they are ripe they split open and when they are touched, or even blown by the breeze, huge clouds of spores shoot out. Puffballs are good to eat when they are young.

Turban Fungus

Chanterelle

Morel

Amanita caesarea

Horse Mushroom

St. George's Mushroom

Parasol Mushroom

Shaggy Cap

Grisette

Boletus

Clitocybe geotropa

Lactarius volemus

Fly Agaric

Wood Blewit

Oyster Mushroom

Lichens

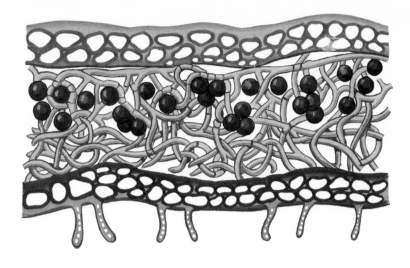

Old walls, especially in country areas, are often encrusted with grey or yellow patches more or less circular in outline. These patches, despite their rather lifeless appearance, are growing plants and they are called lichens (pronounced like-ens). They are extremely hardy plants—hardy enough to grow on exposed rocks—and their hardiness results mainly from their curious construction. A lichen is, in fact, two plants rolled into one. The two plants are a fungus and an alga, each kind of lichen having its own combination of fungus and alga species.

The bulk of the lichen plant is made up of fungus threads, but there is a great difference between the tough, hardy lichen and the soft body of a normal fungus. This is one of the many mysteries still surrounding the lichens. The tiny algae are bound up among the fungus threads and are normally concentrated near the upper surface where they can receive the maximum amount of sunlight.

The fungus threads of the lichen absorb and store water when it is available. They then pass it on to the algae. The algae contain chlorophyll and they can carry out photosynthesis, combining the water with carbon dioxide to make sugar. Mineral salts are obtained partly from the dust that falls on the plants and partly from the rock or soil: the fungus threads produce acids that disolve the rock and release the minerals. The minerals and sugars are combined by the algae to make food for themselves and for the fungus. The lichen is therefore

The bulk of the lichen body is formed by the tangled fungus threads. The algal cells are found mainly in the upper layers, where they get enough light to make food.

Right: Foliose lichens are often leaflike. Fruticose lichens are often branched.

Below: Patches of crutose lichens growing on an old wall.

very much a partnership, with each partner having certain jobs. It is a good example of a *symbiotic association* (see page 134).

There are three main groups of lichens, separated according to the shape of the thallus or body. *Fruticose* lichens are often branched and they look like miniature trees or bushes. *Foliose* lichens consist of spreading 'fingers', often rather leaf-like. *Crutose* lichens cover surfaces with a delicate crust lacking any distinct lobes or 'fingers'.

Reproduction takes place in two ways in the lichens. The fungus produces spores

in brightly colored patches. These tiny spores are scattered in the wind but they contain no algae and will not develop into lichens unless they meet the right kind of alga. The lichen fungi cannot live alone. The other method of reproduction involves the production of *soredia*. These are little granules that break off from the lichen surface. They contain fungus threads and algae and they can grow directly into new lichens.

Lichens are so hardy that they can grow almost everywhere, from the hottest deserts to the coldest polar regions. Large areas of the Arctic Circle are covered with lichens known as reindeer moss because they are the main summer food of the reindeer. As well as growing on the ground, especially where the soil is poor, lichens grow on rocks, walls, and tree trunks. The one thing that they cannot survive, however, is air pollution and they are rarely found in large cities.

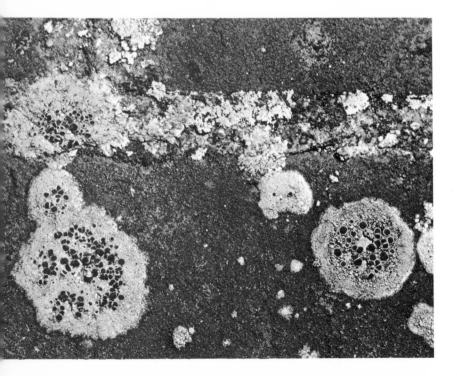

Mosses and Liverworts

Most people can recognise mosses when they see them, but few bother to look closely at these interesting little plants. Their slender stems are rarely more than a few inches long and they carry thin, almost transparent leaves. There are no proper roots, although short hair-like outgrowths anchor the plants to the ground. Mosses usually grow in dense clusters, forming bright green mats or cushions on the ground. They also grow on old walls and tree trunks.

Mosses are not flowering plants, but at certain times of the year dense clusters of leaves develop at the tops of the stems. Among these leaves there are the reproductive organs. The male organs are club-shaped and the female organs, which are usually on separate branches, are like tiny flasks. In damp weather, when the plants are covered with a film of moisture, the male cells are released. They possess minute hairs called *flagella* and these enable them to swim about in the film of water. The female 'flasks' produce a jelly which attracts the male cells, and they then pair up with the egg cells in the flasks. The new cell formed by this union begins to grow and eventually forms a little oval box or capsule. This has its own stalk and it grows up above the rest of the moss plant. It contains the spores—minute dust-like particles that can grow into new moss plants. The spore capsule usually has a pointed cap to start with, but this comes off when the spores are ripe and reveals a complicated system of 'teeth'. When the air is dry these 'teeth' curl back and allow the breeze to scatter the spores. The capsule closes up again in damp weather, for the spores would not float away very well if they were wet.

When the moss spore germinates it sends out a thin green thread. This is called the *protonema* and it soon grows several branches. Buds develop on the branches and each bud grows into a new moss plant. As a result, the young mosses are clumped together right from the start. New threads and buds can grow from the base of the moss at any time, and so it can cover a large area of ground. This habit of forming mats and cushions is very important for the mosses because they do not have waterproof coverings. By clumping together, they retain a good deal of moisture around themselves. Mosses always grow best in moist, shady places.

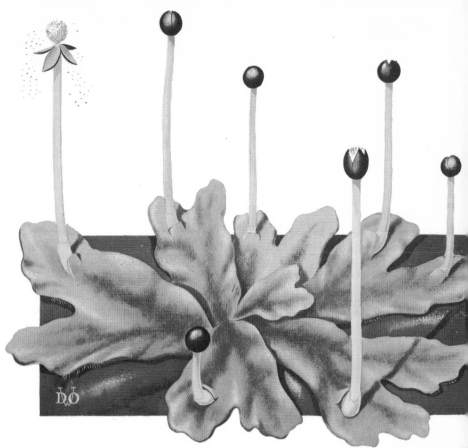

Pellia, one of the flat liverworts, showing the tough thallus and the simple spore capsules. Some liverworts, but not *Pellia*, possess little gemmae cups containing detachable buds (below).

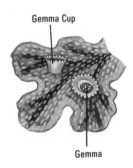

Gemma Cup

Gemma

Some common European mosses of damp grassland.

Liverworts

The liverworts are less well known than the mosses, although many of them are just as common. There are two main groups: flat and leafy. The flat liverworts grow in very damp places, such as river banks and woodland paths. They are especially common by waterfalls, where the water is always splashing them. They look rather like tough green seaweeds creeping over the surface, and it is possible that the earliest land plants were of this type. The leafy liverworts are very much like the mosses, and are readily distinguished only when they are carrying spore capsules.

The life histories of both kinds of liverworts are very similar to those of the mosses, but the spore capsules are very much simpler. They are round and black to start with, but when the spores are ripe they split open and become star-shaped. The spores are scattered by the wind and, when they reach suitable places, they grow into new liverworts. There is no protonema. Many liverworts can also reproduce vegetatively by producing *gemmae*. These are tiny detachable buds, produced in little cups on the liverwort surface. They are scattered by raindrops and quickly grow into new liverworts.

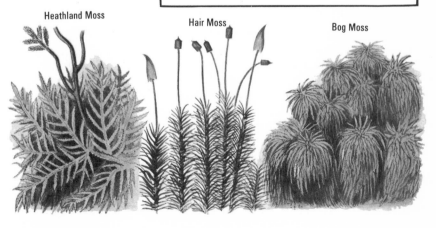

Heathland Moss

Hair Moss

Bog Moss

Ferns and Horsetails

The ferns and their relatives belong to a very ancient group of plants known as the Pteridophyta. This group was in existence at least 400 million years ago and its members included some of the earliest of all land plants. The pteridophytes reached their peak about 250 million years ago, when they formed great forests. Most of our coal consists of the remains of these plants, and so we know quite a lot about them. Some of them reached heights of more than 100 feet. These huge plants have long since died out, and today's pteridophytes rarely exceed a few feet in height. Only the ferns make a significant contribution to today's plant life, although there are several other types of pteridophyte still in existence.

Ferns and the other pteridophytes normally have roots and stems. Most of them also have leaves, but none of them ever has any flowers. They reproduce by scattering tiny spores, rather like those of the mosses and liverworts. The possession of proper roots is an advance over the lowly mosses, but even more important is the development of a system of tubes carrying water and food from one part of the plant to another. When once the plants had evolved such a system they could begin to increase in size, and we have already seen that some of the fern-like plants reached great heights.

Ferns

Most of the ferns possess roots, stems, and leaves. Except in the tropical and subtropical tree ferns, which may reach a height of 70 feet, the stems are small and usually remain under the ground. Fern stems are normally unbranched, but the bracken and some other ferns have branching underground stems called rhizomes. These spread through the soil and throw up leaves all over the place. It is the rhizomes that make the bracken such an invasive plant and so difficult to get rid of. Fern leaves, often called fronds, may be quite large and they are frequently divided into many small leaflets.

The spores are normally produced on the undersides of the leaves, and if you look at fern fronds in late summer you will see hundreds of little brown patches. These are patches of spore capsules. In some species they are covered by protective flaps. Each capsule has a little stalk and looks like a tiny club. When it is ripe it bursts open and the spores are scattered in the wind.

When a spore falls on suitable ground it begins to grow, but it does not grow into

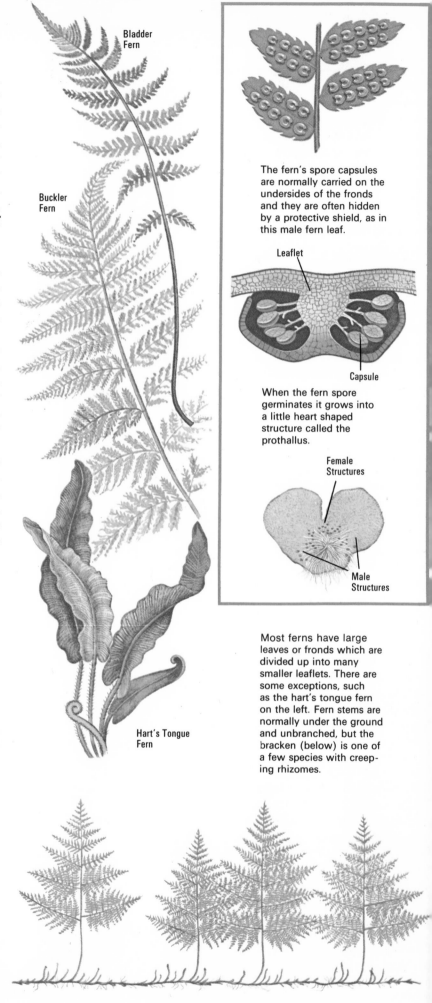

The fern's spore capsules are normally carried on the undersides of the fronds and they are often hidden by a protective shield, as in this male fern leaf.

When the fern spore germinates it grows into a little heart shaped structure called the prothallus.

Most ferns have large leaves or fronds which are divided up into many smaller leaflets. There are some exceptions, such as the hart's tongue fern on the left. Fern stems are normally under the ground and unbranched, but the bracken (below) is one of a few species with creeping rhizomes.

34

a new fern plant. It forms a little heart-shaped plate up to 10 mm across. This plate is called the prothallus. It is green and it has little root-like hairs on its underside. After a while the prothallus produces reproductive cells. These are of two kinds: egg cells and male cells. The male cells are released in wet weather and they swim about by means of tufts of flagella (movable hairs). They are attracted to the egg cells and they join up in pairs. The combined cell then starts to grow into a new fern plant. It is supported by the prothallus at first, but it soon starts to make its own food. The first fronds are quite small and simple, but later ones take on a more mature appearance. It may be several years before the young fern plant begins to produce its own spores.

Many fern plants *can* grow in dry places but, because the life cycle depends on damp conditions during the prothallus stage, the ferns are found mainly in damp places. Shady woods are probably the best places in which to look for them.

Horsetails

About 250 million years ago the horsetails played a very prominent part in the world's plant life. There were many different species and some of them were more than 100 feet tall. Today there are only about 25 species alive and they are rarely more than about three feet tall. Many of them live in marshy places and ponds, although some live in much drier places.

Horsetails are rather spiky plants without proper leaves. Some species have whorls of thin branches, but others have no branches at all and the stems look rather like long

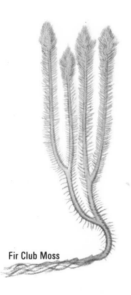

Scale Leaves

Horsetail stems are rigid and jointed. Each joint is surrounded by a collar of tiny scale leaves.

Fir Club Moss

There is a considerable variety of ferns, and in warm damp countries some grow very tall.

pencils sticking up out of the ground or the water. The stems are always somewhat ridged and they bear a number of little 'collars'. These 'collars' are made up of tiny scale leaves and they occur at intervals all the way up the stem. They mark the position of the nodes or joints, and the stems snap very easily at these points.

Under the ground, the horsetails have creeping rhizomes, like those of the bracken fern, and they are difficult plants to get rid of in a garden. The stems usually die down every year, but the rhizome lives on to produce the next year's stems.

The spores are produced in little 'cones' at the tips of the stems. Some horsetail species produce special spore-bearing stems which are brown and which come up before the green stems, but other species carry their 'cones' at the tips of ordinary stems. The spores are green and they grow very quickly when they have been scattered. They grow into prothalli rather like those of the ferns, although they usually have upright 'fingers' on the upper surface. Egg cells and male cells are produced and, after pairing up, they grow into new horsetail plants.

Club Mosses

Living club mosses are all quite small plants and, as their name suggests, they look rather like mosses. They generally grow more or less flat on the ground and their leaves are very small. They are, however, more closely related to the ferns and they have the same sort of life history. The spores are borne in 'cones' at the tips of upright shoots, and the cones are rather club-shaped—hence the common name of these plants. The prothallus of the club moss is something like a small sausage. It normally lives under the ground and gets its food by forming a partnership with a fungus.

Conifers

Seeds are produced only by the two most advanced groups of plants—the cone-bearers and the flowering plants. The ferns and the other more primitive groups of plants all reproduce by scattering tiny spores.

Apart from the strange, palm-like cycads, the cone-bearers are generally large trees. They are often called conifers. Examples include the pines, spruces (Christmas trees), cedars, larches and monkey puzzle trees. The latter are South American trees which form forests on the slopes of the Andes. The conifers nearly all have small narrow leaves called needles, and most of them are evergreens. The needles are very tough and coated with thick cuticles. The breathing pores are sunk well below the surface and water loss is therefore kept to a minimum. This is very important, for the conifers generally live in regions where water is in short supply—at least for part of the year.

Conifers make up a large part of the vegetation in the cooler parts of the world. They stretch in an unbroken belt from Norway to eastern Siberia and also right across Canada. The soil in these regions is frozen during the

Giant Fir   Silver Fir   Sitka Spruce   Douglas Fir

Lodgepole Pine   Western Red Cedar   Norway Spruce

Above: Some well known conifers.

Coniferous forests cover vast areas in northern lands. They are our main source of timber.

winter and water is not readily available to the trees. But the winds are strong and tend to remove water by evaporation from the leaves. Without some means of cutting down water loss, the trees would soon die. Conifers also flourish in the Mediterranean area and in other regions with moist winters and hot, dry summers. The winter rains provide sufficient water for growth, and the needle-

36

Spores and Seeds

There is a big difference in the reproductive processes of the ferns and the conifers. The ferns produce spores, while the conifers produce seeds with baby plants in them.

Most ferns produce only one kind of spore, and all the prothalli (see page 34) are alike. Some ferns, however, produce separate male and female spores, which develop into male and female prothalli. This is what happens in the conifers, but the prothalli are very much reduced and do not lead a separate existence.

The pollen grains are really the male spores. The male prothalli are simply the pollen tubes and the few cells inside them. The ovules are the female spore capsules and the female spores are represented by cells inside the ovules. These spores are never released from their capsules. Only one spore grows in each ovule, and it grows into a bunch of cells representing the prothallus. One cell becomes the egg cell—as it does on the fern prothallus—and is fertilised by a male cell.

In ferns, the male cell has to swim towards the egg cell. Free water is therefore needed, and a drought can prevent fertilisation. By retaining the female spore and prothallus inside the ovule, the conifers avoid the need for free water and reproduction becomes a much less chancy affair. There is always a breeze to carry the pollen to the female cones, and the growth of the pollen tube completes the journey to the egg cell. The early growth of the young plant takes place inside the parent, where it is protected. It is well supplied with food and it is a fairly sturdy young plant by the time it has to look after itself.

Cycads

The cycads are strange plants with many features part way between ferns and conifers. They have seeds, however, and clearly belong with the conifers. They grow in the warmer parts of the world. Most of them have short, stout trunks crowned with tufts of large fern-like leaves. The pollen and seeds are carried in large cones, although the female cones of one species are little more than bunches of ordinary leaves. Male and female cones are carried on separate plants, and the wind blows the pollen from plant to plant. When the pollen grain reaches the ovule it puts out a pollen tube but it does not grow towards the egg cell. It bursts and releases the male cell which actually swims to the egg cell.

like leaves allow the trees to survive the summer droughts.

Coniferous trees grow more quickly than the broad-leaved or flowering trees and this is why the conifers are planted so widely for timber. Their wood is softer than that of most broad-leaved trees, and coniferous timbers are generally known as softwoods. Most of them are less durable than the hardwoods and they rot more quickly, although the western red cedar (not a true cedar) contains natural preservatives that make it resistant to nearly all wood-rotting organisms. This timber is often used for making greenhouses because it does not mind the damp atmosphere.

The Conifer's Life History

Reproduction in the conifers centres on the cones. There are two types of cones—male cones and female cones—and they start to grow in the spring. Both types are normally found on one tree. The male cones are small and yellowish. They are usually rather inconspicuous, but they are easily seen in the pine because they are grouped into large clusters at the bases of the new shoots. The female cones develop near the tips of the new shoots and they are red to start with.

Each cone consists of a series of overlapping scales which are really very special kinds of leaves. The scales of the male cone carry pollen sacs, which are full of yellowish pollen. The female cone scales each carry a pair of ovules which will eventually become the seeds. In dry weather the scales of the cones separate a little and pollen escapes from the male cones. Blown by the wind, the pollen grains enter the female cones. The cones then close up again. The male cones have finished their jobs and they soon wither, but the female cones begin to grow. They become green and then brown and woody. The pollen grains trapped inside the female cones send out fine tubes which grow into the ovules. Inside each ovule there is an egg cell, and a tiny cell from the pollen grain joins with this egg cell to form an embryo plant. Food material is deposited around it and it gradually forms a root and a little shoot with several leaves. While this is going on the ovule wall becomes hard. The whole thing—embryo plant plus food reserve plus coat—is now a seed. The seed coat has a thin 'wing' growing out from one side and, when the cone opens, the seed can float away on the wind. If it lands in suitable soil it will grow into a new conifer. The whole process, from cone formation to the release of the seeds, takes just over a year in most conifers. Pine trees, however, take two years for their cones and seeds to ripen.

White Pine

Spanish Fir

Scots Pine

Lawson

Cedar

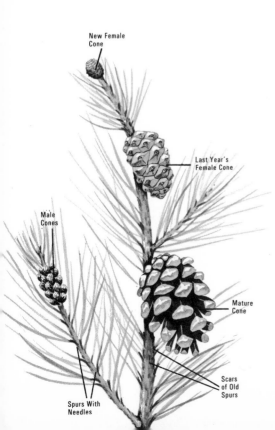

New Female Cone

Last Year's Female Cone

Male Cones

Mature Cone

Scars of Old Spurs

Spurs With Needles

37

Flowering Plants

The flowering plants are the most advanced of all plants. Just as the mammals dominate the animal world, so the flowering plants dominate the vegetable world. Nearly every plant you see around you is a flowering plant. There are about 250,000 different kinds, and they include all the most useful plants as well as the most attractive ones.

Flowering plants range in size from the tiny floating duckweeds to huge forest trees. Their structure varies a great deal but, with the exception of a few parasitic and saprophytic species (see page 47), they all contain chlorophyll and they all make food by photosynthesis. This process is the same in all green plants. Flowering plants also grow in much the same way as the non-flowering plants. It is their methods of reproduction that make the flowering plants so different from the mosses, ferns, and conifers.

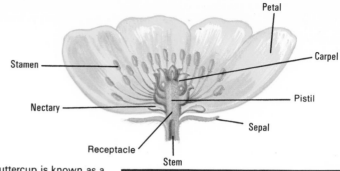

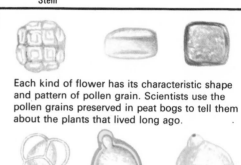

The buttercup is known as a perfect flower because it has all four kinds of floral organs in it.

Most flowers are pollinated by insects. The insects are attracted by nectar, scent, and brightly colored petals as in this cactus flower

Each kind of flower has its characteristic shape and pattern of pollen grain. Scientists use the pollen grains preserved in peat bogs to tell them about the plants that lived long ago.

The previous chapter describes how the conifers have overcome the need for free water during reproduction. This advance has been maintained in the flowering plants, and other improvements have led to better and less wasteful methods of reproduction.

The Structure of the Flower

Conifers produce 'naked' seeds on the scales of their cones. Flowering plants, however, enclose their seeds in containers called fruits. Both the seeds and the fruits are produced by the flowers, and each part of the flower has a particular job to do.

Flowers are composed of very special leaves of which there are four main kinds. Around the outside of a flower there are usually a number of greenish leaves called sepals. In some flowers these are joined together. In others they are separate or free. When the flower is still in bud the sepals surround the other parts and protect them. When the flower opens the sepals bend

Right: Some plants, such as tulips, have only one flower at the top of the stem, but most have 'heads' of flowers with several flowers on one main stem. These heads are called inflorescences. There are a number of different types. The flower on a thistle or a dandelion plant is composed of hundreds of small flowers.

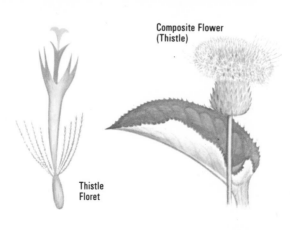

back, or they may even fall off in some flowers. Just inside the sepals we find the petals. These vary a great deal in color, shape and arrangement. They are often joined together. The petals are normally large and brightly colored, and their job is to attract insects to the flowers. Inside the petals there are normally a number of pin-like objects. These are the stamens—the male parts of the flower—and their job is to produce pollen. The pollen is a yellow dust and it is made in the little anthers at the tips of the stamens. Right in the center of the flower we find one or more carpels. These are the female parts of the flower. Each carpel contains one or more ovules which will eventually become seeds. The carpels may be joined together or they may be free. Each one has a sticky top called a stigma, which may be raised up on a slender stalk.

Most flowers contain stamens and carpels, but there are some which contain only one or the other. Marrow flowers, for example, are either male or female—they have stamens or carpels, but not both. There are even some species which have male and female flowers on different plants. Examples include willows and holly trees. Only the female trees will bear fruit and seeds, so many holly trees will never carry berries.

Pollination

Before a flower can form seeds it must be pollinated. This means that pollen of the right kind must fall on to the stigmas of the carpels and begin to grow. Some flowers are self-pollinated, that is pollen falls from the stamens to the stigmas of the same flower. Most flowers, however, have some way of preventing or discouraging self-pollination. The stamens may ripen and scatter their pollen before the stigmas are ready, or the stigmas may ripen before the stamens are ready. The flowers may also be self-sterile, which means that the pollen will not grow on the stigmas of the same flower. These flowers must be cross-pollinated with pollen from another flower of the same kind.

Wind pollinated flowers, such as these plantains, are generally drably colored and they usually have long hanging stamens which scatter pollen in the breeze.

Some flowers can be pollinated only by certain insects. The sweet pea can be pollinated only by an insect heavy enough to weigh down the keel petal and get into the flower.

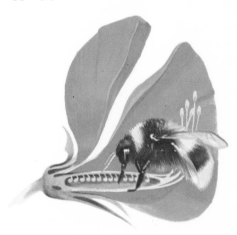

The petals are the most obvious parts of the flower at first. Their job is to attract insects. After pollination, however, the petals wither and the carpels become obvious as they swell up and become fruits.

Some plants, including grasses and many trees, rely on the wind to carry their pollen from flower to flower. Such wind-pollinated plants generally have small and inconspicuous flowers, often without any petals. The stamens and carpels are often in separate flowers, and the male flowers often hang in long clusters called catkins. These scatter clouds of pollen when blown by the wind. The stigmas of the wind-pollinated plants are usually branched and feathery, and they therefore have a better chance of trapping pollen.

Carpels
Containing
Ovules

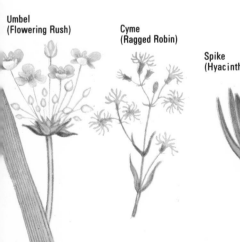

Umbel
(Flowering Rush)

Cyme
(Ragged Robin)

Spike
(Hyacinth)

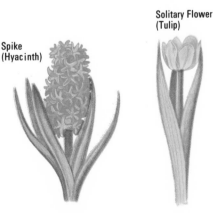

Solitary Flower
(Tulip)

Wind pollination is a very wasteful method because large amounts of pollen never get to other flowers at all. Most flowering plants rely on insects to carry their pollen from flower to flower. The insects are attracted to the flowers by the brightly colored petals and also by the scent and the sugary nectar that many flowers produce. Most of the insects are interested only in the nectar, and the flower's stamens and stigmas are cleverly placed so that the insects must brush against them as they reach for the nectar. The insects thus get dusted with pollen

39

Oats

Strawberry

Blackberry

Poppy

Horse Chestnut

Pea

Tomato

Rose

Bean

Apple

A selection of fruits. The apple, strawberry, and rose hip are false fruits because they are not formed from the carpels alone. A fruit may contain only one seed, but most fruits contain more than one.

The seeds of the poppy and snapdragon are scattered as the capsules are rocked by the wind. The seeds of the stock and violet are shot out as the fruits dry out and shrink.

down the tube and enter an ovule. One of them pairs up with the egg cell in the ovule. The egg cell is now said to be fertilized, and it rapidly grows into an embryo. This is a tiny plant, with root, shoot, and either one or two leaves. Food materials are deposited in or around these leaves, and the wall of the ovule becomes tough. The whole thing is now a seed.

While the seed or seeds have been growing, the carpel around them has been growing and changing too. It has become the fruit. Fruits come in many shapes and sizes. Some are hard and woody, some are soft and juicy, some are tough and leathery. They all contain one or more seeds. Most fruits are formed from the carpel or carpels alone, but some are formed from other parts of the flower. Apples and pears, for example, are formed mainly from the flower stalk. The

Many juicy fruits are eaten by birds. The seeds are then scattered far and wide.

which they then carry to the stigmas of another flower. Because the pollen is carried directly from flower to flower, the plants do not have to make so much. Very little is wasted.

The Formation of Fruits and Seeds

When a pollen grain falls on to a ripe stigma of the right kind it begins to grow. It sends out a thin tube which forces its way down through the stigma and into the carpel. Two minute cells from the pollen grain move

carpels of the apple flower are concealed inside the flower stalk, below the petals. After fertilization the flower stalk swells up to form the fruit. The original carpels become the core of the fruit, with the seeds inside them.

Scattering Fruits and Seeds

When the seeds are ripe the parent plant must release them so that they can grow. It would not be much good if the seeds simply fell beneath the parent plants, because they would be overcrowded there and the parent plants would take all the water and minerals from the soil. Most plants have some method for ensuring that the seeds are carried some way away. The fruits play a large part in this scattering or dispersal of the seeds.

Some fruits, including the pods of peas and beans, dry out as they ripen. When com-

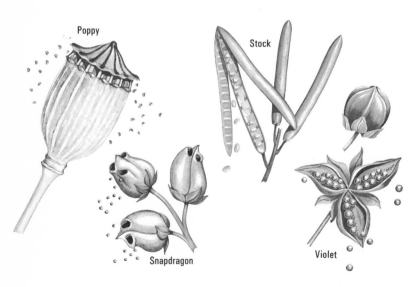

Poppy

Stock

Snapdragon

Violet

pletely ripe, the pods burst open and throw out their seeds. Some make quite a noise in doing so, and the seeds may travel several yards. Many plants rely on the wind to scatter their seeds. Very small seeds are blown about by the wind quite easily. Larger fruits and seeds may have feathery outgrowths, such as the parachutes of dandelions and thistles. Many trees carry fruits which have wing-like flaps. When the fruits fall they are borne away on the wind and come to earth some way away from the parent tree.

Juicy fruits are attractive to birds and other animals, which eat the fleshy parts and spit out the seeds. The smaller seeds may pass right through the birds' food canal and come out unharmed with the droppings. In both examples, the seeds are scattered far away from the parent plant. Many fruits have small hooks on them and the hooks catch into the fur of animals. By the time the animal shakes off the fruit it might be several miles away.

Germination

When the seed has fallen on suitable ground, and when the temperature is high enough, it will begin to grow, or germinate.

Composite flowers, such as thistles and dandelions, produce fruits with little 'parachutes' of hairs. These carry the fruits away very efficiently.

The fruit of corn and other grasses consists of nothing more than a thin case around the starch-filled seed. The whole thing is called a grain. When it reaches a warm and moist place it will start to germinate. Like all seeds, it contains enough food to last the seedling until it can make food for itself. The root is the first to grow out from the seed. It begins to absorb water and then the shoot grows up above the ground. When the leaves open out the young plant can make its own food.

It will absorb water and swell up, the seed coat will burst, and out will come the little root. No matter which way up the seed was lying in the ground, the root will always grow downwards and anchor the young plant in the ground. The young shoot soon follows the root out of the seed and it grows upwards to the light. The seed leaves may grow up as well, although they often stay inside the seed. While the seed is germinating it relies on the food stored up in it by the parent plant. Then, when its own shoot and leaves have grown up into the light, it starts to make food for itself and becomes an independent plant. In time it will produce flowers and seeds of its own and start the whole cycle off again.

Annuals, Biennials and Perennials

Many plants live for only one season. They germinate as seeds in the spring, flower in the summer, and then die. They are called annual plants. Some of them actually have several generations in one year, taking only a few weeks to grow and produce their own seeds. Most of the annual plants are rather small. Biennial plants take two seasons to complete their life histories. They grow from seed in the first season and store up supplies of food. These food stores keep them going through the winter and enable them to produce flowers and seeds in the following spring or summer. After flowering the plants die. Perennial plants go on living for several years—hundreds of years for some species of trees. Most of them flower many times during their lives. All trees are perennials.

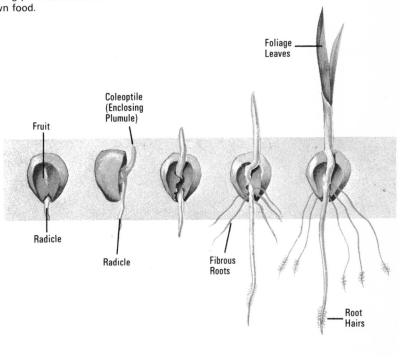

Flowering Trees

Trees are plants, just as much as the daisies and daffodils and cabbages are plants. The main difference is that trees are much taller than other plants and therefore need strong woody trunks to support them. The trunk takes a long time to grow, however, and young trees are much like any other young plants when they first leave their seeds.

All modern trees produce seeds, but there are two main kinds of trees—the conifers and the flowering trees. The conifers carry their seeds in cones, and they are described on page 36. All the other trees are flowering trees, although their flowers may not always be conspicuous.

Flowering trees belong to a wide range of families. Some trees, such as the elms, have families to themselves. Others share their families with small herbaceous plants. The violet family, for example, contains many large tropical trees as well as the familiar violets. The pea family contains the herbaceous peas and clovers, the shrubby gorse and broom, and the attractive but poisonous laburnam tree. These plants all look very different, but you can see the connection if you look at the flowers.

Some flowering trees have very attractive flowers. Examples include apple, almond, lilac, horse chestnut, magnolia, and pussy willow. Some of these are cultivated in parks and gardens. Many other trees, however, have small and rather dull flowers which are easily overlooked. Examples of these trees include oak, ash, elm, maple, holly, and poplar. It might surprise some people to learn that these trees have flowers at all.

Reproduction

Most of the inconspicuous flowers are pollinated by the wind, although the lime or linden tree is a notable exception. Its hanging bunches of yellowish flowers produce abundant scent and nectar and the trees literally buzz with bees in June and July. The wind-pollinated flowers usually open early in the year, often before the leaves have started to open. The hazel is a good example of an early flowering tree.

Right: A selection of flowering trees. 1. Japanese Maple, 2. Lombard Poplar, 3. Gray Birch, 4. Beech, 5. Common Ash, 6. White Oak, 7. White Poplar, 8. Goat Willow, 9. Purple Beech, 10. Sycamore Maple.

42

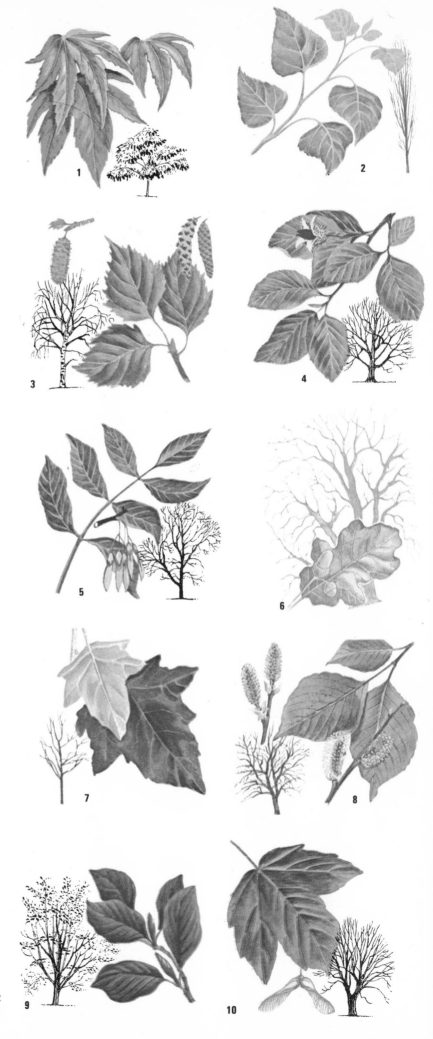

All flowering trees carry their seeds in fruits of some kind or other. Many trees bear juicy fruits which are brightly colored. These are eaten by birds, which then scatter the seeds over a wide area. Other trees may bear woody fruits. These may have 'wings' which help them to drift away, or they may simply fall to the ground to be carried away by squirrels and other rodents. Poplars and willows have hairy seeds which drift away on the wind.

Trees and Shrubs

The gorse, the gooseberry, and the rhododendron are clearly not herbaceous plants because they have woody stems. But we can hardly call them trees either. Trees are woody plants which *normally* have one main stem or trunk carrying branches some way above the ground. The gorse and the other plants mentioned above have several stems of more or less the same size, all arising at about ground level. Plants which *normally* grow in this way are called shrubs.

A bush is any woody plant with several stems arising at about ground level. It may be a true shrub, or it may be a tree which has been cut down and allowed to send up side shoots from the base.

Right: 1. London Plane, 2. Tulip Tree, 3. Linden, 4. Spanish Chestnut, 5. Walnut, 6. Sugar Maple.

Below: The attractive rhododendron is a typical shrub.

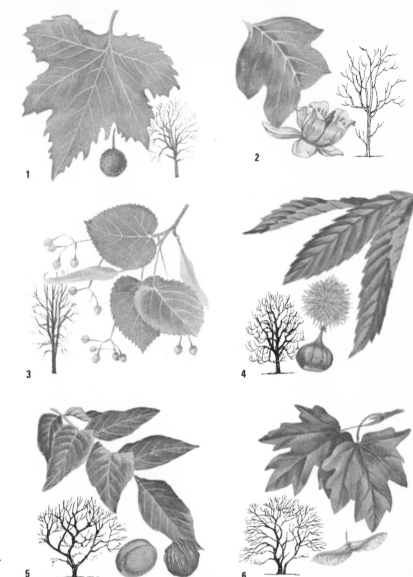

Grasses

The grass family, known as the Gramineae, is the most important of all plant families as far as man is concerned. From this one family come all our cereal crops. Corn, wheat, oats, barley, rice, and many other cereals have all been developed from various wild grasses. These cereals provide flour and they are the staple foods in most parts of the world. About two-thirds of the world's sugar comes from the tall grass that we call sugar cane. As well as feeding us, the grasses also feed all our important meat-producing animals such as cattle and sheep and rabbits. Pigs, too, are fed largely on cereal grains.

If all the world's grasses were to disappear we would lose a very large proportion of our food materials, and we would also lose our lawns, parks, and playing fields. Much of the land surface would become a barren waste because the grasses play a very important part in binding the soil together and preventing it from being blown away. Marram grass and some other species are often planted on sand dunes to stabilize them and stop them from moving.

The grasses belong to the group of flowering plants called monocotyledons (see page 25) and, like the other members of the group, they have narrow leaves with parallel veins. They are nearly all small herbaceous (non-woody) plants, although some of the bamboos are very woody and reach heights of 100 feet. Sugar cane is also a large grass, and so are some of the reeds whose tough stems are still used for thatching houses in some places.

The stems of most grasses are normally very short and close to the ground. Only the leaves grow upwards at first, and if you look at a lawn or a meadow in the spring you will see that it is nearly all leaf. This is what enables the grasses to stand up to repeated mowing or grazing. The stems are not normally removed and they can easily send up a new crop of leaves. The stems also send out a number of side shoots at ground level. These are called tillers and they enable the grass to cover the ground and form a turf quite quickly. If the main stem is damaged or removed the plant will produce even more tillers, and so mowing, trampling, and grazing actually help the grass to form a thick turf. Many grasses also produce runners or rhizomes (see page 24) and spread even more quickly over the ground. A dense mat of roots develops underneath the turf and this helps to bind the soil together.

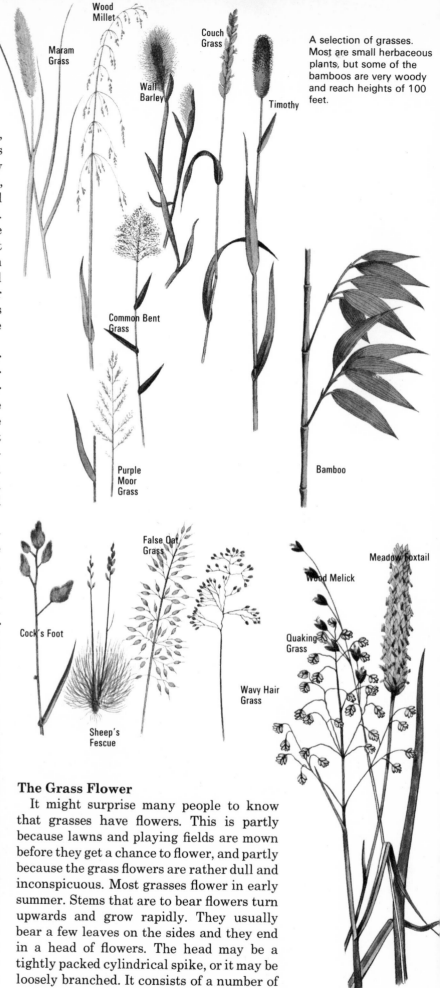

A selection of grasses. Most are small herbaceous plants, but some of the bamboos are very woody and reach heights of 100 feet.

The Grass Flower

It might surprise many people to know that grasses have flowers. This is partly because lawns and playing fields are mown before they get a chance to flower, and partly because the grass flowers are rather dull and inconspicuous. Most grasses flower in early summer. Stems that are to bear flowers turn upwards and grow rapidly. They usually bear a few leaves on the sides and they end in a head of flowers. The head may be a tightly packed cylindrical spike, or it may be loosely branched. It consists of a number of small oval structures called spikelets, each of which contains one or more flowers.

The flowers have no petals, no bright

44

colors, and no scent. They consist simply of the stamens and carpels, concealed among the greenish scales that make up the spikelets. When they are ripe the stamens hang out of the spikelets and scatter their pollen to the wind. Some of the pollen reaches the feathery stigmas of the grasses but, as with all wind-pollinated flowers, much of it is wasted. Grass pollen blowing about in the air is mainly responsible for producing hay fever. Many people are sensitive to the pollen blowing about in the air and it affects their noses, giving them all the symptoms of a nasty cold.

After a grass flower has been pollinated its ovary develops into the grain. The grain contains a food reserve composed mainly of starch and this is what makes the grasses so important as human food. Several wild species have been cultivated to produce bigger and better yields of starch (flour). The plant breeders are also interested in the type of flour produced as well as the quantity, because flour that is good for making bread is not good for making biscuits or macaroni. There are therefore many different varieties of wheat and other cereals in existence today.

Cultivated grasses are far and away the world's most important crops. In temperate countries wheat is the chief cereal crop. The great wheat-fields of the North American and Russian plains often stretch unbroken to the horizon. Harvesting the crop is a highly mechanized affair carried out by teams of combine harvesters.
Rice is the chief cereal crop in much of Asia. Often it is planted, gathered and threshed by hand, and the farmers can produce hardly enough grain for their own families.
Both wheat and rice have been grown for many centuries and through selective breeding the grain yields are now far greater than the original wild strains.

45

Insect-Eating Plants

Although most plants are able to make their own food from simple substances in the air and the soil, a number of plants supplement their food supplies by catching and digesting insects. The insects provide additional nitrates and other minerals, and so plants with insect-catching habits are often able to survive on poor soils. Most of the insect-eating plants actually live in peat bogs and similar wet places. There are relatively few soil bacteria in such places and the plants do not decay very rapidly. The partially decomposed remains of mosses and grasses accumulate to form the peat, and very little mineral matter becomes available for growing plants.

Insects are caught by specially modified leaves or parts of leaves, and the plants then digest them with digestive juices not unlike those found in our own intestines. All of the insect-eating plants can survive without catching insects. They possess chlorophyll and they can make their own food. If they are grown in good soil they will grow and flower quite normally even if they cannot catch insects. When grown on their normal soil, however, they need the extra minerals and they do not flower well if they are unable to catch insects.

Pitcher plants trap insects in various kinds of 'jars' which are formed from the leaves. The 'jars' contain fluids which attract the insects and then digest them when they fall in. The digested materials are absorbed by the plant. Some pitcher plants have lids on their pitchers. The lids prevent rain from getting in and diluting the juices.

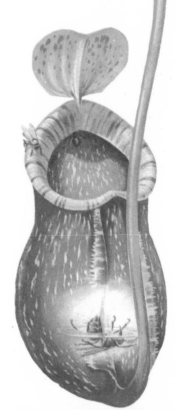

The sundew attracts insects with the shiny droplets on its leaf tentacles. The insects are trapped by the sticky fluid and the leaf closes up while the insect is digested.

Parasitic Plants

The mistletoe is best known as a Christmas decoration, although in some parts of Europe it is also believed to protect homes from fire and lightning. Its name in Switzerland actually means 'thunder broom'. But apart from its various legendary properties, the mistletoe is interesting because of its strange way of life. It has no roots of its own and it grows on the branches of other trees, especially apples and poplars. It sends suckers into the branches and these suckers take in water and mineral salts. The mistletoe is therefore a parasite—a 'lazy' organism that gets its food from another plant or animal without having to work very hard for it. The mistletoe is actually only a partial parasite, because it does contain its own chlorophyll and it can make its own food when once it has 'stolen' water and minerals from its host plant.

Another very common partial parasite is the eyebright, a little plant with nettle-like leaves and pink and white flowers. The eyebright has roots of its own, but they are not very efficient and they obtain most of their water and minerals by sending suckers into neighboring grass roots. The eye-

Looking like pink strands of cotton, the dodder twines around a heather plant and sends suckers into it to obtain food. The dodder's flower clusters are clearly seen.

bright can grow without the help of the grasses, but it then makes only a small plant and it does not flower well.

A number of flowering plants have taken the parasitic habit a stage further and have become complete parasites. They have no chlorophyll and they can make no food for themselves: they take everything they need from other plants. The only thing these parasites have to do for themselves is produce flowers and seeds. All their energies are concentrated in this direction and all other parts of the plant are reduced.

The dodder is a well known parasite whose tangled red stems often smother low growing shrubs and grass in the summer. The thread-like stems twine around the host stems and send suckers into them. The suckers absorb food materials and pass them into the parasite. The upper parts of affected host plants often become discolored and they may die. The dodder's leaves are reduced to tiny scales on the stem, and the plant does not even have a root after the first few days of its life. Bunches of pale pink flowers cluster around the stems and each flower produces a vast number of tiny seeds. Each seed contains a minute embryo and a small store of food. When conditions are suitable the seed germinates and the thread-like stem emerges together with a

temporary root. The stem lengthens rapidly and creeps over the ground. If it finds a suitable host it will coil around it and send in suckers. The root will then die, leaving the dodder completely dependent upon its host. If the seedling does not find a host within a few days its food reserve will give out and it will die.

The World's Biggest Flower

One of the most amazing plants of all is the Rafflesia, which grows in Malaya and neighboring countries. It has no roots, stems, or leaves but it produces the world's biggest flower. Rafflesia is a parasite of various vines. A seed becomes wedged in the bark and it sends a number of hair-like threads into the wood. These threads take food from the host and they soon produce a little flower bud at the surface. The bud takes about nine months to grow and then it opens to produce a huge flower, sometimes more than two feet across and weighing about 20 pounds.

Parasites and Saprophytes

A parasite is a plant or an animal that lives in or on another living creature, taking food from it without giving anything in return. The creature attacked is called the *host* and it may be a plant or an animal. The parasite often kills the host.

A saprophyte is a plant that takes food from dead and decaying organisms. Saprophytes do not have any chlorophyll. Most fungi are saprophytes, taking food from decaying leaves and so on.

Mistletoe is an attractive plant frequently used for decorative purposes. It is only a partial parasite because it does make some of its own food. But it sends suckers into its host plant for its water supply, as shown lower left.

Right: The toothwort is a flowering plant without any chlorophyll. It gets all its food by sending suckers into the roots of trees such as the elm and hazel.

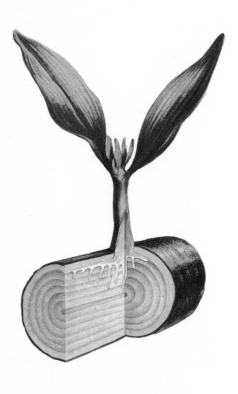

Right: Diagram showing how the dodder sends out suckers (haustoria) which penetrate the host plant's tissues and suck up food materials.

Climbing Plants

The taller a plant grows, the more chance it has of getting up above its neighbors and getting the full benefit of the sunlight. Tall plants, however, need sturdy stems to support them and this means that they grow rather slowly as a rule. But some plants have overcome this problem by climbing over their neighbors. These climbing plants do not need strong stems of their own and they can put all their efforts into upward growth. They grow very quickly, therefore. Climbing plants belong to many different families and they have developed many different methods of climbing. A few of the different methods are illustrated on this page.

The ivy is one of the best known climbing plants. The young stems creep over the ground and when they find a wall or a tree trunk they turn upwards. Small roots grow out from the stems and take a hold in little cracks and pores. They hold the slender ivy stems tightly against the support and allow them to continue their upward growth. The ivy does not take food from the tree on which it grows, but a lot of ivy certainly weakens a tree.

Many plants climb by simply twisting or twining themselves around their neighbors. Honeysuckle and bindweed are common examples, and so are the runner beans or stick beans that we grow in our gardens. Some species twine clockwise and others twine anticlockwise.

Peas and several other plants possess slender outgrowths called *tendrils*. They are often specially modified leaves. They coil tightly around any twig they happen to touch and they hold the climber firmly in position.

Right: A selection of climbing plants. Clematis climbs by twining its leaf stalks (petioles) about supports. Honeysuckle is the best known of the 'clockwise' twiners and Convolvulus or bindweed one of the most common 'anticlockwise' climbers. Ivy climbs by means of aerial roots developed on the lower part of the stems. The tendrils of the Virginia Creeper may develop into adhesive discs which enable the plant to climb.

Below: Ivy is a common climber of both trees and walls. Its aerial roots secrete a sticky fluid which glues them to the support.

Most people have scratched themselves on the curved prickles of roses or blackberries at some time or other, but these prickles are not basically for the protection of the plant. They are climbing aids. The prickles hook over the twigs of other plants and support the briar, or bramble.

Protection in Plants

Some plants are carefully avoided by man and other animals because they are unpleasant to touch. Many of them have sharp thorns or prickles capable of inflicting painful wounds. Others possess poisonous hairs that produce irritating rashes.

Thorns and Spines

When picking blackberries or gooseberries we always try to avoid the prickles and we do not pay any particular attention to them. A careful look at them, however, will show that they are not all alike and that there are in fact several different types of prickles.

The hawthorn and the blackthorn or sloe tree both have very unpleasant prickles, which you will probably have felt if you have tried to pick their attractive flowers or fruits. These prickles are actually dwarf branches. They develop from buds just like any other branch does, but they soon stop growing and they become sharply pointed. Prickles of this kind should really be called *thorns*. We can see that they are formed from special stems because they often carry leaves.

In some other plants the leaves themselves form the prickles. Prickles formed from or growing on the leaves are called spines. The barberry has some of its leaves modified as long, slender spines. Gooseberry and acacia spines are outgrowths from the leaf bases. Gorse bushes have both thorns and spines: the leaves are needle-like and short, pointed stems grow in their axils. Another form of spine develops when the edges of leaves are drawn out into sharp points. This occurs in holly and thistles, for example. The spines of cacti are not really understood. They might be special leaves, they might be dwarf stems, or they might simply be out-growths from the skin. Whatever they are, they

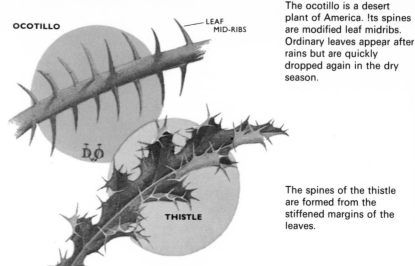

OCOTILLO — LEAF MID-RIBS

The ocotillo is a desert plant of America. Its spines are modified leaf midribs. Ordinary leaves appear after rains but are quickly dropped again in the dry season.

THISTLE

The spines of the thistle are formed from the stiffened margins of the leaves.

always arise on little 'cushions' called *areoles* and these will always distinguish a cactus from any similar prickly plant. No other kind of plant has areoles.

The curved prickles of roses and blackberries are neither stems nor leaves. They are outgrowths from the skin and they are often called *emergences*. Although they certainly do protect the plants, their main job is to help the plants climb (See page 49).

The Function of Prickles

Most prickly plants live in dry or semi-dry regions, or else in places where water drains readily from the soil. The plants must not lose too much water by evaporation, and one

The prickles of the blackberry are climbing aids. The function of the cactus spines is unknown.

BARBERRY

The spines of the gooseberry and acacia are outgrowths from the leaf bases. Those of the barberry are actually modified leaves, while the gorse has both thorns and leaf spines.

TRUE LEAF (SPINOSE)

SPINOSE LEAVES ON DWARF BRANCH

DWARF BRANCH

GOOSEBERRY

GORSE

FALSE ACACIA

of the commonest ways of cutting down water loss is by reducing the size of the leaves. The leaves tend to become spiny and, the drier it is, the more woody they become. Prickles therefore seem to be mainly concerned with reduction of water loss, but they certainly protect the plants from animals as well. Some spines, however, are probably basically for protection. The holly, for example, has very spiny leaves on its lower branches—where animals can browse—but its higher leaves are often quite smooth.

Stinging Hairs

Most plants have some sort of hair on their stems and leaves. Some of these hairs are most attractive structures when looked at under a lens or microscope. They come in many shapes and sizes. Long coats of hair may keep a plant warm in a cold climate, or they may keep the sun off it in a hot climate. The hairs also help to cut down the loss of water. The goosegrass or cleavers has tiny hooked hairs which help it to climb, just like the prickles of a wild rose.

Many hairs secrete oils and other fluids,

The thorns of the hawthorn are really dwarf branches, shown by the fact that they often carry leaves.

DWARF BRANCH

HAWTHORN

Left: The curved prickles of roses both protect the plant and help it to climb.

and the stinging hairs come under this heading. The stinging hairs of the stinging nettle grow on the stem and leaves. They are strengthened with a glassy material called silica and they end in little spherical swellings. These break off when something brushes against the nettle and the sharp needle-like point sticks into the animal. Poison then flows into the wound from the cavity of the hair and it sets up an irritating rash. The poison is a complicated mixture of substances and there is no real antidote.

Several other plants have stinging hairs. The fever nettle of West Africa has a very strong poison that can make a man quite ill. Even elephants are said to avoid this plant.

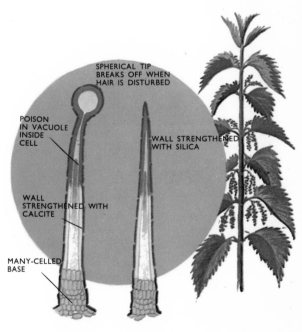

The stinging hairs of the nettle end in a little round swelling. This breaks off when the plant is touched, leaving a sharp needle which injects poison into your skin.

SPHERICAL TIP BREAKS OFF WHEN HAIR IS DISTURBED

POISON IN VACUOLE INSIDE CELL

WALL STRENGTHENED WITH SILICA

WALL STRENGTHENED WITH CALCITE

MANY-CELLED BASE

Dictionary of Flower Families

Arrowhead family (Alismataceae) A family of monocotyledons living in water or wet places. The plants have erect and/or floating leaves and some of them have submerged leaves as well. The flowers, usually carried on whorls of slender branches, have three rounded petals and three sepals. The petals are usually pink or white, often with a darker patch at the base. There are about 80 species, mainly in the Northern Hemisphere. They include the arrowhead, named from its leaves, and the water plantains.

Arum family (Araceae) A family of monocotyledons containing over 1,000 species. Most are herbs, but some are woody climbers. The flowers are normally small and they are crowded together on a fleshy stalk called a spadix. This is very often surrounded by a leaf-like sheath called the spathe. There are often separate male and female flowers, the male ones being above the female ones on the spike. The family has its headquarters in the forests of South America, but it is widely distributed. One of the commonest European species is the cuckoo pint, also known as lords-and-ladies. It grows in shady woods and hedgerows and is most obvious in the autumn when its bright red fruits are ripe. The flowers have an unpleasant smell which attracts flies. The insects crawl down to the base of the spathe and become trapped there by a ring of hairs. The female flowers open first and they are pollinated by the flies. The male flowers then open and dust the flies with fresh pollen. The hairs wither and the flies escape, only to be lured into another arum where they pollinate another batch of female flowers.

Balsa family (Bombacaceae) This is a family of trees found in many parts of the Southern Hemisphere, although the balsa itself is found only in South America. The balsa is an evergreen tree growing in the forests of Ecuador and Peru. It has large white flowers, but it is best known for its very light wood. Another famous member of the family is the strange baobab, found in Africa and Australia. The trunk is quite short but it may be 30 feet across. It bears a crown of branches at the top and looks for all the world as if it has been uprooted and thrust into the ground again upside down. The trunk is full of water. The large white flowers are usually hidden among the leaves, but the trees are deciduous and when the leaves fall they reveal the large gourd-like fruits. These contain a pleasant-tasting pulp. The durian tree of South East Asia is another member of the family with edible fruits.

Balsam family (Balsaminaceae) A family of dicotyledonous herbs with succulent stems. There are over 400 different kinds, most of them living in tropical Africa and Asia. Most of the species belong to the genus *Impatiens,* so named because of the 'impatient' way in which the fruits burst when touched. When they burst they throw their seeds out. The flowers are frequently pink or orange and they are very irregular. There are usually five petals and three sepals, although some of the sepals look like petals. Many of the species live along river banks and in other damp places.

Banana family (Musaceae) A family of monocotyledons found mainly in the tropics. There are about 200 species in the family, in-

Sweet chestnuts showing the spiny cups surrounding the fruits.

Arrowhead, showing the characteristic leaf shape.
An arum lily cut away to show the clusters of male and female flowers at the bottom of the spike.

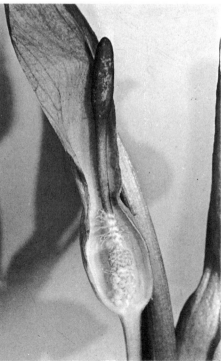

cluding the striking bird-of-paradise flowers of South Africa. Although we often talk of banana 'trees', the banana plant is not really a tree at all. Cultivated banana plants may reach a height of 40 feet and they may be a foot across, but the 'trunk' is composed entirely of leaf stalks wrapped around a very slender stem. The leaf blades may be 12 feet long and they form a large crown at the top of the plant. The real stem emerges from the crown and carries the flowers. The flowers start out in a bunch of purple bracts about the size of a football. The bracts turn back one by one to reveal clusters of tubular flowers, each with six yellow petals. Each cluster of flowers produces a hand of bananas, and there are dozens of hands on a stem. There is only one flowering stem per 'tree' and the whole shoot is finished when it has produced its fruit. New shoots are always being produced, however, from the underground rhizome of the plant.

Bedstraw family (Rubiaceae) A large family of dicotyledons including both herbs and woody plants. The leaves are either in opposite pairs or else in whorls around the stem. The flowers are small, often white or yellow, and have four or five petals. Most of the 5,000 species live in the tropics, but some of the bedstraws (*Galium* species) are common in cooler regions. One of the best known of these is the goosegrass or cleavers. The coffee plant also belongs to this family.

Beech family (Fagaceae) A family of trees and shrubs including oaks and sweet chestnuts as well as beeches. There are over 800 species, more than 500 of them being oaks. They are found mainly in the Northern Hemisphere, although the southern beeches (*Nothofagus*) are found in South America, Australia and New Zealand. There are both evergreen and deciduous species. There are separate male and female flowers. The male flowers are usually in catkins or in tassel-like clusters. They are not very conspicuous and their pollen is scattered by the wind. The female flowers are arranged in small groups and they give rise to the nuts. The latter are partly surrounded by scaly or spiny 'cups'

Beet family (Chenopodiaceae) A family of dicotyledons consisting mainly of herbs and shrubs. The leaves are often covered with swollen hairs and they appear downy. Several species have succulent leaves. The flowers are always small and green. Most of the species live in open situations, often in quite dry places. Many live on sand dunes by the sea, and several are common weeds in cultivated land. Beetroot and sugarbeet are both cultivated varieties of the wild beet that grows on coastal dunes and cliffs.

Begonia family (Begoniaceae) A family of succulent dicotyledons found mainly in the tropical forests of South America and the Pacific islands. Their large and colorful flowers have made them popular subjects for growing indoors. The male and female organs are carried in separate flowers. The female flowers have between two and five petal-like parts, formed from both sepals and petals. The male flowers have two petals and two larger, petal-like sepals. The many stamens also add to the appearance of the flower.

Bellflower family (Campanulaceae) A large family of dicotyledons whose members nearly all have bell-shaped flowers. The five petals are all joined together to form the bell, but their

52

outer ends are separated to form five lobes. The majority of the flowers are blue or purple. Nearly all the 700 species are herbaceous plants and they are found in most parts of the world. Examples include the harebell (the Scottish bluebell), the clustered bellflower, and the cultivated Canterbury bells.

Bindweed family (Convolvulaceae) A family of dicotyledons including herbs, shrubs, and some trees. Many are climbers and the plants often contain a white, milky juice. The flowers are regular, with the five petals joined to form a funnel or a bell. There are five separate sepals. There are about 1,000 different species in the family. Most of them live in the tropics, but there are some common ones in cooler regions. As well as the familiar bindweeds, the family contains the parasitic dodders, the morning glory, and the sweet potato. The latter comes from South America, where its sweet, starch-filled roots are an important food.

Birch family (Betulaceae) A family of deciduous trees and shrubs containing the birches and alders. They are found mainly in the temperate regions of the Northern Hemisphere. Some of the species, such as the dwarf birch, reach far into the tundra. The male flowers are carried in hanging catkins and their pollen is scattered in the wind. The female flowers are carried in little upright catkins. These are quite slender in the birch, but they are cone-like in the alder. The fruits have little wings and they are carried away by the wind. The female cones of the alder remain on the trees long after the fruits have gone.

Black bryony family (Dioscoreaceae) A family of monocotyledons, although its members have broad leaves more like those of the dicotyledons. Most of the 170 species are climbing plants. Some are woody and some are herbaceous. Most of them have tuberous underground stems. The flowers are small and pale, with male and female flowers usually on separate plants. Each flower has six petal-like parts, united at the base to form a short tube. Most of the species live in the tropics, but the black bryony is very common in Europe. It

Aubretia flowers showing the four petals characteristic of the cabbage family.

A birch twig showing male catkins (hanging) and female catkins.

The cacti have some of the most attractive flowers in the world, but many of the flowers last for only a few hours.

climbs over hedges and its bright berries make a pleasant sight in the autumn. Several species are cultivated in tropical lands. Their starch-filled root tubers and rhizomes are called yams.

Box family (Buxaceae) A small family of dicotyledons, with about 40 species scattered throughout the world. Most of the species belong to the genus called *Buxus* and they are nearly all evergreen shrubs. The flowers are small and green, consisting of four sepals without any petals. Male and female flowers are separate, but both kinds grow on one plant. The box is quite common on the limestone and chalk of southern England.

Broomrape family (Orobanchaceae) A family of dicotyledons whose members are all root parasites. They have no chlorophyll and they obtain their food by sending suckers into the roots of other plants. The aerial shoots are scaly and they are cream, pink, or brown in color. They carry dense clusters of flowers in

The lesser celandine, a member of the buttercup family.

which the petals are joined to form a twisted tube. Like all parasites, these plants produce huge numbers of seeds. Many of the species are pollinated by bumble bees. There are about 130 species, mainly in the Old World. Common species include the toothwort, which attaches itself to the roots of hazel and other trees, and the various kinds of broomrapes. The broomrapes parasitize clovers, thistles, and many other herbaceous and shrubby plants.

Buckbean family (Menyanthaceae) A small family of dicotyledons whose members are all marsh-loving plants. There are less than 40

These cactus flowers bloom briefly in the North American deserts.

Lady's smock, or cuckoo flower, showing the four petals arranged in the form of a cross. This is typical of the cabbage family.

Traveller's joy or wild clematis, a climbing member of the buttercup family.

species. The buckbean is widespread in the Northern Hemisphere. It grows in bogs and marshes and also around the edges of lakes and ponds. Its large shiny leaves each have three leaflets and they look like huge clover leaves sticking up out of the water. The flowers are pink or white, densely packed on tall stalks. The edges of the petals are divided up to form dense fringes. The family also contains the fringed waterlily, whose small yellow flowers are also fringed at the edges.

Buckthorn family (Rhamnaceae) A family of dicotyledonous trees and shrubs with about 500 members. The flowers are generally small and inconspicuous, often being greenish in color. The sepals form a little tube, with four or five lobes at the top. The petals, if there are any, are attached to the rim of this tube. The fruits are often fleshy and brightly coloured. The buckthorn and the alder buckthorn are common in Europe. The buckthorn is a thorny shrub, found mainly on limestone soils. Its fruits are black. The alder buckthorn is not thorny and its fruits are more of a purple-black.

Buddleia family (Loganiaceae) A family of dicotyledons found mainly in the tropical regions. Most of the 600 species are trees and shrubs. The four or five sepals are united into a tube. The four or five petals also form a tube, although their outer parts fold back to form a star-shaped flower. Many of the flowers are small, but they are grouped into dense heads. Members of the family include the cultivated buddleia—a native of China—and various plants from which we obtain the drugs strychnine and curare.

Buttercup family (Ranunculaceae) A large family of dicotyledons containing herbs, shrubs, and woody climbers. There are about 1,400 species, most of them found in the cooler regions of the Northern Hemisphere. Most of the plants in this family have a bitter taste, and many of them are very poisonous. The flowers are usually regularly arranged, although there are some irregular flowers in the family. All the parts of the flower are separate. Distinct sepals and petals are normally present, although some flowers have sepals that look just like petals. There are many stamens in the flowers. Among the many members of the family there are: buttercups, lesser celandines, anemones, marsh marigolds, delphiniums, columbines (granny's bonnets), and clematis.

Butterwort family (Lentibulariaceae) A small family of dicotyledons whose members all increase their food supplies by catching insects. There are about 250 species. Many live in marshes or in water, and some live as epiphytes on trees. The petals are united to form a hood-shaped tube. The flowers are often carried singly on slender stalks. The butterworts grow in bogs and wet grassland. They have rosettes of leaves which are covered with sticky secretions. Insects get stuck on the leaves and the leaf margins then role in a little until the insects have been digested. The bladderwort also belongs to this family. It grows in water and catches small water creatures in little balloon-shaped traps. There are about 200 kinds of bladderwort, most of them living in the tropics.

Cabbage family (Cruciferae) A large family of dicotyledons with about 2,000 species. Most of them live in the Northern Hemisphere. The majority are herbs, but there are some woody plants in the family. There are four petals, arranged in the form of a cross. This gives the family its name because crucifer means 'to carry a cross'. There are four sepals

55

flowers are pink, with pink sepals as well. The petals are united at the base. The flowers are followed by large fruits rather like footballs. Each fruit contains numerous seeds which are the cocoa beans. Cocoa is obtained from the roasted beans. Another member of the family is the kola tree of West Africa. The fruits, borne among the foliage in the normal way, yield seeds which are used to make refreshing drinks.

Cucumber family (Cucurbitaceae)   A family of dicotyledons with about 700 species, mainly found in tropical regions. Most of the species are herbaceous, although some are shrubs. Many of the species are climbers. The leaves are broad and their veins spread out like the fingers of a hand. The leaves and stems are often covered with coarse hairs. Male and female flowers are normally separate, but they often grow on one plant. The flower is trumpet-shaped, although the petals are not completely joined. The fruit is a berry or a rather large and thick-skinned object called a pepo. The cucumber is a pepo. As well as the cucumber, the family contains melons, marrows, pumpkins, gourds, and the white bryony. The latter is a common wild plant in the hedgerows of Britain and Europe. The flowers are small and greenish. Male and female flowers are carried on separate plants.

Currant family (Grossulariaceae)   A family of shrubs with about 150 species, all in the

as well, and usually six stamens. Two of the stamens have short stalks and the other four have long stalks. There are two carpels joined together, but all the other parts of the flower are separate. Most of the flowers are white, yellow, or pink. Among the members of the family are wild cabbage (and its many cultivated varieties), turnip, candytuft, honesty, lady's smock, water cress, wallflower, and aubretia. All of them contain strongly-flavored juices, particularly noticeable in the mustard plant.

Cactus family (Cactaceae)   A family of dicotyledons with about 1,700 members. With one possible exception, they all come from the New World. The stems are all thick and fleshy and modified for water storage. Leaves are generally absent, but the stems bear a number of large or small spines. These spines arise on little 'cushions' called areoles. Cactus flowers are very showy, with numerous petals and sepals all grading into one another. Most of the cacti live in dry places, but not all of them. The Christmas cactus and related species grow in the rain forests of South America.

Carrot family (Umbelliferae)   A large family of dicotyledons in which the flowers are carried in large heads called *umbels*. The individual flower stalks all leave the main stem at one point and, although the individual flowers are small, they form large circular patches. There are nearly 3,000 species, most of them in the cooler parts of the Northern Hemisphere. Nearly all of them are herbaceous plants, with ridged, hollow stems. The leaves are usually greatly divided and fern-like, and most of them are pleasantly scented. Among the members of the family are: carrot, hog-weed, parsnip, celery, and ground elder. Many herbs used in cooking also come from this family. They include parsley, dill, fennel, chervil, and coriander. Their value lies in their scented leaves.

Cocoa family (Sterculiaceae)   A small family of trees growing in the warmer parts of the world. The cocoa tree itself, often called the chocolate tree, is a native of South and Central America, although most of today's cocoa is grown in West Africa. The tree reaches a height of about 30 ft and carries large leathery leaves. The flowers are carried not on the branches, but on little 'cushions' on the main trunk. The

A cactus flower with numerous petals, greenish and sepal-like on the outside but brilliantly white in the center of the flower. The cactus below is a prickly pear covered with juicy fruits.

56

The daffodil family, illustrated by this sectioned snowdrop, is separated from the lily family because the seed box or ovary is outside the petals.

The sunflower is a composite, made up of many small florets. Only the outer florets have the broad yellow rays.

genus *Ribes*. The leaves are broad and often palmately lobed—like the fingers of a hand. The flowers are often small and inconspicuous, but they are followed by prominent berries. Several species are cultivated. They include red, white, and black currants as well as the gooseberry.

Daffodil family (Amaryllidaceae) A family of monocotyledons with about 500 species, nearly all of which have bulbs. Most of them come from temperate regions. The flowers, which are often solitary, are enclosed in a leafy or papery sheath before they open. There are six petals, joined to form a bell in the snow-

flakes but separate in daffodils and snow-drops. The cup or trumpet of the daffodil is just an outgrowth from the petals. The members of this family are very similar to the lilies, but the ovary (seed box) lies outside the petals. In the lilies the ovary is inside the petals. The family also includes the large spiky agaves of the American deserts.

Daisy family (Compositae) With about 15,000 species of dicotyledons, this is the largest of all flowering plant families. There are

The deadnettle family can be recognised by its rather square stems and the whorls of irregular flowers.

Tansy

Cornflower

The dandelion is another composite. Each floret produces its own fruit, which is borne away by a 'parachute' of hairs. Some more composites are shown below and to the right.

Sneezewort

Common daisy

trees, shrubs, herbs, and climbers in the family. They inhabit all possible places from ponds and streams to the coldest mountains and the driest deserts. But the flowers are remarkably alike. What we normally call the flower is actually a collection of many tiny flowers packed tightly into a single head. Each tiny flower is called a *floret* and it is made up of five little petals joined up to form a tube. It has five stamens, and a carpel at the base. The sepals are never leaf-like and they are usually represented by a ring of little hairs. After flowering they grow up to form the 'parachute' that carries away the fruit. The florets are of two main types, called *ligulate* and *tubular*. The petals of a ligulate floret grow out on one side to form a flat blade or ligule. Dandelion florets are all of this kind. Tubular florets do not have a ligule. Thistle florets are all tubular. Sunflowers, daisies, and many other composite flowers have both types of floret. The ligulate florets are on the outside, helping to make the flower heads conspicuous, and the tubular ones are in the center. As well as the flowers already mentioned, the family includes knapweed, ragwort, hawkweed, colts-foot, and chicory. Many species are cultivated for their flowers. They include asters, dahlias, zinnias, cornflowers, and marigolds. Lettuce, endive, and globe artichokes are among the species cultivated as vegetables.

Deadnettle family (Labiatae) A family of dicotyledons with about 3,000 species, nearly all of them herbaceous. The stems are rather square and they carry the leaves in opposite pairs. The flowers usually form whorls around the stem. The five sepals form a toothed tube, and the five petals are joined to form a helmet-shaped flower. The four stamens are concealed in the upper part of the helmet. There are two carpels, but each is divided into two so that there appear to be four little 'nutlets' at the base of the sepals when the petals fall. Many members of the family contain fragrant oils. As well as the deadnettles, the family contains lavender, rosemary, mint, and thyme.

The dandelion flower head is a collection of tiny flowers or florets at the top of a leafless stem. The green objects surrounding the flower are bracts, not sepals.

The blue funnel-shaped flowers of the stemless gentian are typical of the gentian family. The cushions of tough leaves are well suited to the alpine conditions in which the plant lives.

Dock family (Polygonaceae) A family of dicotyledons with about 800 species, mainly in the temperate parts of the world. There are herbs, shrubs, and climbers, together with a few trees. The flowers are often small and dull, but they are grouped into dense clusters in many species. Many of the plants, including the docks and knotgrasses, are troublesome weeds. Among the few useful plants in the family are rhubarb and buckwheat.

Dogwood family (Cornaceae) A family of about 100 dicotyledons, mainly trees and shrubs. The small flowers usually hang in clusters and they are normally followed by small stone fruits (drupes). The flowering dogwood is one of North America's most decorative trees. The white petaled flowers erupt in the early spring. These are followed by shiny red berries. In the autumn, the leaves turn a beautiful dark red. The wood of the dogwood is extremely hard and was once used for making skewers and wooden spikes used to hold rope lashing.

Duckweed family (Lemnaceae) This family of monocotyledons contains the smallest of all flowering plants. They are all tiny floating plants, consisting of a little green disc only a few millimetres across. There may or may not be a root hanging down into the water. The minute flowers are borne in 'pockets' on the disc. There are separate male and female flowers and they consist simply of stamens or a carpel. Some species rarely flower but reproduce by budding—small pieces become detached and then grow into new plants. They often form continuous green carpets over ponds in the summer. There are about 20 species of duckweeds, found in fresh waters throughout the world.

Ebony family (Ebenaceae) The hard black wood known as ebony can be obtained from the heartwood of several tropical trees, but most of it comes from the true ebony trees of India and Africa. The family is a small one and its members are all trees and shrubs. The flowers are of two types: solitary female ones, and male flowers hanging in large clusters towards the tips of the branches. Some of the stamens may be modified into bright petal-like structures. The flowers produce abundant nectar and they are very attractive to bees. The persimmon is another member of the family. This is a small tree that produces a pleasant tomato-like fruit.

Eelgrass family (Zosteraceae) A small family of monocotyledons containing some of the few flowering plants that live in the sea. The plants have creeping rhizomes, slender stems, and grass-like leaves. There are separate male and female flowers, borne on the sides of the stem and more or less enclosed by a sheath. They are pollinated by water and have no petals. The plants more often reproduce vegetatively by breakage of the rhizome. There are about a dozen species, found around the coasts and estuaries of most of the temperate seas. Most of them grow round about low tide level.

Elm family (Ulmaceae) A family of trees with about 150 species, mostly in the Northern Hemisphere. The elms themselves are the most important members of the family. They are large deciduous trees and their clusters of tiny reddish flowers are produced very early in the year, long before the leaves appear. The leaves are often unevenly developed at the base. The fruits have little circular wings around them and they are scattered by the wind. Elms have

Flax has been cultivated for many years for its fibers and oil.

The forget-me-not, showing the partial closure of the throat by a ring of scales.

very tough wood, suitable for furniture, but they are attacked by bark beetles and by a serious fungus disease. They are not planted as often as they were in the past. Another member of the family is the hackberry. This small tree grows in the warmer regions and produces small plum-like fruits.

The grass family provides us with all our cereals. The left hand picture shows ripe ears of wheat. On the right is the flower head of annual meadow grass, a very common weed in gardens.

Flax family (Linaceae) A family of dicotyledons containing herbs, shrubs, and trees. There are nearly 300 species. The flowers are regularly arranged, with four or five petals, and they contain both stamens and carpels. The flax itself is probably a native of the Mediterranean region and it has been cultivated for a very long time. The fibers from its stems are used to make linen, while the seeds yield linseed oil. The plant is also grown in gardens for its attractive blue flowers.

Flowering rush family (Butomaceae) A small family of monocotyledons closely related

to the arrowhead family. The only species in Europe is the flowering rush, a tall plant with broad strap-like leaves and pink flowers arising in a cluster from the top of the stem. There are six petals, three large and three small. The plant is found in ditches and along the edges of ponds and streams. There are only about 10 species in the family.

Forget-me-not family (Boraginaceae) A large family of dicotyledons with over 1,500 species. Most of them live in Europe and Asia. Most of them are herbs, but there are some woody plants in the family. The five sepals are joined to form a tube, and the five petals are also joined. They sometimes form a bell-shaped flower, sometimes a funnel-shaped flower, and sometimes they open out to form a saucer-shaped flower. The throat of the flower may be partly closed by scales or hairs. Many

of the flowers are blue. The leaves and stems are often very hairy. The family also includes lungwort, borage, comfrey, and viper's bugloss.

Frogbit family (Hydrocharitaceae) A family of monocotyledons whose members are all aquatic herbs. They live in fresh water or in the sea. Some species float freely at the surface, others are rooted and completely submerged. The frogbit is a floating plant with palette-shaped leaves and conspicuous white flowers. There are separate male and female flowers, but each has three large petals and three smaller sepals. Several flowers normally arise from a cluster of bracts on the stem. The water soldier is a similar kind of plant, but it has large spiky leaves. Both float in ponds and ditches, especially in chalky districts. Other members of the family include vallisneria and Canadian pondweed. Both are commonly grown in aquarium tanks. They are rooted plants.

Gentian family (Gentianaceae) A family of dicotyledons famous for the brilliant blue colors of many of its members. They are all herbs, usually with hairless, shiny leaves. There are about 800 species, mainly found in the temperate regions. The petals are joined together and are generally funnel-shaped, although the outer regions may turn back to form a four or five armed star. The petals do not fall when they wither, but remain around the developing fruit. Many of the most famous gentians are alpine plants and they are widely grown in rock gardens. Other members of the family include the pink centaury and the yellow-wort.

Geranium family (Geraniaceae) A family of dicotyledons with about 700 species, most of them herbaceous. The leaves are broad and lobed or deeply divided. The flowers, which are usually some shade of pink or purple, are more or less regular. They have five free petals, 10 stamens, and several joined carpels which are prolonged into a long 'beak'. This is especially prominent when the fruits swell up after flowering. As well as the cultivated geraniums, the family includes the cranesbills, storksbills, and the very common herb Robert.

59

Rye grass, showing the large hanging stamens which scatter pollen. This is one of the most nutritious grasses and is widely sown in hay meadows.

The hazel, like many other wind-pollinated trees, produces its flowers early in the year before the leaves appear and get in the way of the pollen.

Grass family (Gramineae) A very large family of monocotyledons of great importance to man and other animals. There are about 10,000 species, nearly all of them herbaceous. The leaves are strap-shaped and they sheath the stems at their lower ends. The flowers have no petals and they are borne in little spike-lets. Each spikelet consists of two leaf-like scales which enclose one or more flowers. Each flower normally has three stamens and a carpel. The spikelets are arranged in various ways in tight or loose heads. The stamens hang from the spikelets when they are ripe and they scatter their pollen to the wind. It is picked up by the feathery stigmas of other flowers. Grasses grow in all possible places. As well as all the wild grasses that grow in our open spaces, the family includes all the cereals, sugar cane, and bamboos.

Hazel family (Corylaceae) A small family of deciduous trees and shrubs found mainly in the northern temperate regions. The male flowers are carried in hanging catkins, while the female flowers are carried in little bud-like catkins. The fruits are nuts and they are partly surrounded by leaf-like bracts. The two best known members of the family are the hazel and the hornbeam. There are about 50 species altogether.

Heather family (Ericaceae) A family of dicotyledonous shrubs with about 1,600 species. Most of them live in the cooler parts of the world, including mountain slopes. The flowers are more or less regular, with the petals attached to the edge of a fleshy disc. As a rule the petals are partly or completely joined to each other, forming bell-shaped or lantern-shaped flowers. About half of the members of the family belong to the *Rhododendron* genus. These include both evergreen and deciduous shrubs. The heathers and heaths are well known plants on sandy soils. Another well known member of the family is the bilberry or whortleberry whose juicy black fruits make delicious pie fillings.

Holly family (Aquifoliaceae) A family of about 300 kinds of trees and shrubs, most of them belonging to the genus *Ilex* which in-cludes the holly itself. Some species are deciduous, while others are evergreen. The small white flowers appear in the summer months. There are separate male and female flowers, and they normally grow on different trees. This is why some trees never bear the familiar red fruits.

Honeysuckle family (Caprifoliaceae) A family of dicotyledons nearly all of which are shrubs. There are about 400 species, most of them living in the northern temperate regions. The five petals are joined together to form a long or short tube. The tube may have two distinct lips. The stamens are usually five in number and they are attached to the petals. Many species produce attractive berries and stone fruits. As well as the honeysuckles, the family contains the elder, the guelder rose, the wayfaring tree, and the snowberry. Many species are cultivated for their flowers or fruits.

Hop family (Cannabinaceae) A small family containing only four species of dicotyledonous herbs. Hops are climbing herbs, with male and female flowers on separate plants. The male flowers are small and brownish and they are carried on branching stalks. The female flowers develop in cone-like clusters. These clusters are used in making beer. They give the beer its bitter taste. Hops are widely cultivated for this purpose. There are three species of hops. The other member of the family is the hemp, from which we get valuable rope-making fibers. The plant also produces the drug known as marijuana.

Horse chestnut family (Hippocastanaceae) A small family with about 25 species of trees and shrubs in the Northern Hemisphere. The leaves are large and have several leaflets radiating out like the fingers of a hand. The flowers are grouped into large pyramidal

Honeysuckle

bunches at the ends of the twigs. The large nut-like seeds (conkers) are carried in leathery capsules which are often prickly. The common horse chestnut, which is widely planted for its attractive flowers, comes from Turkey and Eastern Europe. The seeds were once thought to be useful in treating sick horses but, like most other parts of the plant, they are poisonous. Both the horse chestnut and buckeye occur in the United States.

Hydrangea family (Hydrangeaceae) A family of trees and shrubs with about 200 species. Most of them live in the northern temperate regions. They are closely related to the saxifrages and often regarded as part of that family. The flowers are regularly arranged and they usually contain both stamens and carpels. Some flowers, however, especially around the outside of a cluster, are sterile. They have large and showy petals which attract insects to the smaller fertile flowers in the center. As well as the cultivated hydrangeas, which come from China and Japan, the family includes the syringa or mock orange. This is another shrub that is commonly planted in gardens.

Iris family (Iridaceae) A family of monocotyledons with about 1,000 species. They are all herbs with corms, bulbs, or rhizomes. The leaves are slender and grass-like. The flowers are usually regular and they contain both male and female parts. There are six petals, not always alike, and they are usually joined together near the base to form a tube. There are three stamens. Many of the species are cultivated. The family includes the irises, crocuses, and gladioli. It also includes the little sisyrinchium or blue-eyed grass, a common garden plant which also grows wild in Ireland and in North America.

Garden irises

Right: lily of the valley showing the berries (red) which develop from the fertilized flowers.

Below: Forsythia is a popular garden shrub. Its bright yellow flowers appear before the leaves.

Lily of the Valley

Ivy family (Araliaceae) A family consisting mainly of trees, shrubs, and woody climbers. Most of the 700 species live in the tropics. The flowers are small but they often grow in large bunches. They are followed by berries or by stone fruits (drupes). The common ivy grows throughout most of Europe and climbs on trees and rocks by means of special little roots. Its greenish flowers open late in the year, usually in October, and they are very attractive to flies, wasps, and other insects.

Lilac family (Oleaceae) A family of trees and shrubs with about 400 species, most of them growing in Asia. The flowers are carried in large bunches. Each flower is quite small and usually consists of four petals joined to form a tube. There are rarely more than two stamens. Many of the flowers are sweetly scented, and several of the shrubs are cultivated. As well as lilacs, the family includes the ash, privet, winter jasmine, forsythia, and olive. The ash and the olive have no petals, and the ash produces its flower clusters long before the leaves are out.

61

The snake's head fritillary has nodding flowers of a tulip-like shape.

Lily family (Liliaceae) A family of monocotyledons with about 2,500 species. Most of them are herbs, but there are a few shrubby plants in the family. Woody plants are very rare among the monocotyledons. The flowers are regular and usually have six petals. The petals may be free or they may be joined together. There are six stamens. The three carpels are joined together and they are inside the petals. This last feature separates the lily family from the daffodil family. Many of the plants grow in desert climates and survive the dry season as bulbs. Many are cultivated for their pretty flowers. The family includes the lilies, bluebells, hyacinths, colchicums, tulips, and onions. The woody plants in the family include the yuccas, the aloes, and the dragon

Butcher's Broom

trees. These all produce dense clusters of spiky leaves, either close to the ground or high up on bare trunks and branches. They are very similar to the American agaves of the daffodil family.

Linden family (Tiliaceae) A family of trees, shrubs, and occasional herbs. There are about 300 species. Linden trees (sometimes called limes in Europe, or basswood to avoid confusing them with the citrus limes), are deciduous trees growing in the Northern Hemisphere. The small greenish or yellowish flowers hang in bunches, each bunch bearing a leaf-like 'wing'. Each flower has five sepals, five petals, and many stamens. The flowers are very fragrant and they attract plenty of insects. The fruits are later carried away by the 'wing' The inner bark of the trees is very fibrous and it was once used as a source of raffia or bast. Another very important member of the family is the jute plant. Grown in India and Pakistan, this plant provides fibers for most of the world's sacks.

Lobelia family (Lobeliaceae) A family of dicotyledons including both herbs and small trees. There are about 400 species, found mainly in the warmer parts of the world. Several species are cultivated for their flowers. The petals are joined to form an irregular tube. The anthers (pollen sacs) of the stamens are also joined together to form a tube around the stigma. Most of the lobelias are some shade of blue, although there are red and white ones as well. Many lobelias, including those grown in gardens, are quite poisonous plants.

Loosestrife family (Lythraceae) A family of dicotyledons with about 500 species. It contains herbs, shrubs, and trees. The flowers are usually regular and they contain both male and female parts. The sepals, petals, and stamens are joined together at the base to make a tube surrounding the ovary. The purple loosestrife, one of the best known members of the family, is a slender plant growing in marshy places. It has long spikes of purple flowers. The yellow loosestrife is quite different and belongs to the primrose family.

These fine clivias, together with the other plants on this page, belong to the lily family. The butcher's broom (bottom left) is a strange plant because it has no proper leaves. What look like leaves are in fact flattened stems—shown by the fact that they carry flowers.

Solomon's seal, a woodland perennial plant, showing the small black berry which develops from a fertilized flower.

Magnolia family (Magnoliaceae) A small family containing about 20 species of trees and shrubs. The magnolias are among the most beautiful of the flowering shrubs. Many of them produce their large star-shaped or cup-shaped flowers before the leaves open in the spring. There are between six and twelve petals and many stamens. The carpels are clustered on a sausage shaped receptacle in the center of the flower. Most of the magnolias come from the Far East. Some are evergreen and some are deciduous. Another member of the family is the tulip tree, which comes from North America. There are three sepals which form a 'saucer' for the petals. The latter are six in number and they form a deep cup.

Mahogany family (Meliaceae) A small family of trees famous for their hard red wood which is excellent for cabinet making. Several species live in Central America and the West Indies, and related species live in Africa. They all have compound leaves, the leaflets of which are arranged in unequal pairs. The small greenish flowers have their stamens joined together. Much of the timber sold today as mahogany actually comes from other types of trees, although the wood is just as good. Many cigar boxes, especially those made in the West Indies, are made of wood from another tree in this family. It is often known as the Spanish cedar. Its strong scent appears to discourage insect pests.

Mallow family (Malvaceae) A family of dicotyledons containing trees, shrubs, and herbs. There are nearly 1,000 species. The leaves are usually palmately veined (like the fingers of a hand) and may be sub-divided. The leaves and stems are often covered with star-shaped hairs. The flowers are usually regular and they normally contain both stamens and carpels. There are normally five sepals and five petals. The latter are free but they usually form a trumpet-like tube. Besides the mallows themselves, the family contains the hollyhock, the hibiscus, and the cotton.

Mangrove family (Rhizophoraceae) A family of trees that grow in coastal swamps in the warmer parts of the world. They produce great tangles of roots that trap mud and debris and gradually build up dry land. The trees have small flowers, followed by leathery fruits. The seeds germinate in the fruits while they are still on the trees. They each produce a dagger-like root and when they fall from the trees the roots dig into the mud and get a firm hold. The name mangrove is also given to trees of various other families which live in similar situations and which have developed the same kind of root system.

Maple family (Aceraceae) A family of trees and shrubs with about 120 species, mainly in the northern temperate region. All but two belong to the genus *Acer*. The leaves are often palmately veined and some are compound (made up of several leaflets). The flowers are small and greenish and they hang down in bunches. They are followed by the familiar double winged fruits which children call 'airplanes'. Many oriental species are grown for their attractive foliage. This is especially colorful in the autumn when the leaves turn many shades of red, orange and yellow. The sycamore maple is one of the largest of the maples, reaching a height of about 100 ft. It comes from Southern and Central Europe and it seeds so freely that it is becoming a weed in many woodlands.

Milkwort family (Polygalaceae) A family of herbs, shrubs, and trees with about 700 species. About 500 of these belong to the genus *Polygala* and are called milkworts. These are herbs or small shrubby plants with irregular flowers. There are five sepals but the inner ones look like petals. There are three petals, joined together and also joined to the eight stamens. The common milkworts of Europe are small plants usually found growing in grassland. The flowers are blue, pink, or white.

Mistletoe family (Loranthaceae) A family of shrubby plants, most of which are partial parasites of trees. There are about 600 species. The flowers are regular, with sepals and bright petals in many species. In other species the sepals are missing and the petals are then small and green. This is the case with the mistletoe itself. The mistletoe has separate male and female flowers, and these normally develop on different plants. The female flowers are followed by the sticky white berries.

Nasturtium family (Tropaeolaceae) A family of 80 dicotyledons all belonging to one genus and all coming from America. They are herbaceous plants and many of them climb. The leaves are more or less circular and they have a spicy flavor. They make a pleasant addition to salads. The bright orange and yellow flowers are favorites in many gardens. Some of the wild species possess underground tubers which are a good source of starch.

Orange family (Rutaceae) A family of trees and shrubs with about 1,300 species ir the warmer parts of the world. Several species are used for making furniture, but the most important members of the family are the citrus

A fine large-flowered magnolia.

trees that provide us with oranges, lemons, limes, grapefruits, and similar fruits. These trees all came originally from South East Asia, although they are now grown in nearly all warm regions of the world. They have shiny, evergreen leaves and most species have white flowers with five petals and many stamens. The fruits are berries in which the outer skin becomes thick and leathery.

Orchid family (Orchidaceae) A family of monocotyledons containing some of the world's most fascinating flowers. The flowers have six petal-like segments, but one of these is always very different from the others. There is often a long spur on one of the petals and the nectar is produced there. Most of the orchids are found in the tropics, but many attractive species grow in cooler regions. Some orchids grow as epiphytes, perched high up on the branches of tropical trees. Others are saprophytes, lacking chlorophyll and getting food from decaying leaves in the soil. Many orchids are named after animals because the petals often take on the shapes of animals. The lower petal of the bee orchid, for example, looks very much like a bee. It is thought that this attracts other bees to the flower and thus helps pollination. Some of the other resemblances are pure chance. The man orchid has its lower lip drawn out to look like the body of a man, complete with arms and legs. The lower lip of the monkey orchid is drawn out in similar fashion, but the 'limbs' are relatively longer and more bent. There are about 7,500 species of orchids.

Many tropical orchids grow up in the tree tops and, like this one, they send sprays of flowers arching down from the branches.

Many orchids resemble insects. The bee orchid looks very much as if it has a large bee sitting on it.

Palm family (Palmae) A large family of monocotyledons whose members are all woody plants, either slender trees or climbers. The leaves are large and greatly divided. They usually spring from near the top of the plant. The flowers arise in large bunches and hang down from the tops of the plants. There are about 1,500 species of palms and they provide a great many useful materials. Among the most important are the date palm, the coconut palm, the oil palm (which yields valuable oil used in making margarine and soap), and the raffia palm from which we get raffia. Only the grass family is more important to man than the palm family.

Pea family (Papilionaceae) A large family of dicotyledons with very characteristic flowers and fruits. There are more than 6,000 species, ranging from tiny herbs to large trees. The leaves are usually compound (made up of several leaflets). The flowers have five petals, one of which stands up at the back of the flower and is called the standard. There is a wing petal at each side, and the other two are partly joined to form the keel. The five sepals normally form a tube, and the 10 stamens are also normally joined together. There is a single carpel which develops into the pod. The pod, which is characteristic of the family, contains several seeds and it normally bursts open as it dries. It throws the seeds out. Another characteristic of the family is the possession of root nodules. These are little swellings on the roots and they contain special bacteria which are able to make nitrates (plant foods) from the nitrogen in the air. Members of the pea family, often called legumes, enrich the soil by means of these nodules. Members of the family include: peas, beans, clovers, lupins, gorse, vetches, laburnum, and acacias. Many species are cultivated for their flowers or their fruits and seeds.

Pepper family (Piperaceae) A family of dicotyledons with over 1,500 species in the warmer parts of the world. Many are woody climbers. One of the most important species is *Piper nigrum*, from which we get the pepper in our pepper pots. It is a woody climber that clings to supports by means of special roots. It has broad, shiny leaves and little clusters of flowers without petals. The flowers are followed by little orange berries which are harvested when almost ripe. They are then dried and ground up to make pepper. The plant came originally from South East Asia. The red and green peppers that we eat as vegetables are quite different and they belong to the potato family.

Dyer's greenweed (below) and vetch, two members of the pea family showing the characteristic flower shape with a standard petal forming a sort of hood at the back.

The fragrant orchid, a common species in chalk grassland.

Broom is found on dry, acid soils.

The wild carnation grows in southern parts of Europe and Asia. It is from this plant that the garden carnation has been developed.

Vetch

Periwinkle family (Apocynaceae) A family of dicotyledons containing just over 1,000 species. Most of them are woody plants and many of them are climbers. The majority live in the warmer parts of the world. The petals are joined to form a tubular, funnel-shaped flower. The stems normally contain a milky juice and several species yield useful forms of rubber. Others provide medicinal drugs and arrow poisons. The periwinkles are creeping evergreen plants with blue or white flowers. They grow in shady places and are commonly planted in gardens. The oleander is another member of the family.

Pineapple family (Bromeliadaceae) A family of monocotyledons with about 1,500 species, all hailing from the New World. They produce clusters of spiky leaves around fairly short stems. The flowers are more or less regular and densely packed on thick spikes. Some of these flower heads are many feet long. Many members of the family, commonly called bromeliads, live in the rain forests. Some live as epiphytes, perched on the branches of trees (see page 196). The leaf bases hold a great deal of water and many animals make their homes in these 'ponds'. The pineapple itself is a small plant which is cultivated in fields. It has purple flowers. The fruit is formed by the fusion of the flowers and the leaves and stalks of the spike. Commercial pineapples have no seeds, the plants being raised from cuttings.

Pink family (Caryophyllaceae) A family of dicotyledons containing about 1,500 species. Most of them are herbs and they live mainly in the northern temperate regions. The leaves are generally narrow. The flowers are regular as a rule and normally contain both male and female parts. There are four or five sepals, sometimes joined at the base. There are four or five petals, but they are often deeply notched so that there appear to be twice as many. All the petals are free. There are commonly twice as many stamens as petals, surrounding between two and five stigmas. The family includes the pinks (and the cultivated carnations), stitchworts, chickweeds, campions, ragged robin, sweet williams, and gypsophila.

Ragged robin

Cultivated primrose

The rose family contains a very wide variety of flower types. There are normally five petals, although this blackthorn has six.

Pitcher plant family (Sarraceniaceae) A family of dicotyledons in which the leaves are modified for catching insects. There are about a dozen species, found mainly in the eastern parts of North America. They live in peat bogs and in other poorly drained places where normal food materials are in short supply. The leaves of the pitcher plants are like slender beakers or ice cream cones and they contain water. Insects and other small creatures that fall in gradually decay and the plant absorbs the resulting material. The rim of the pitcher is often brightly colored and slippery. Insects are attracted to it and then they cannot get out. Some of the species have little 'lids' on the pitchers. These prevent too much rain from entering. Another family of pitcher plants lives in South East Asia. It is called the *Nepenthaceae.* There are about 60 species. The pitchers are formed by the tips of the leaves and they are provided with 'lids'. Some of these pitchers hold as much as a pint of water.

Plantain family (Plantaginaceae) A family of dicotyledons, although their leaves are often narrow and have parallel veins like those of monocotyledons. They are herbs and their leaves normally form a rosette on the ground.

Deadly nightshade

The small flowers are carried in dense spikes at the top of leafless stalks. Each flower has four pale, scale-like petals, all joined to form a tube. There are both stamens and carpels in each flower, the stamens having very long stalks. The flowers at the bottom of the spike open first, and the stigmas ripen before the stamens in each flower. A flowering spike will therefore have several regions: unopened flowers at the top, a ring of spiky stigmas further down, a ring of hanging stamens below that, and a region of developing fruits at the bottom. The flowers are pollinated by the wind. There are about 200 species. They are very common in grassland.

Pondweed family (Potamogetonaceae) A family of monocotyledons. There are about 90 species, all in the genus *Potamogeton.* They are all aquatic herbs living mainly in fresh water. They nearly all have rhizomes buried in the mud. The leaves are generally strap-shaped and all submerged, although some species have oval floating leaves. The flowers are clustered on loose or dense spikes. Each has four pale segments, with four stamens and usually four carpels. Many of the flowers never rise above the surface and their pollen is transferred by the water. Other flowers emerge from the water and are pollinated by the wind.

66

Poppy family (Papaveraceae) A family of dicotyledons with about 120 species, mainly in the Northern Hemisphere. Most of them are herbs and they all contain a thick milky juice. The flowers are regular, usually with four petals. There are normally two sepals, which fall off as the flower opens. The petals are often crumpled when they first open. There are numerous stamens, and the joined carpels are top-shaped. They scatter their tiny seeds through small pores as they sway in the wind. As well as the poppies themselves, the family includes the greater celandine. Many species are cultivated as garden plants.

Potato family (Solanaceae) A family of dicotyledons with about 2,000 species. The flowers are usually regular, with five joined petals. These sometimes form a bell-shaped flower, but they are more often spread out to form a star. The stamens are often joined together and they form a little cone sticking out of the center of the flower. The fruits are usually berries, although some species produce tough capsules. The family includes herbs, shrubs, and small trees. Many of them are poisonous. The members include henbane, deadly nightshade, woody nightshade (bittersweet), potato, and tomato.

Above: plantain heads showing the long hanging stamens. The flowers at the bottom of the spike open first, and so the oldest spike here is in the middle: the uppermost flowers are open and seeds are already well formed at the bottom.

Above left: The oxlip, a member of the primrose family restricted to the chalky boulder clay in England. Left: polyanthus a cultivated member of the primrose family.

Left: The white rockrose, a rather rare plant found only in certain limestone areas.

Below: a cultivated rose being visited by two hover-flies. Plant breeders have greatly increased the number of petals in cultivated species.

Cowslip

Primrose family (Primulaceae) A family of dicotyledons with about 350 species, mainly in the Northern Hemisphere. Most of them are herbs. The flowers are regular, usually with five sepals and five petals. Both the sepals and the petals are joined and they often form tubes. Many of the species are cultivated for their flowers. The family includes primroses, cowslips, oxlips, creeping jenny, and yellow loosestrife. It also includes the cyclamen and the many cultivated forms of primula and polyanthus.

Reedmace family (Typhaceae) A small family of monocotyledons with only nine species. They are stout herbs which grow in shallow water. They are often wrongly called bulrushes. The leaves are broad and sword-like. The flowers are carried in dense cylindrical spikes. The male flowers are in narrow spikes at the tops of the stems, while the female flowers are below them in thicker spikes which become a rich brown when mature. Each flower is surrounded by a few hairs, but there are no petals. The hairs carry the fruits away when they are ripe.

Rockrose family (Cistaceae) A family of dicotyledons with about 200 species, most of them found in the Mediterranean region. They include both herbs and shrubs. The flowers are regular, with three or five sepals and usually with three or five petals as well. The parts are

Dog's mercury, a very common member of the spurge family which grows in shady woods and hedgerows.

all separate and the petals are easily detached. There are many stamens. The rockroses and their relatives favor warm, dry soils. Several species are common on chalk slopes and grassland.

Rose family (Rosaceae)   A very large family of dicotyledons including herbs, shrubs, trees and climbers. There are about 3,000 species, many of which provide us with valuable fruit. Many more are cultivated for their flowers. The leaves, which may be simple or compound, almost always bear little outgrowths called stipules at the base. The flowers are regular. There are usually five sepals, partly joined together, and five free petals. There are many stamens and a variable number of carpels

which may or may not be joined. The plants produce a very wide variety of fruits. Among the many members of the family are: roses, blackthorn, blackberry, hawthorn, mountain ash, meadowsweet, cinquefoil, and geum. Important fruits provided by the family include apples, pears, plums, cherries, peaches, apricots, raspberries, and strawberries.

Rush family (Juncaceae)   A family of herbaceous monocotyledons with grass-like leaves or with leaves reduced to little brown scales. There are about 350 species, most of them growing in damp places. Many species form thick tufts. The stems are long and slender, often round in section and filled with soft pith. They are pointed at the top and many stems bear clusters of flowers emerging from one side near the top. The flowers are small but perfectly formed with six brownish or greenish petal-like segments arranged in a star shape. They contain both stamens and carpels and they are pollinated by the wind. Members of the genus *Luzula* are more like grasses than the other rushes. Their leaves are more blade-like and they are fringed with soft hairs. The plants are often known as wood rushes and they grow in drier places than the other rushes.

St John's-wort family (Hypericaceae)   A family of dicotyledons with about 300 species, mainly found in the temperate regions. The family includes herbs, shrubs, and trees. More than 200 of the species belong to the genus *Hypericum* and are known as St John's-worts. They have yellow flowers, with five sepals and five petals. There are many stamens, often joined together in bunches. The flowers produce abundant pollen, but no nectar. One of the unusual features of the plants is the possession of translucent dots on the leaves of many species. These are easily seen if the leaves are held up to the light: it looks as if the leaves have been pierced all over with a pin. The pale regions are glands and they secrete a poison which is harmful to animals. Several species are weeds in farmland, but others are cultivated for their attractive heads of flowers. One of the best known is a low-growing shrub called the rose of Sharon.

Saxifrage family (Saxifragaceae)   A family of dicotyledonous herbs with about 500 species, mostly found in the cooler regions. The flowers are usually regular, with four or

Right: A sedge in flower, showing the male flowers in the upper spike and the female flowers in the lower spikes.

Below: Speedwell, a member of the snapdragon family.

A wood rush, quite a common plant in grassland.

five petals. They are very similar to the flowers of the rose family, but they have fewer stamens as a rule and their fruits are rather different. Many species grow in cold regions, especially alpine areas, and they form cushions of small, tough leaves. Plants of this type are widely grown in rock gardens. The garden plant called London pride is a member of this family.

Sedge family (Cyperaceae)   A family of monocotyledons which are often confused with grasses and rushes. Many of them actually have common names including the words grass or rush. Many sedges can be identified at once by their triangular stems and the triangular arrangement of the leaves. There are about 3,000 species of sedges and most of them grow in damp places, although many can be found in dry grassland. The flowers have no sepals or petals but each is surrounded by a

little scale. Male and female flowers are separate in many species and are carried on separate spikes of the flower head. The male spikes are usually nearer the top and they can be recognised by the long anthers hanging from them. The family includes the cotton grass, the bulrush, and a wide variety of true sedges belonging to the genus *Carex*. Many of them grow in large tufts, and several are used for thatching.

Snapdragon family (Scrophulariaceae) A large family of dicotyledons with more than 2,500 species. Most of them are herbs, although there are some trees and shrubs. The petals are joined together and the flowers are usually highly irregular. In many species they are tubular, and they often have two distinct 'lips' which are held closed around the stamens and stigmas. Such flowers can be pollinated only by insects heavy enough to weigh down the lower lip and enter the flower. The four or five stamens are usually in the upper part of the flower, and so is the stigma. They are in the right position to touch the head and back of an insect as it enters the flower. As well as the snapdragons, the family includes foxgloves,

Mullein, a member of the snapdragon family.

mulleins, speedwells, and many garden plants such as nemesias and calceolarias. There are also many partial parasites in the family. These include the eyebrights and the yellow rattle, all plants which take water and dissolved salts from the roots of other plants (see page 47).

Spindle family (Celastraceae) A family of about 450 species of trees, shrubs, and woody climbers. Many of the lianas that drape the trees in the tropical rain forests belong to this family. Better known is the European spindle tree. Its pale greenish flowers are inconspicuous, but its gay fruits make it very attractive in the autumn. The fruits are pink and they hang from the branches like little lobed cherries. When ripe, they split open and reveal the bright orange seeds. The leaves turn purple in the autumn, and a related American tree is often called the burning bush, or wahoo, because of its autumn colors. The spindle is a deciduous shrub, but many of its relatives are evergreens.

Spurge family (Euphorbiaceae) A family of dicotyledons with more than 7,000 species. Most of these live in the tropical regions and they inhabit a wide variety of habitats. The plants are therefore very varied in their appearance. There are herbs, shrubs, trees, climbers, and aquatic plants in the family. Many of the species contain a milky juice. Most of the plants have no petals, their place being taken by modified leaves which often assume bright colors. These leaves may be taken for a single flower, but they actually surround a cluster of several flowers. The actual flowers are very tiny. Male and female flowers are separate and the male flowers each consist of just a single stamen. The petty spurge is a very common bright green weed in gardens. It pours out its latex when the stem is broken. Other members of the family include the crown-of-thorns and the poinsettia, both popular house-plants. Two useful members of the family are the rubber tree and the castor oil plant. The family also includes the dogs' mercury, a very common plant in European

Crown-of-thorns, a very prickly red-flowered spurge from Madagascar.

Orpine, a large member of the stonecrop family is a favourite feeding place for autumn butterflies.

woodlands. It differs from the spurges themselves in having no milky juice. It also has a different kind of flower.

Stinging nettle family (Urticaceae) A family of dicotyledons, most of which are herbs or small shrubs. Many, but not all of the 500 species have stinging hairs (see page 51). The flowers are small and greenish, with stamens and carpels usually in separate flowers. The common stinging nettle carries its flowers in drooping spikes. The plant is a common weed throughout the temperate parts of the world. The leaves can be eaten when they are young. They are cooked like spinach and they taste very nice.

Stonecrop family (Crassulaceae) A family of dicotyledons which usually have succulent (juicy) leaves. They are all herbs or small shrubs with small star-shaped flowers which are born in large flat heads. There are over 1,200 species, mostly found in the warm and dry regions of the world. They are especially prominent in rocky places, including sea cliffs, and they are popular rockery plants. Many are particularly attractive to butterflies. The family includes the house-leeks as well as the stonecrops and the orpine.

Sundew family (Droseraceae)   A family of dicotyledons whose members are all adapted for trapping insects. There are about 100 species, all more or less herbaceous. The sundews themselves have glandular hairs all over their leaves and these hairs secrete a sticky fluid which attracts insects and then traps them. The leaves then fold over and pour digestive juices on to the insect. The leaves absorb the resulting material. Sundews live in acid bogs and other places with poor soil. Their flowers are often small and insignificant. The family includes the Venus fly trap, which traps insects with a gin-trap arrangement. The end portion of the leaf is hinged and it snaps shut when an insect alights on it.

Tea family (Theaceae)   A family of small evergreen shrubs with its headquarters in South East Asia. The flowers are often fragrant, with five white petals and many stamens. As well as the tea plant, the family contains the beautiful camellias which are grown in gardens

all over the world. Many varieties of these plants have been produced, with a wide range of colors.

Teasel family (Dipsaceae)   A family of dicotyledonous herbs with about 150 species, mainly in the Mediterranean region. The flowers are small and are grouped into dense heads surrounded by sepal-like bracts. The petals are joined to form a small and irregular tube. The teasel head bears numerous stiff bristles between the bluish flowers. These bristly heads were once used for combing out wool. The leaves and stems of the plant are also very spiky. The family also includes the various species of scabious.

Thrift family (Plumbaginaceae)   A family of dicotyledonous herbs and shrubs with about 250 species. Many of them live on sea shores,

cliffs, and mountains. The flowers are regular. There are five sepals, united into a tube at the base, and five petals. The petals may be free or joined to form a tube. There are also five stamens and five stigmas. The flowers are often tightly bunched into heads. The sea lavender and the thrift or sea pink are well known members of the family.

Vine family (Vitaceae)   A family of about 600 species of climbing shrubs found throughout the tropical and temperate regions. The plants climb by means of tendrils. The leaves are palmately veined and often broken up into several leaflets. The flowers are small and greenish and they are followed in many species by the familiar berries which we know as grapes. Several species are grown for their attractive leaves, especially colourful in the autumn. One of the best known of these decorative plants is the Virginia creeper. Its tendrils are branched and each branch ends in a little sticky disc. These attach the plant to trees and walls and allow it to climb up.

A sundew plant showing the sticky hairs. The club-shaped objects are the developing fruits.

The spiky heads of the teasel were once widely used for combing out wool and other cloth.

Most willowherbs can become garden weeds since they seed very easily.

The thrift is found mainly in coastal areas in Britain and is here seen growing on a cliff top in company with dwarf gorse. On the opposite page it is seen growing in a rock crevice.

Violet family (Violaceae) A family of dicotyledons containing about 800 species of trees, shrubs, and herbs. About half of these belong to the genus *Viola*, which includes the familiar violets and the pansies. The flowers of this genus are irregular, with one of the five petals forming a spur. The flowers are followed by rounded capsules which split into three parts when ripe and force the seeds out as they shrink. The family has its headquarters in South America, where some species are large trees.

Walnut family (Juglandaceae) A family of about 50 trees, mainly in the Northern Hemisphere. The leaves are large and compound. Male and female flowers are separate, the male ones being in hanging catkins and the female ones solitary or in small spikes. The petals are very small or missing altogether. The walnuts that we buy in the shops for Christmas are not the complete fruits. The complete fruit is a stone fruit, rather like the plum except that the outer layer is green and leathery. This outer layer is removed as a rule before the 'nuts' go to the shops. The woody part is the equivalent of the plum stone and the seed is inside it. The

The large floating leaves of the water lily provide shelter for many small water dwelling animals. They are also favorite resting places for damselflies and other insects.

Violet

Right: Thrift

family also contains the hickory trees of North America. These provide valuable timber and one species also provides pecan nuts.

Waterlily family (Nymphaeaceae) A family of water-dwelling or marsh-dwelling dicotyledons with about 60 species. They usually have sturdy rhizomes buried in the mud and these throw up long-stalked leaves which are usually heart-shaped and which float at the surface. There may also be some thin submerged leaves. The flowers arise singly and usually float, although the flowers of the yellow waterlily or brandy-bottle stand up above the surface. The flowers are regular and they often have numerous petals. There are usually many stamens as well.

Willow family (Salicaceae) A family of deciduous trees and shrubs with about 340 species—300 willows or sallows and 40 poplars. Most of them live in the northern temperate regions and some extend right into the Arctic. The flowers have no petals and they are carried in upright or hanging catkins. Male and female catkins are borne on separate trees. The willows and sallows generally have rather narrow leaves and their catkins are erect. They are pollinated by insects and many have

sweetly scented flowers. The male catkins are laden with golden pollen in the spring. Poplars are all fairly large trees. Their leaves tend to be triangular and their flowers are carried in hanging, wind-pollinated catkins.

Willowherb family (Onagraceae) A family of dicotyledons with about 500 species. Most of them are herbs, but the fuchsias and a few other members of the family are shrubs. The flowers are usually regular, with two or four free sepals and two or four free petals. The willowherbs themselves have four petals. Willowherbs bear long pod-like fruits which split into four parts and release the hairy seeds. The rosebay willowherb is a very common plant of waste land. It is also called fireweed because it springs up so readily on burnt

ground. Other members of the family include the evening primrose, the enchanter's nightshade, and the clarkia.

Wood sorrel family (Oxalidaceae) A family of dicotyledons with nearly 1,000 species. Most of them are herbs, but there are some shrubs and trees. About 800 species belong to the wood sorrel genus *Oxalis*, in which the leaves are divided into several leaflets rather like clover leaves. The flowers are regular, with five petals and usually with 10 stamens. The wood sorrel itself is a delicate plant with little white flowers whose petals are streaked with pink or mauve. It is common in the woodlands of Europe and North America. Many related species are cultivated in gardens. Several of them fold their leaves down at night.

71

PART 3

Skeletons and Muscles

Most animals have a skeleton of some sort or another. Our own skeletons, like those of all other backboned animals, are inside our bodies. They are called *endoskeletons* and they are made of bone. But some animals have skeletons on the outside of their bodies instead. These animals include insects, crabs, and snails. Their skeletons are called *exoskeletons*. They are hard and chalky in crabs and snails, but insect exoskeletons are made of a horny material called *chitin*.

Are Skeletons Necessary?

Imagine what would happen if our skeletons were suddenly spirited away. We would sink to the floor in shapeless heaps. Skeletons are clearly necessary. The skeleton takes the weight of the body and all the other parts of the body are normally suspended from it. The skeleton also gives shape to the body. This is especially obvious in crabs and other animals with exoskeletons, but it holds true for people as well.

The skeleton also enables animals to move. All but the simplest animals move by means of muscles. The muscles shorten and pull various parts of the body. Such a system can work only if the muscles are attached to something firm. The skeleton provides this firm support. The other main job of the skeleton is to protect the animal. Again, this is most obvious in animals with exoskeletons, but our own skeletons protect our bodies to some extent. Our skulls protect our brains, and our ribs protect our hearts and lungs.

Joints

Skeletons are not rigid structures. We could not move if they were. The skeleton is made up of numerous sections which meet at *joints*. The joints of an exoskeleton are quite simple, usually little more than thin and flexible regions of the outer covering or cuticle. The joints of endoskeletons are a little more complicated. When two bones meet at a joint they are held together by very tough straps called *ligaments*. The ends of the bones are covered with smooth cartilage and they are often enclosed in a little bag of liquid. Both of these features help the bones to move smoothly against each other.

There are several kinds of joints in the vertebrate body, especially in the body of a mammal. One important type of joint is called

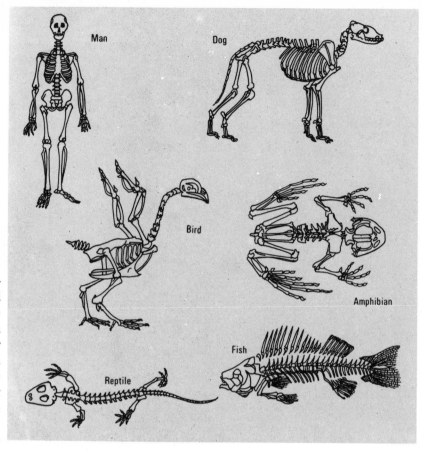

Vertebrates skeletons all have the same basic pattern, but it is adapted for different ways of life.

The crab does not have a backbone: its skeleton consists of hard plates on the outside of its body.

Insect wings are made to vibrate very rapidly by the contraction of two sets of muscles in the thorax.

a *hinge joint*. The two bones meet to form a hinge and they can move only in one plane. The best examples are found in our elbows and knees. Hinge joints are also very common in insects and crabs. The legs of these animals are made up of several segments, each joined to the next by a hinge joint. Another important type of joint in our own bodies is the *ball and socket joint*. This is found in our hips and shoulders. The upper ends of the leg and arm bones are ball-shaped and they fit into cup-shaped hollows in the hip and the shoulder. The balls can move round in any direction, and so we have a very wide range of movement at the hip and the shoulder.

Other joints in the vertebrate body allow rather less movement. The head can swivel round on the upper two vertebrae, but the rest of the vertebrae can move only slightly against each other, but there are a lot of them and all the little movements added together allow us to bend our backs. Some of the tightest joints are found in the skull. The skull bones are not tightly joined in young mammals because they have to be able to grow, but in mature animals they are tightly knitted together. They have interlocking edges rather like a very complicated jigsaw puzzle.

72

Muscles for Movement

Muscles form the flesh of an animal and they generally make up a large proportion of its bulk. They consist of thousands of little tapering cells collected into bundles. A muscle, such as the biceps of the arm, contains several bundles of fibres. Most muscles, are connected to the skeleton at each end. Vertebrate muscles are attached to bones by means of very strong cords called *tendons*.

When a nerve stimulates a muscle the muscle fibres contract rapidly. The whole muscle gets shorter and fatter and, because it is attached to a bone at each end, one of the bones must move. Contraction of the biceps muscle in the arm pulls the forearm up and closes the elbow joint. A muscle can relax and it can then be stretched out again, but it cannot expand itself and push a bone. Muscles therefore have to work in pairs. The biceps works with the triceps on the back of the arm: the biceps bends the arm and the triceps stretches it again.

The muscles are not always very close to the bones which they pull. The bones of a bird's leg and foot are worked by muscles high up in the leg. These muscles are joined to the foot bones by long tendons—the sinews that you have to remove when you prepare a chicken for the oven. The lower part of the leg of a horse or an antelope contains very little flesh or muscle. The bones are worked by large muscles at the top of the leg The legs themselves are quite light and this helps the horse to run swiftly.

The muscles that help us to move—the *skeletal muscles* as they are called—can contract very strongly but they get tired quickly. The powerful contractions enable us to run very fast for a short distance, but they soon get tired and we cannot keep it up. Similarly, we can lift heavy weights for a short time, but we soon have to drop them because our muscles get tired.

Involuntary Muscles

The skeletal muscles are also called *voluntary muscles*, because we can move them when we want to. There are other muscles, however, which go on working without our knowing about them. They include the muscles that push food along the digestive canal and the muscles that control the size of the pupil in the eye. Unlike the skeletal muscles, these *involuntary muscles* do not get tired. They go on working all the time.

Both the frog and the cheetah have very strong muscles in the upper parts of their legs. Contraction of these muscles straightens the legs and sends the animal hurtling forward. The cheetah is the fastest of all land animals and can run at 70 mph for short distances.

The Senses of Animals

Animals use their eyes, ears, and other sense organs to find out what is going on in their surroundings. Signals are sent from the sense organs to the brain and the brain then instructs the body to react accordingly. Most of the reactions concern escape from danger or the finding and catching of food, so the senses are very important for an animal. An animal that does not detect changes in its surroundings, or an animal that does not react correctly to these changes, will soon perish.

The simplest animals, such as Amoeba, have no special sense organs. The whole body surface of these creatures is sensitive and they react to unpleasant sensations by moving away. The more advanced animals have a number of special sense organs, each constructed for picking up different kinds of signals from the surroundings. There are also nerves, which carry signals from one part of the body to another, and a brain which is the control center. Signals are received in the brain and the brain then sends signals to the body telling it to take the correct action.

We tend to think that we have only five senses: sight, hearing, taste, smell, and touch. These are the most obvious senses that we use to tell us about our surroundings, but there are several more. The senses of balance and temperature, for example, are quite important. And there are senses that tell us about the conditions inside our bodies. The sense of pain tells us when something is wrong, and other nerves in our bodies tell us when we are hungry or thirsty. Similar kinds of sense organs are found in other animals, although they are not always made up in the same way and they are not always on the same part of the body. The senses are not equally developed, and some of the sense organs may be absent. Many animals living in the dark, for example, have no eyes.

The Eyes

The eyes are the organs of sight. They are normally found on the head and they are sensitive to light rays. The eyes of vertebrate animals are all more or less alike. The basic parts are the *lens* and the *retina*. The lens gathers the light rays and focuses them on to the retina. The latter consists of millions of light-sensitive cells and, when light falls upon them, they send signals to the brain. The brain then interprets the signals as a picture.

Insect eyes are constructed on a very

Hunting animals need to be able to judge distances well. Two eyes at the front (instead of on the sides) help them to do this. The owl's large eyes help it to see well at night.

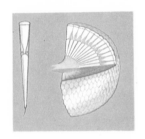

Insect eyes are made up of lots of small cone-shaped lenses. The surface of the eye is often brilliantly colored, as in the horsefly above.

Sounds reaching the human ear drum set up vibrations in a chain of tiny bones. These carry the sounds to the inner ear, where the vibrations are picked up by nerves and transmitted to the brain.

different pattern. They are known as *compound eyes* because each one is made up of numerous tiny units. Each unit is cone-shaped and they all fit together so that their lenses form a patchwork pattern on the surface of the eye. Each lens sends its own signal to the brain, so that the insect sees a rather blurred mosaic image. Although the image is not very clear, the insect easily spots movements in its neighbourhood. You will know just how easily if you have ever tried to swat a fly or, better still, catch a dragonfly. The dragonflies feed on smaller insects, which they catch in mid air. They rely entirely on sight to do this, and their eyes have up to 30,000 units. Insects which rely less on sight have smaller eyes with fewer units.

The Ears

Sounds play a large part in the lives of many animals. They are used as warnings and as signals to attract mates. The sound-receiving organs, or ears, are normally little

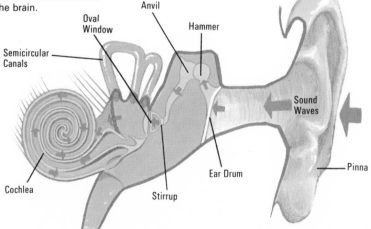

membranes stretched like drumskins across an opening. They vibrate when sound waves hit them and then send signals to the brain. Some insects pick up sounds with sensitive hairs. The sound receiving organs are normally on the head, but crickets have their ears on their front knees.

Chemical Senses

The senses of taste and smell are the chemical senses. The sense of smell tells us about chemicals in the air. Most substances give off some sort of vapor or scent and this consists of minute particles called molecules which affect our smelling organs, or olfactory organs to give them their proper name. Our own sense of smell, centered in our noses, is not very good but some other animals have extremely good senses of smell. Insects smell with their feelers, and some male moths have such sensitive smell receptors that they can smell female moths more than a mile away.

The sense of taste tells about the substances we are actually touching. Minute

The impala must always be alert to danger. Even while drinking, its eyes are scanning the area and its ears are pricked up waiting for the faintest sound of danger.

amounts of the substance dissolve and affect sense organs. We can taste things only with our tongues, but other creatures can use other parts of their bodies. Flies, for example, have taste organs on their feet and they actually taste things just by landing on them.

The Sense of Touch

Most of the body surface is sensitive to touch, although some parts are more sensitive than others. Hairs normally have tiny nerves wrapped around their bases, and the nerves tell the brain when the hairs are touched. Many animals use hairs as special organs of touch. Mice and cats, for example, have very sensitive whiskers which help them to find their way in the dark. If the whiskers touch anything the animal draws back.

Right: The arrangement of the nervous system is very similar in man and the frog, although man's brain is very much larger and more efficient.

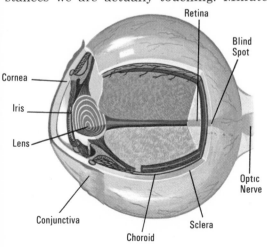

Left: Man's eyes are his most important sense organs and they tell him a great deal about his surroundings. Light is gathered in by the lens and focused on to the retina at the back. Nerve cells there are stimulated and they send the picture to the brain.

Retina

Blind Spot

Cornea

Iris

Lens

Optic Nerve

Conjunctiva

Sclera

Choroid

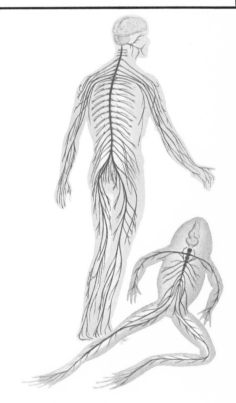

Feeding and Digestion

All animals must have food, and most of them must eat regularly if they are to survive. Their food provides energy and it also provides the materials they need for building up their own bodies. Green plants can make energy-containing foods from water and carbon dioxide by the process of photosynthesis (see page 17). They get the energy from the sun. Animals cannot do this and they have to rely on green plants to provide their energy. They eat plants or other animals, but the energy always comes from a plant to start with. If you follow any food chain back a stage or two you will always find a plant at the beginning. To take the simplest of examples, fox eats rabbit and rabbit eats grass.

Some animals feed entirely on plants. They are called herbivores. Some species feed entirely on other animals. They are called carnivores or flesh-eaters. Other species feed on both plants and animals. They are called omnivores.

The waterbuck eats grass and other leaves and is a typical herbivore.

The toad is a carnivore, feeding on slugs and insects which it catches with its long sticky tongue.

Herbivores

Herbivorous or plant-eating animals include voles, rabbits, deer, sheep, horses, snails, caterpillars, and many others. Most of them feed on the leaves and other juicy parts of the plants, but some animals feed on wood. Examples include the termites and many beetles. One large group of insects feeds by sucking the sap from plants through very specialized mouths equipped with 'hypodermic needles'. These insects include greenfly and scale insects and they do great damage to crops.

Brachiopods are marine creatures resembling scallops in some ways. Tiny hairs on their 'arms' draw a current of water into the body and the animals then swallow any food particles that are brought in.

Carnivores

Carnivorous or flesh-eating animals include cats, dogs, weasels, owls, snakes, frogs, sharks, ladybirds, starfishes, sea anemones, and many other creatures. Their methods of catching food vary a great deal and some are described in the following pages. Some ani-

mals chase their prey, others lie in wait until something comes along. Some spiders and some caddis fly larvae actually set traps for their prey by spinning silken webs or snares. Some snakes and other animals use poison to kill or paralyze their food before they eat it.

Most of the carnivores catch their food while it is alive, but some feed on dead animals, or carrion. Carrion feeders include vultures and hyenas. These creatures may catch living prey, especially young or sick animals, but they normally wait until an animal dies before attacking it. This seems a rather unpleasant habit, but the carrion feeders do a good job in keeping the countryside clear of rotting bodies. Many insects also feed on dead animals, including the famous burying beetles or sexton beetles. These little creatures bury dead animals by dragging soil out from under them. They then feed on them or on the fly maggots that also eat the rotting flesh.

Various blood-sucking creatures, such as fleas and bed bugs can also be classed as carnivores.

Omnivores

Many herbivorous animals will eat flesh sometimes. For example, voles and other rodents will often eat insects or even their own babies. Carnivores may also eat vegetable food from time to time. But there are some animals which regularly eat a mixture of plant and animal materials. These are the omnivorous animals. They include most bears, the badger, the rat, and the cockroaches. Man is also an omnivorous animal.

Filter Feeders

Many animals living in water feed on tiny particles which they filter out from the water. These particles might be tiny plants or animals, or else they might be decaying matter falling to the bottom of the sea or the pond. The filter feeders are therefore omni-

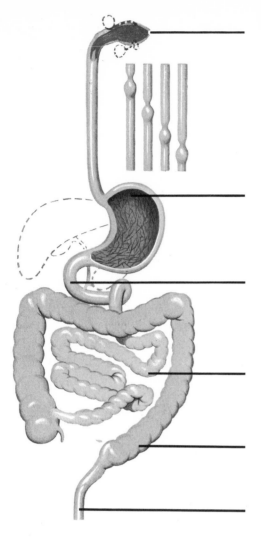

Digestion begins in the mouth, where the salivary glands pour saliva on to the food. The chewed and moistened food then passes down the gullet and into the stomach.

In the stomach the food is churned up with dilute acid and certain digestive juices. Food particles are reduced in size and the chemical breakdown of protein begins. After a few hours the food is reduced to a liquid 'soup' and it is ready to pass on.

In the duodenum the food meets other digestive juices, including some from the liver and pancreas. Chemical breakdown is accelerated.

Chemical breakdown is completed in the small intestine and the food materials are taken into the bloodstream.

Undigested material passes into the large intestine or colon and water is reabsorbed from it there.

The remaining material passes into the rectum and is then periodically passed out of the body.

particles held back are passed to the mouth. Freshwater mussels are useful in fish tanks because they filter out a lot of the microscopic algae which would otherwise make the water green. The animals obtain oxygen from the water current at the same time as they strain out their food.

Some marine worms are also filter feeders. They have crowns of feathery arms or tentacles around their heads and these strain out particles falling from above.

Digestion

When an animal has eaten its food it must convert it to some usable form. This conversion is carried out in the digestive system. The food is broken down into simple substances which can be carried around the body and used to provide energy or to build new cells. The detailed structure of the digestive system varies a great deal, but it is basically a tube which begins at the mouth and ends at the anus. Food is taken in through the mouth, and the front part of the system is concerned mainly with breaking down the food into smaller pieces and simpler substances. The next part of the system is concerned with absorbing the digested food into the blood, and the last part is concerned with getting rid of the undigested material or waste. Sea anemones, jelly fishes, and some other simple creatures have only a mouth and this must serve for taking food in and for passing waste out.

vorous animals. Bivalve molluscs are among the most efficient. These are the shellfish with two parts to their shells. They include scallops, mussels, and razor shells. Two tubes or siphons stick out from the body, one to suck in water and the other to squirt it out. While the water is flowing through the animal it is strained or filtered and the

GETTING RID OF WASTE

The food that is taken into the body through the blood-stream is used to provide energy and building material. Worn out parts of the body must be replaced. These processes all produce waste products which must be removed. The removal of these waste products is called excretion. The main organs involved in excretion are the kidneys, which lie near the back of the body. Each kidney consists of about a million little tubes and each tube is surrounded by a tiny blood vessel. Blood coming into the kidney is filtered through these little tubes. Unwanted materials, including excess water, are retained by the kidney tubes and they flow into a large collecting duct called a ureter. This carries the waste materials (now called urine) from the kidney to the bladder. From the bladder the urine passes to the outside.

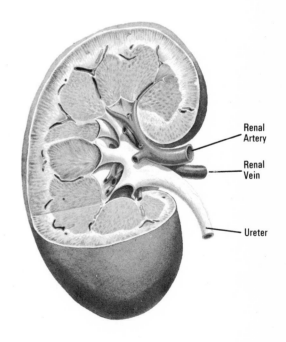

Renal Artery

Renal Vein

Ureter

Animal Teeth
Only the backboned animals have proper teeth and jaws, and even then they do not all have teeth. The teeth of fishes, amphibians, and reptiles are normally simple spikes, although the snake's fangs are very special teeth. Birds have no teeth at all today, although some extinct birds had them. The most complex teeth are found in mammals. Their job is to catch, cut, and chew the food. The number and arrangement of a mammal's teeth depend to some extent upon its food. Flesh-eaters, such as dogs, have large stabbing eye teeth (canines) near the front. These stab the prey and rip the flesh. The back teeth of flesh-eaters have sharp edges and they slice the meat into pieces. Grazing animals do not need stabbing teeth, and they have gaps where the eye teeth would normally be. Their back teeth have broad flat surfaces suitable for grinding up the grass.

How Animals Breathe

Breathing, like feeding, is one of the characteristics of all living things. Breathing involves taking in oxygen, which is then used to 'burn' the food in the cells of the body and to release energy. The whole process of obtaining oxygen and using it in the body is called *respiration*. One of the waste products of respiration is a gas called carbon dioxide. The breathing organs get rid of carbon dioxide as well as obtaining oxygen for the animals.

The simplest animals, such as the protozoans and the jellyfishes, have no special breathing organs. They are quite small creatures and they are not very active. They do not use up much energy and they do not need very much oxygen. Enough oxygen seeps into their bodies from the surrounding water without the need for special breathing organs.

Larger and more active animals need special breathing organs and they also need some sort of transport system to carry the oxygen from the breathing organs to the rest of the body. The blood stream is the transport system. It carries all sorts of materials around the body, not just oxygen and carbon dioxide.

Breathing under Water

Many aquatic animals come to the surface every now and then to breathe air. Examples include whales, seals, and some water snails. Many others, however, can remain

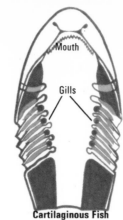

Cartilaginous Fish

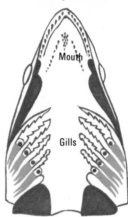

Bony Fish

The fish's gills lie in pouches at the back of the throat. In the shark-like fishes (upper picture) each gill has its own separate opening on the side of the body. The gills of the bony fishes (lower picture) are covered by the operculum and there is only one opening on each side.

Birds are very active animals and they need a good supply of oxygen. Their ribs and keel move backwards and forwards, forcing air in and out of the lungs and air sacs.

permanently under the surface and rely on oxygen dissolved in the water. The smaller creatures absorb oxygen all over their body surface, as already explained, but the larger ones have special breathing organs called *gills*. In common with all breathing organs, the gills have a large surface area through which to absorb oxygen. They also have thin walls and a good blood supply to carry the oxygen away to other parts of the body.

Fish gills consist of columns of little finger-like organs situated in cavities connected to the throat. The cavities also open through a number of slits on to the sides of the body. The fish takes in a mouthful of water and then, by shutting its mouth and raising the floor of its throat, it forces the water over the gills. The blood arriving at the gills has relatively little oxygen in it, and so oxygen tends to pass into the blood from the water.

Other animals breathing by means of gills include the crabs and lobsters, the bivalves (cockles and mussels), and many water snails. The gills of the crabs and lobsters are rather feathery outgrowths from the body wall and from the bases of some of the legs. They are covered by the shell. One of the other limbs wafts a continuous current of water through the gill chamber. This brings a continuous supply of oxygen to the gills.

The gills of bivalves are very large and they serve to trap food particles as well as to absorb oxygen. The animals have two breathing tubes or siphons. One brings a current of water into the gill chamber and the other carries the water away. Many water snails, including most of those that live in the sea, breathe by means of gills. These are feathery structures lying in a space between the animal's body and the 'cloak' of skin that encloses it. Water is pumped in and out of the cavity and the gills absorb oxygen from it. Land snails have lost their gills, but they still have the gill cavity. It is lined with blood vessels and acts as a lung with the snails pumping air in and out of it. Some snails returned to the water after a period of life on land. Most of them still breathe air and they can be seen coming to the surface now and then to renew the air in their lungs.

Air-breathing Vertebrates

The reptiles, birds, and mammals all breathe by means of *lungs*. These are thin-walled sacs, rather like balloons, and they lie in the chest cavity. Air is pumped in and out of them and oxygen is absorbed. Most amphibians also have lungs, although these are rather inefficient organs. The amphibians are moist-skinned animals and some can absorb a good deal of oxygen through the skin. The lining of the frog's mouth and throat is also very thin and well supplied with blood vessels. Much oxygen is absorbed in this region and a frog can often be seen drawing air into its throat and pumping it out again through the nostrils only occasionally taking it down into its lungs. The lungs themselves are small and have smooth walls. This means that they have only a small surface area and they cannot absorb much oxygen from the air.

Reptile lungs are much more efficient than those of amphibians because the reptiles cannot breathe through their skin. The lungs are elaborately folded and they have a fairly large surface area. This folding is taken even further in the lungs of mammals. Each lung is made up of thousands of tiny air pockets, each of which is surrounded by blood vessels. This arrangement provides a very large surface area for the absorption of oxygen, but it does mean that there is a good deal of stale air in the remote parts of the lung. Not all the air can be exchanged at each breath.

Birds are extremely active animals and they need large amounts of oxygen. Their lungs are extremely efficient. Instead of being closed at the end, the lungs lead on to thin-walled air-sacs. When the bird breathes in the air rushes in through the lungs and along to the air-sacs. At each breath, the air in the lungs is completely changed and there is never any stale air there. The air-sacs reach many parts of the body and, although no oxygen is absorbed from them, they do help to buoy the bird up in the air.

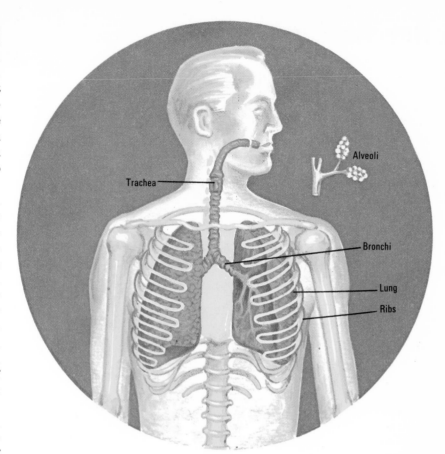

The human lungs shown in position in the chest cavity. The small diagram shows how the lung tubes end in little clusters of thin-walled pouches where oxygen is actually absorbed.

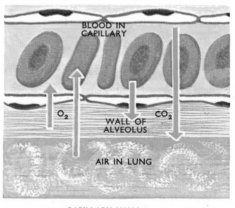

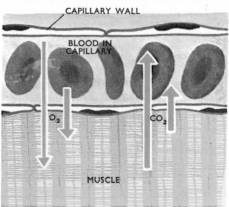

Left: Diagrams showing how oxygen spreads throughout the human body. The oxygen concentration in the lungs is higher than that in the blood. Consequently oxygen spreads, or *diffuses*, through the wall of the alveoli into the blood capillaries. At the same time the blood is richer in carbon dioxide than the air in the lungs, so carbon dioxide passes from the blood into the lungs. At the tissues the situation is reversed. The blood is now richer in oxygen and poorer in carbon dioxide. So oxygen passes from the blood to the tissues and carbon dioxide from the tissues to the blood. The blood then flows back to the lungs and the cycle is repeated.

Right: Part of an insect's breathing system. The fine tubes reach all parts of the body and carry oxygen to all the tissues.

Insect Breathing Systems

The insects, together with the centipedes and a few other arthropods, have a breathing system quite different from that of other animals. The body is penetrated by a maze of tiny tubes called *tracheae*, which open on to the body surface at holes called *spiracles*. Air seeps along the tracheae and carries oxygen to all parts of the body. The inner ends of the tubes normally contain liquid, but this is reabsorbed when the insect is very active. The air can then get further along the tubes and provide the extra oxygen needed for the activity. The blood plays no part in carrying oxygen to the cells, although it does carry some of the carbon dioxide away.

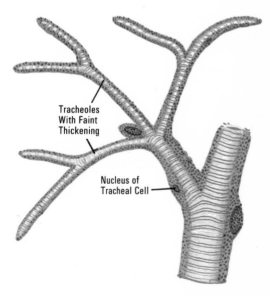

The Blood System

All but the simplest animals have some sort of blood system. It is the body's transport system, carrying materials from one part of the body to another. It carries food from the digestive organs to all parts of the body. It also carries waste products from the tissues to the excretory organs, which pass them out of the body. In most animals the blood carries oxygen from the breathing organs to the tissues. Hormones—chemical 'messengers'—are also distributed by the blood. A less obvious function, but still a very important one is the distribution of heat around the body. The blood plays a very important part in keeping the bodies of birds and mammals at constant temperatures. Yet another function of the blood in some animals consists of fighting infections and destroying germs.

Mammalian Blood

Mammalian blood, like that of the other vertebrates, consists mainly of a liquid called plasma. Floating in the plasma there are millions of tiny cells called corpuscles. The plasma is a pale yellowish color. It is mainly water, but it contains a very complicated mixture of proteins, sugars, and dissolved minerals. This is in addition to the materials that it carries from place to place in the body.

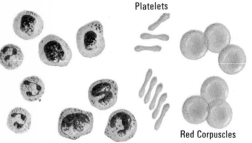

White Corpuscles

Platelets

Red Corpuscles

The blood corpuscles are of two main types: red and white. Red corpuscles, which give the blood its red color, are minute disc-shaped cells. They are so numerous that a thimble full of human blood would contain almost 10,000 million of them. Their job is to carry the oxygen around the body. The red coloring matter, haemoglobin, takes up the oxygen in the lungs and releases it again as the blood flows around the body (see page 79). Mammalian red corpuscles are made in the bone marrow and, when they are mature, they contain no nuclei. They do not live very long—about 100 days in man—and new ones are always being produced. The red cells of other vertebrates do have nuclei and they last much longer than mammalian cells.

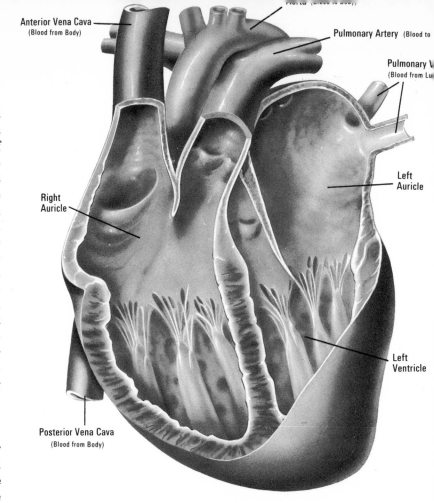

Anterior Vena Cava
(Blood from Body)

Pulmonary Artery (Blood to

Pulmonary V
(Blood from Lu

Right
Auricle

Left
Auricle

Left
Ventricle

Posterior Vena Cava
(Blood from Body)

Above: A drawing of the human heart showing the various chambers and the valves which ensure that the blood flows in the right direction.

Left: Red and white blood corpuscles. The red ones have no nuclei.

White blood corpuscles are much less numerous than the red ones. They are concerned with the control of infection. In other words, they fight germs that get into the blood. Some of the white cells are rather like amoebae (see page 86) and they actually swallow up the germs. Other white cells produce *antibodies*. These are proteins which are specially designed to kill the germs. Each kind of germ causes the white cells to produce a special kind of antibody. As well as killing germs, the white cells help to heal wounds. They 'eat' the damaged cells and clean up the area so that new cells can grow properly.

In addition to the red and white corpuscles, the blood contains some smaller bodies called *platelets*. These break down when they come in contact with the air and they cause the blood to clot. They are thus very important, because they help the blood to form a scab which stops the blood flowing from an injury.

The Heart and the Blood Vessels

The insects and other arthropods have what is known as an open blood system. The blood lies in large cavities and surrounds all the internal organs. Most animals, however, have a closed blood system with the blood running in narrow vessels.

The blood has to circulate round the body, and so the blood system has to have a pump. The heart is the pump. The shape and structure of the heart vary from one animal group to another, but the heart is always a muscular bag and it is normally divided into two or more chambers. The blood vessels that carry blood away from the heart are called *arteries*. The main arteries have thick, muscular walls. As they get further from the heart they divide and send branches to all parts of the body. When these blood vessels reach their destination they divide up into lots of tiny vessels called *capillaries*. These have very thin walls and materials can pass very easily between the blood and the tissues. The capillaries then join up again to form the *veins* which carry the blood back to the heart. The veins have thinner and less muscular walls than the arteries and the blood pressure in them is much lower.

The diagrams below show the arrangement of the major blood vessels in various vertebrate animals.

Warm-blooded or Cold-blooded?

The birds and mammals can be distinguished from all other animals because they are able to keep their bodies at constant high temperatures. For this reason, we call them warm-blooded animals. Fishes, frogs, snakes, and all the invertebrate animals are unable to keep their temperatures up and we call them cold-blooded animals. But this does not mean that they are always cold. Their bodies are always more or less at the temperature of their surroundings. A snake basking in the sunshine will actually be quite warm—warmer than a warm-blooded animal sometimes. Birds and mammals have to use up a lot of energy to keep their temperatures up. This is why they need regular food. Cold-blooded creatures do not need so much energy and they can go without food for long periods. Some large snakes, for example, have been known to go without food for a year.

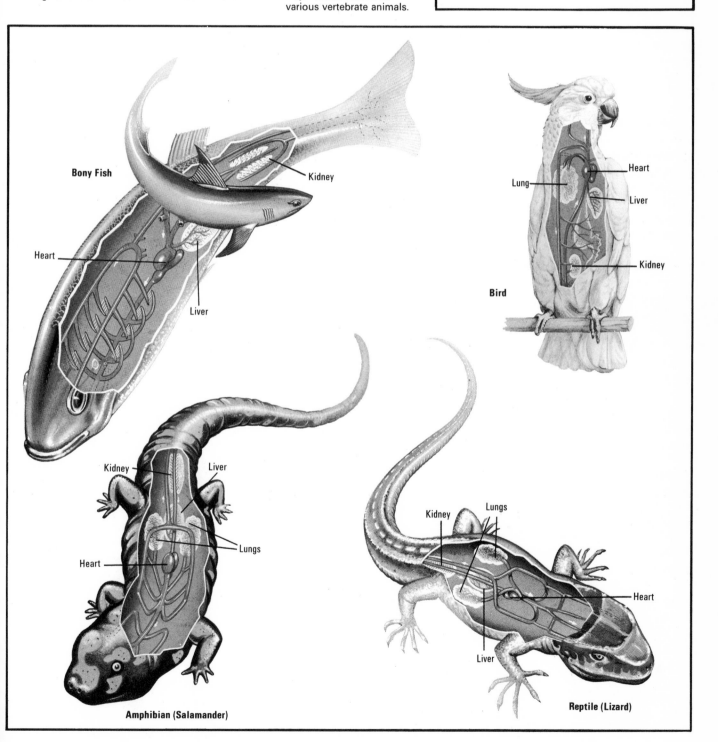

Bony Fish

Kidney

Heart

Liver

Bird

Heart

Lung

Liver

Kidney

Kidney

Liver

Lungs

Heart

Amphibian (Salamander)

Kidney

Lungs

Heart

Liver

Reptile (Lizard)

Reproduction

Reproduction is one of the characteristics of living things. Not every animal actually reproduces itself but, with the exception of worker ants and other similar examples, all normal animals *can* reproduce when they are old enough. Reproduction is essential if the species is to continue in existence. There are two main methods of reproduction—sexual and asexual. Sexual reproduction, which is the more important of the two, involves the joining together of special cells called gametes or sex cells. Asexual reproduction takes place without any joining of cells and often involves little more than the breaking of an animal into two parts. Almost all animal species reproduce sexually at some time or other, but only the simpler animals can reproduce asexually.

Splitting in Two

The simplest form of reproduction takes place in some of the protozoans, such as Amoeba. These creatures consist of only one cell. The cell goes on growing until it reaches a certain size—perhaps until it gets too big for its nucleus—and then it divides into two. Each half then starts growing again. In a sense, these creatures have eternal life because they do not die from old age, but large numbers are eaten by other creatures. Splitting into two halves, or binary fission to give it its scientific name, is a form of asexual reproduction. But Amoeba can also reproduce sexually at times.

Budding

Some simple animals, notably the corals and their fresh-water relative Hydra, produce new individuals by a process called budding. A 'bud' develops somewhere on the body and gradually grows to form a branch.

Female toads take no interest in their offspring. They lay large numbers of eggs, but only about two will ever reach maturity.

Earthworms all contain both male and female organs, but they have to pair up before they can lay eggs. Some species pair on the surface of the ground at night, joining themselves together with bands of slime.

Below: Hydra has two methods of reproduction. New individuals can grow out as buds (upper pictures) or the animal can reproduce sexually by producing eggs and sperm. The fertilised egg grows into a new animal (bottom pictures).

This branch is a smaller version of the parent. A branch of Hydra normally separates from the parent and begins an independent life. In many other creatures, however, including most of the corals, the branches do not separate. The animals form colonies. Each individual normally looks after itself, but some colonies contain special individuals which are concerned only with sexual reproduction.

Sexual Reproduction

This method of reproduction involves the joining together of special cells called gametes. The joining together of the gametes is called fertilization. In Amoeba and some of the other simple animals all the gametes are alike, but the more advanced animals have two kinds of gametes. One is called a sperm and the other is called an egg or ovum. The two kinds of gametes are usually, but not always, produced by separate individuals. Sperms are produced by the male sex and eggs are produced by the female sex. A sperm must join with an egg before a new animal can be produced.

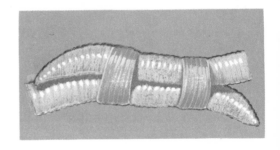

Some water-dwelling animals merely release their sperms and eggs into the water and leave fertilization to chance. In most animals, however, the two sexes have to meet before reproduction can take place. Some animals give out a scent which attracts the opposite sex. Others rely on elaborate displays to bring the sexes together and to ensure that eggs and sperms are produced at the right time. The eggs may be fertilized inside the female's body or outside it.

When an egg has been fertilized it starts to develop into the new animal. The egg contains enough food material for all the early stages of development. Eggs that have been fertilized inside the mother's body may be laid soon afterwards, or they may remain in the mother's body. The young completely develop before they leave the mother's body. The mother then gives birth to young that are small but fully formed editions of their parents. In most of the mammals, the group of animals to which we ourselves belong, the young animal remains inside its mother, developing for many months. It obtains food from the mother's own body.

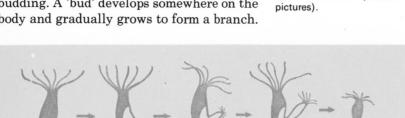

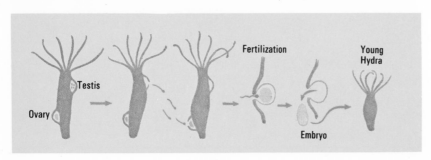

Testis

Ovary

Fertilization

Young Hydra

Embryo

Parental Care

Some animals produce large numbers of eggs or young, while others may have only one or two babies. The number of eggs or young depends very much on the dangers that they face during their lives. Young animals are easily caught and eaten by other creatures. Animals that merely scatter their eggs and abandon them suffer the most, and these animals survive only by producing huge numbers of eggs. The female cod, for example, may produce about 9 million eggs, but only two of them can be expected to reach maturity. Animals that look after their offspring do not face the same problem and they can get by with far fewer young. Birds, for example, do not rear huge families. Most mammals have even smaller families.

Hermaphrodites

The earthworms and many snails do not have separate sexes. Each individual possesses both male and female reproductive organs and can produce both sperms and eggs. Such animals are called hermaphrodites. But, although they can produce sperms *and* eggs, the animals have to pair before they can reproduce. When they pair they exchange sperms. Each then goes away and lays its own eggs, fertilized by sperms from the other animal. This arrangement is very useful for slow-moving creatures such as worms and snails. It means that they can pair with *any* adult of the same kind. The chances of finding a mate are therefore greatly improved.

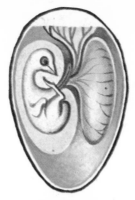

A bird's egg contains just the right amount of food for the development of the young bird. The bird completely fills the egg when it has used up all the food.

Emperor penguins, like many other sea birds, breed in large colonies. The male looks after the single egg for about six weeks, then the female returns to look after the chick while the male collects food. This is a very high degree of parental care.

Parthenogenesis

Many stick insects, greenflies (aphids), and other insects can lay eggs which hatch without ever being fertilized. The male sex is very rare or even unknown in some species and the females reproduce almost entirely by laying these special eggs. This method of reproduction is called parthenogenesis. It is an asexual method really, because there is no joining of cells. But, on the other hand, we can think of it as a special kind of sexual reproduction because it does involve the production of eggs in the normal reproductive organs of the insect.

Regeneration

A number of the simpler animals are able to re-grow or regenerate parts of their bodies that have been lost or damaged. In some animals this ability amounts to a form of asexual reproduction. Hydra, for example, will not normally die if it is cut into two parts. Each part will regenerate the missing sections and become a complete animal. Flatworms are also very good at regenerating themselves. These are the little flat grey or black creatures that glide about in ponds. They can be cut into several small pieces and each piece will normally grow into a new animal. What is more, the new head will always develop on the front end of the section. Starfishes can grow new arms if any are lost, and the broken arms can sometimes grow new bodies! Young insects can grow new legs and antennae if these are broken or lost, but the new ones are usually a bit smaller than the originals. Adult insects cannot grow new parts.

Animals with backbones can repair cuts and wounds. A few amphibians, like the newt, can regenerate lost limbs. Many lizards have tails which break off when seized. The piece broken off continues to wriggle for a while and distracts an enemy while the lizard makes its escape. A new tail will grow, but it is normally shorter than the old one and it contains no real backbone.

Animal Classification

The animal kingdom can be divided quite conveniently into two large 'parcels'—animals with backbones, and animals without backbones. These groups are properly called vertebrates and invertebrates, but they are very unequal groups. To divide the animal kingdom into vertebrates and invertebrates is rather like splitting up a town's shops into those that sell cakes and those that do not sell cakes. The vertebrates belong to just one of about 20 major groups into which the animal kingdom is divided.

These major groups are called *phyla* (singular *phylum*), and most of them are further divided into *classes*. The main phyla and classes are shown on this 'family tree' of animals. Some of the phyla are obviously related to each other and have clearly descended from a common ancestor. Others are more difficult to fit in because they are not clearly related to other groups.

MOLLUSCS

Molluscs are soft-bodied animals and many have shells. The young stages of some snails suggest that they are related to the annelids.

BRACHIOPODS

Brachiopods are shelled animals, some of which are like cockles at first sight. But they are very different inside. They were very common in earlier times but only a few exist today.

POLYZOANS

Polyzoans are tiny colonial creatures living in the sea and freshwater. Some are called sea mats. This group is rather a mixed bag but some of its members are related to brachiopods.

NEMATODA

Roundworms (Nematoda) are nearly all minute creatures. Many are parasites. Their needle-like bodies are not segmented and they are not related to earthworms.

PLATYHELMINTHES

Flatworms (Platyhelminthes) are rather flat, worm-like creatures. We do not know how they are related to other groups. They include tapeworms, liver flukes and planarians.

OCTOPUSES

BIVALVES

SEA UR

COELENTERATE

Coelenterates, represent jellyfishes and sea anem are simple many-celled tures. They branch from other groups very

PORIFERA

Sponges (Porifera) are many-celled animals but they are quite unlike any other animal group. They have no close relatives. Sponges feed by filtering tiny particles from the water.

PROTOZOA

The protozoa are the simplest of all animals, consisting of just one cell each. There are several groups but we do not know how they are related.

THE EARLIEST FORMS OF LIFE

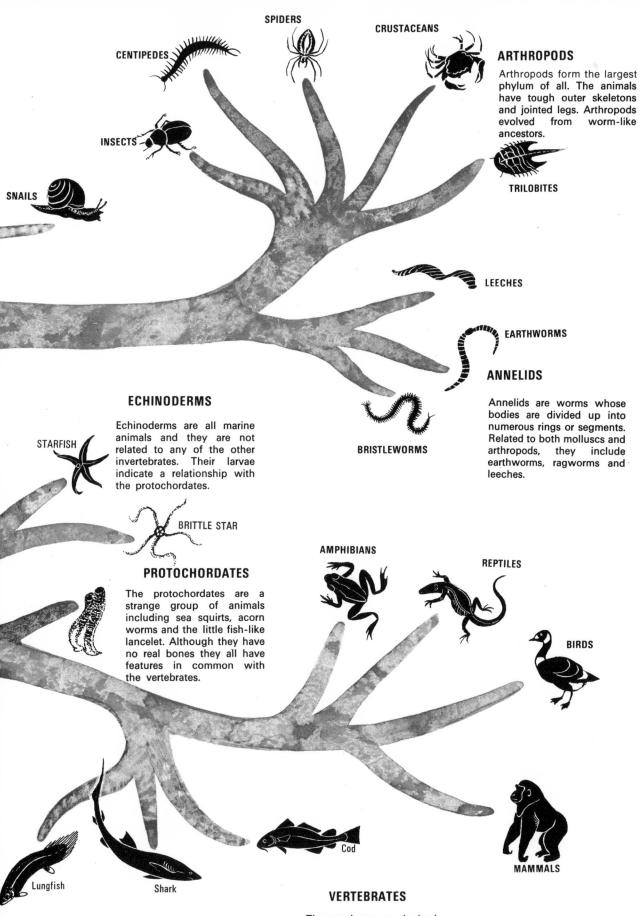

CENTIPEDES

SPIDERS

CRUSTACEANS

ARTHROPODS

Arthropods form the largest phylum of all. The animals have tough outer skeletons and jointed legs. Arthropods evolved from worm-like ancestors.

INSECTS

TRILOBITES

SNAILS

LEECHES

EARTHWORMS

ANNELIDS

Annelids are worms whose bodies are divided up into numerous rings or segments. Related to both molluscs and arthropods, they include earthworms, ragworms and leeches.

ECHINODERMS

Echinoderms are all marine animals and they are not related to any of the other invertebrates. Their larvae indicate a relationship with the protochordates.

STARFISH

BRISTLEWORMS

BRITTLE STAR

AMPHIBIANS

REPTILES

PROTOCHORDATES

The protochordates are a strange group of animals including sea squirts, acorn worms and the little fish-like lancelet. Although they have no real bones they all have features in common with the vertebrates.

BIRDS

Cod

Lungfish

Shark

MAMMALS

VERTEBRATES

The vertebrates are the back-boned animals and, with the protochordates, they make up the phylum chordata.

The Microscopic World

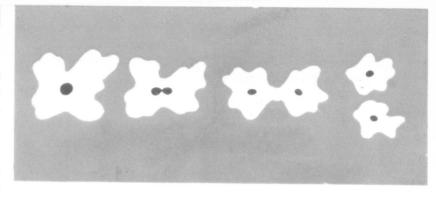

A jar of dirty pond water looked at with the naked eye does not look very interesting, but if some of the water is looked at under a microscope it becomes a busy highway, with animals of many kinds moving to and fro. There will probably be various algae too.

Most of the animals will be tiny single-celled creatures called *protozoa*. This name means 'first animals', for it is likely that the earliest animals were quite similar to some of today's little pond creatures. One of the animals that you are likely to see, especially if your sample contains some mud from the bottom of the pond, is *Amoeba*. This is popularly regarded as the most primitive form of life and it is often described as a shapeless blob of jelly. But there really is more to it than that. Inside the little speck of jelly, no more than half a millimeter across, all the processes of life take place.

To say that Amoeba is shapeless is not quite accurate. It would be more correct to say that the animal's shape is always changing. Finger-like projections are continually being pushed out from the body and withdrawn and, as they move forward, the animal's body moves too. As one of these 'fingers' is pushed out, living material from the rest of the body flows into it. The whole animal therefore flows along. It can change direction merely by putting out a 'finger' in another direction.

The animal also feeds by putting out these 'fingers'. They flow around small particles of

Amoeba merely splits into two halves when it gets too big. Each half then begins life as a new animal.

Euglena, one of the flagellates which are on the border between plants and animals.

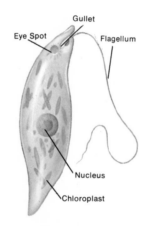

Gullet
Eye Spot
Flagellum
Nucleus
Chloroplast

Left: Amoeba has no problem getting oxygen or getting rid of waste. Materials simply pass through the thin walls of the animal.

food and completely engulf them. The particles then move around inside the body and are gradually digested. Any indigestible material is simply left behind as the animal flows forward.

Amoeba finds its food by a chemical sense. The whole body surface is sensitive to chemicals in the water, and the animal reacts by moving towards food materials and away from harmful materials such as acids. The animal is also sensitive to light and it moves away from bright light. This sort of reaction ensures that the animal remains in suitable surroundings.

Breathing is no problem for the Amoeba because it is so small. Plenty of oxygen seeps in through its thin skin. The thin skin does, however, have one disadvantage: water can pass through it and the animal tends to absorb water from its surroundings. This extra water must be removed. Amoeba collects it in an 'elastic bag' somewhere in the body and then squirts it out through the skin.

Oxygen In
Through Body
Surface
Waste Out
Food is
Engulfed
Water And
Waste Out
Oxygen In

Below: Paramecium, another very common pond dwelling protozoan.

Ectoplasm
Food
Vacuole
Endoplasm
Pseudopodia
Engulfing
Small Plant
Contractile
Vacuole
Nucleus

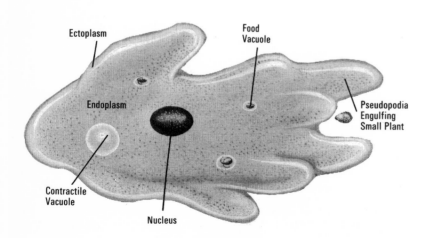

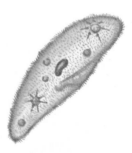

Left: A greatly enlarged Amoeba about to engulf a small plant. The pale sphere on the left is the contractile vacuole—the 'elastic bag' which removes excess water.

86

As it grows, Amoeba gradually gets too big for its nucleus, or brain. When it reaches a certain size the animal simply divides into two halves. The nucleus divides first and one half goes into each new Amoeba.

Plant or Animal?

Some of the simplest creatures in the world are known as flagellates. They have one or more whip-like hairs, called *flagella*, which help them to swim about. Some flagellates, such as *Chlamydomonas*, are clearly plants. Some are clearly animals. But others are right on the border-line between plants and animals. *Euglena* is one of these border-line creatures. There are several different kinds, some of which feed like plants and some of which feed like animals. Some species can even feed like plants in the light and like animals in the dark. No one knows whether these strange creatures are plants or animals. They are, in fact, often called plant-animals. It seems certain that all plants and animals have descended from this sort of ancestor.

A trypanosome, one of the most harmful protozoans. It causes sleeping sickness when it gets into the nervous system.

Some of the small plants and animals (not all protozoans) which can be found in the sea.

Marine Protozoans

Many protozoans live in the plankton of the sea. Two groups are of particular interest. These are the *radiolarians* and the *foraminiferans*. Both groups are related to the amoeba, but they have shells around their bodies. The foraminiferans have chalky shells and the radiolarians have glassy shells, often with very delicate patterns. When the animals die the shells sink to the bottom, and some parts of the sea bed are covered with thick layers of these shells.

Disease-causing Protozoans

Many protozoans live inside the bodies of other animals and cause illness or disease. A relative of Amoeba lives in the food canals of humans and causes dysentery. Sleeping sickness, a dreadful disease carried by African tsetse flies, is caused by flagellate protozoans called *trypanosomes*. These affect the nervous system and usually kill the victim in time. Malaria is another serious disease caused by protozoans. It is carried by mosquitoes.

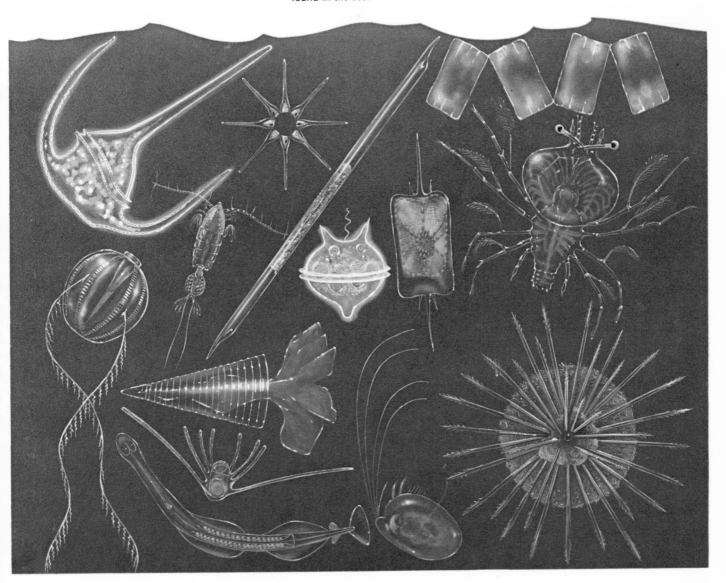

The Reef Builders

Coral reefs occur in many of the warmer seas of the world. The most famous reef is the Great Barrier Reef off the coast of Queensland in Australia. This is hundreds of miles long and several miles wide in places. It is hard to believe that such a massive structure has been built almost entirely from the skeletons of the tiny animals that we call corals.

Corals are coelenterate animals closely related to the sea anemones. The main difference between the two is that the corals form limestone skeletons around themselves. Some corals live by themselves, but the majority live in colonies built up by the repeated branching of one individual and its offspring. These colonies may be several feet across and they are the basis of the coral reefs.

The body of a coral animal, whether solitary or colonial, is just like a tiny sea anemone—a hollow tube with a number of tentacles surrounding the mouth. It is called a *polyp*. Some polyps reach an inch or so across, but the majority are about the size of a grain of rice. They are often brightly coloured. The skeleton is deposited around the base of the polyp in the form of a little cup. More material is added to it as the coral grows. Inside the cup there are numerous vertical plates of limestone. These are easily seen when the coral withdraws into the cup or when the coral dies.

The coral polyps are carnivorous creatures despite their small size. They feed on small crustaceans and other animals, which are trapped by the tentacles. These tentacles are armed with stinging cells just like those of the jellyfishes and sea anemones. They quickly paralyze small animals and the tentacles then carry the food to the mouth. Many reef-building corals contain green

Part of a mature reef, with several different kinds of corals growing on it.

The coral starts off as a single polyp and forms a single cup around itself. New polyps branch off later and the skeleton grows up around them all. The older parts of the polyps die in due course and leave the branches as separate animals, although they are still connected by the skeletons.

algae in their bodies. This is probably a symbiotic arrangement (see page 133) and it seems likely that the algae remove waste products from the polyps.

Corals and sea anemones all reproduce by scattering eggs and sperms into the water. Some of the eggs are fertilized and they settle down to grow into new polyps. After a while the colonial forms begin to develop buds or branches. These grow out from the original polyp just above the rim of the skeleton. The latter then grows up around the branches as well. The new polyps form branches themselves later and the whole colony grows. The older polyps die, but their skeletons remain to support the rest of the colony. Although all coral polyps look much alike, their skeletons are remarkably varied. In some species the branches remain very close together and form a solid, rounded colony. In other species the branches are widely separated and they form delicate branching colonies.

The Distribution of Coral Reefs

Solitary corals are found in most of the seas of the world, but the colonial species are found only in warm waters and the reef-formers are confined to areas in which the sea temperature never falls below 21°C

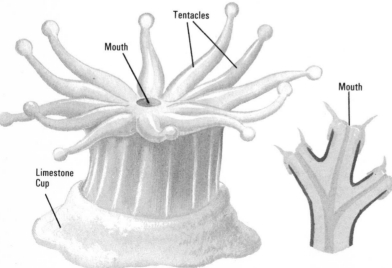

Tentacles

Mouth

Mouth

Limestone Cup

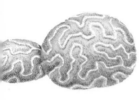

Above: Some of the many forms of coral

(70°F). This means that reefs are rarely found outside the tropics. But they are not found everywhere inside the tropics. Cold currents prevent them from growing along the west coasts of Africa and South America. Reef-forming corals also need clear water, and so they do not grow where rivers discharge large amounts of mud into the sea.

The Formation of Reefs

There are three main kinds of coral reef. *Fringing reefs* grow close to the shore. *Barrier reefs* grow further away from the shore and are separated from it by lagoons. *Atolls* are little coral islands, usually horseshoe shaped with a lagoon in the centre. Fringing reefs and barrier reefs are probably just two stages in reef development. The reef would start to grow close to the shore and it would gradually rise up. Growth of the reef is most vigorous at the outer edge, where the waves bring plenty of food and oxygen. This part would rise to the surface first and the waves might pile broken coral on top of it to form a ridge. This would protect the inner region even more and sediment would begin to settle. The sediment would kill the corals and the inner part of the reef would break down. The outer part would continue to grow outwards and, especially if accompanied by a gradual sinking of the land, it would eventually form a barrier reef. Atolls could be formed in the same way if the central island sank below the waves. Alternatively, atolls could grow up on submerged platforms. They would become ring-shaped because the inner part would not receive enough food and it would die. The same would happen to the sheltered part of the ring, and it would become horse-shoe shaped.

Whatever the type of reef, living corals form only its outer coat. The rest is made of the compacted skeletons and other materials, such as shell fragments, that become cemented together. The whole thing is very hard and it is used for building houses in some parts of the world.

The corals are the main reef-building organisms, but many others contribute their skeletons to the reef. Some algae are particularly important in this respect, while sponges and shellfish also provide material. As you might expect, the reef provides food and shelter for a great many other creatures. Fishes abound around them, feeding on the corals themselves or on the other plants and animals. Starfishes also enjoy a meal of corals, and one species, known as the crown of thorns, appears to be destroying parts of the Great Barrier Reef. When once the living coral is removed, there is nothing to prevent the erosion of the coral rock underneath.

Jellyfishes

Jellyfishes are free-floating coelenterates distantly related to the corals and sea anemones. If you imagine a sea anemone turned upside down and floating freely, you will see the similarity. Jellyfishes normally have numerous tentacles and most of them have powerful stinging cells with which they catch small fishes and other animals. The Portuguese man o' war, however, is not a true jellyfish. It is really a colony of animals more closely related to Hydra.

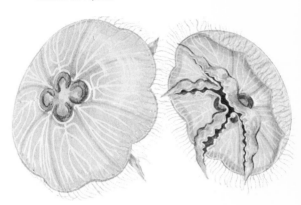

Hydra

Hydra is a small freshwater coelenterate, rather similar to the sea anemones although only distantly related. It feeds on water fleas and other little creatures, which it catches with its tentacles. Hydra grows branches from time to time, but the branches normally separate when fully formed and become separate animals.

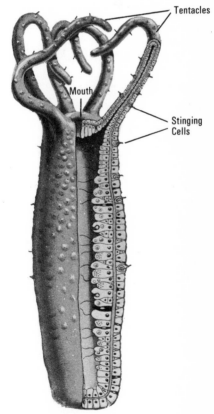

Tentacles

Mouth

Stinging Cells

The Stinging Cells

The stinging cells of the coelenterate animals are nearly all found on the tentacles. Each one consists of a fluid-filled bag with a coiled thread inside it. The outside of the cell is provided with a 'trigger'. When the 'trigger' is touched the cell fires out the thread. There are several kinds of threads. Some simply coil around the hairs of the small creature that touches them. Some are sticky, and some actually penetrate the animal and inject poison. The threads all combine to trap the prey and the tentacles then carry it to the mouth.

Worms and Leeches

We use the name worm for a whole range of long and slender creatures. They belong to many different groups, but the most important are the flat-worms, the round worms, and the segmented worms or annelids.

The Earthworm

The segmented worms or annelids get their name because their bodies are divided up into a number of little rings or segments. There are three main groups of annelids, of which the earthworms are by far the best known.

The humble earthworm might not seem to be a very important animal at first, but it is really an extremely important animal on any farm. Huge numbers live in the soil and their tunnels help to drain and aerate the soil, providing good conditions for plant roots.

The various kinds of earthworms all look more or less alike, although they vary in length and in the number of segments. They are nearly all reddish brown or greyish brown. The adult worm has a thickened region called the 'saddle' somewhere in the front half of the body, but otherwise the segments are all more or less alike. The front end of the animal is normally pointed, and the hind end is often somewhat flattened, but there is no real head. The animal has no eyes, although its whole body is sensitive to light and it quickly burrows into the soil when disturbed.

Each segment apart from the first and last

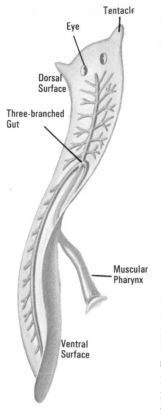

Planarian worms glide about in water, propelled by the beating of thousands of minute hairs on the body. They have simple eyes on the head, but the mouth is near the middle of the body and can actually be pushed out on the end of a tube. The digestive system sends branches to all parts of the body.

Below : Ragworms

Flatworms and Roundworms

The flatworms are all rather primitive creatures whose bodies are flattened. They include the little planarian worms that glide about the ponds, propelled by the beating of thousands of minute hairs called cilia. The flatworms also include the flukes and tapeworms, parasites that live inside the bodies of other animals.

Roundworms, or nematodes, are slightly more complex creatures and, as their name suggests, they have cylindrical bodies. Many of them live as parasites inside other animals or plants. Those living in plants often cause serious diseases and they are known as *eelworms*. Most of the roundworms, however, live freely in the soil or in water. Most of them are minute creatures, visible only with a microscope, but some are quite large. *Ascaris*, a nematode that is found in the human intestine, is sometimes more than 15 inches long.

has four pairs of tiny bristles on the lower surface. You can feel them if you run your fingers along the underside of a worm. They help the worm to move. When the animal starts to move forward the middle and hind regions are anchored by their bristles. The front end stretches forward and anchors itself. The middle regions are then drawn forward and, when these are anchored again, the hind regions are pulled up.

Feeding and Reproduction

Earthworms feed on decaying matter. They will often pull dead leaves down into the soil, but they get much of their food simply by swallowing the soil as they tunnel. Organic material is digested and the undigested soil particles are passed out as *worm casts* on or near the surface of the soil. The worms thus perform a very useful ploughing

90

action, bringing mineral-rich sub-soil up to where the plants can use it. The deposition of worm casts on the surface means that objects are gradually buried by the worms. The Stonehenge monuments, for example, are sinking by six inches every century and most of this is due to the action of earthworms. Worms can swallow only tiny particles of soil and so the material deposited at or near the surface is very fine. Stones gradually sink to lower levels and this is why we do not find stones under old lawns and meadows.

When it is fully grown the earthworm has both male and female organs in its body. The animals have to pair up, however, before they can reproduce. They lie close together, surrounded by a coating of slime, and exchange sperms.

The sperms are stored in special pouches. The worms separate. Eggs develop in each individual and, when they are ripe, they pass back to the 'saddle'. A cylinder of skin breaks away from the saddle and works its way forward with the eggs. It then receives sperms from the other worm and the eggs are fertilized. The cylinder of skin then comes right away and becomes wrapped around the eggs to form a cocoon. The eggs hatch after a few weeks and produce tiny worms.

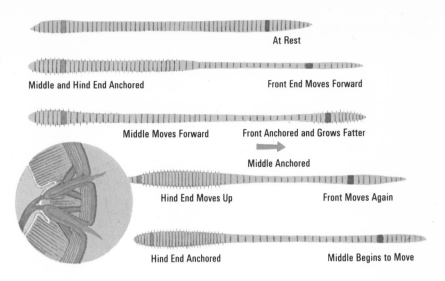

At Rest

Middle and Hind End Anchored

Front End Moves Forward

Middle Moves Forward

Front Anchored and Grows Fatter

Middle Anchored

Hind End Moves Up

Front Moves Again

Hind End Anchored

Middle Begins to Move

The earthworm moves its body forward in stages. While one region is being extended the other parts are anchored by the little bristles sticking out from the underside of the animal. The bristles also help to anchor the worm in its burrow.

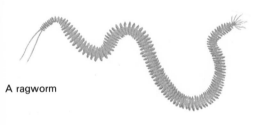

A ragworm

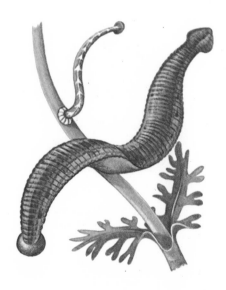

Leeches can be recognised by the large suckers at each end.

Bristle Worms

Bristle worms are marine annelids, many of which have long bristles on most of their segments. Some of them, known as ragworms, are free-living creatures. They are much more active than the earthworms and they have well developed heads with eyes and jaws. Other bristle worms live buried in the mud or sand or else they form tubes of sand or limestone. The lugworm burrows in the mud and feeds in the same way as the earthworms. The peacock worm forms tubes of sand grains. It has a crown of special bristles around the mouth and these trap particles of food floating in the water. The tentacles are rapidly withdrawn if the animal is disturbed. Bristleworms reproduce simply by scattering eggs and sperms into the water and leaving fertilization to chance.

Leeches

Unlike the other annelids, the leeches have no bristles. They are rather flattened worms and they are readily identified by the large sucker at each end of the body. Some live in the sea and some live in damp soil, but most leeches live in fresh water. They are all carnivorous creatures. Some of them eat other small aquatic creatures. Other leeches are blood suckers, attaching themselves to various animals and taking two or three times their own weight of blood for one meal. The blood is stored in special pouches and is digested very slowly. The blood-feeding leeches can go for very long periods without feeding.

Leeches reproduce in the same way as earthworms. They have 'saddles' and they produce egg cocoons. These cocoons are often flat and horny and they are commonly seen on stones and water plants.

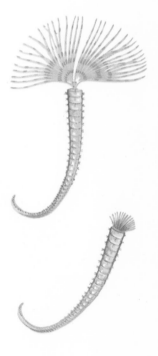

Fan worms are bristle worms that live in tubes. Bristles at the head end form a fan which traps food particles. The fan is quickly withdrawn if the animal is disturbed.

Crabs and their Relatives

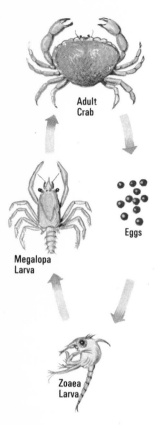

Adult
Crab

Eggs

Megalopa
Larva

Zoaea
Larva

Crabs have a complex
life cycle, often involving
several stages that look
nothing like the adult
crab. The young stages
usually float freely in the
surface waters.

Crabs, lobsters, shrimps, woodlice, and barnacles all belong to the large group of arthropods known as the *crustaceans*. This name refers to the hard, chalky coat which covers most of the animals. The fishmonger calls them shellfish. He also uses this name for winkles and cockles, although the shells of these molluscs are quite different from those of the crustaceans.

Nearly all the crustaceans live in water. This fact distinguishes them from most other arthropods, but there are other differences as well. For example, the crustaceans have two pairs of feelers or antennae, although one pair might be rather small. Many have a 'cloak' covering all or part of the body. It is called the *carapace* and it is formed from a fold of skin which becomes hardened with deposits of lime. It becomes as hard as stone in some animals. There are usually many pairs of limbs, but they are of several different types even in one animal.

The crustaceans range from minute floating creatures to lobsters weighing 30 lb and crabs with legs more than five feet long. Many live in fresh water, but the majority live in the sea. They make up a large part of the animal plankton—the drifting animals in the surface layers. Some of the crustaceans live permanently in the plankton. These include the shrimp-like creatures called krill which form the bulk of the diet of the large whalebone whales. Another important planktonic crustacean is called *Calanus*. It is a relative of the little freshwater *Cyclops* and it is the main food of the

herring and other shoaling fishes. Many other crustaceans live in the plankton only during the early part of their lives. Crabs and lobsters all have floating larvae, and most species have several different larval stages.

Filter Feeders

The simplest of the crustaceans are the little fairy shrimps. These have many pairs of broad, hinged limbs which filter food particles from the water while they 'row' the animal along. The limbs also act as gills and absorb oxygen from the water. Related to the fairy shrimps are the water fleas, the little reddish brown creatures that are very common in ponds. The trunk limbs are enclosed by the carapace but they still vibrate and draw a current of water through the carapace. Food and oxygen are obtained from the water. The animals swim by 'rowing' with their large second pair of antennae. This produces a jerky upward movement, from which the animals get their name.

Swimming in the pond with the water fleas there are usually some little pear-shaped creatures called *copepods*. There are several different kinds, but *Cyclops* is the commonest. The animals have no carapace and they have no compound eyes like those found in most other crustaceans. Some copepods filter food from the water, but *Cyclops* picks up individual pieces with its mouthparts. Like the water fleas, it can swim by beating its antennae, but it also swims more rapidly by paddling with its tiny trunk limbs. Female copepods can easily be recognised by the little egg sacs they carry at the hind end.

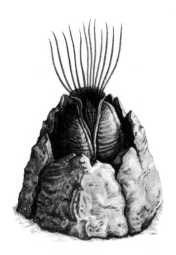

Left: These barnacles coating a glass float look rather lifeless out of the water, but when submerged by the tide the inner plates open up and out come the barnacle's limbs (above) to comb food from the water.

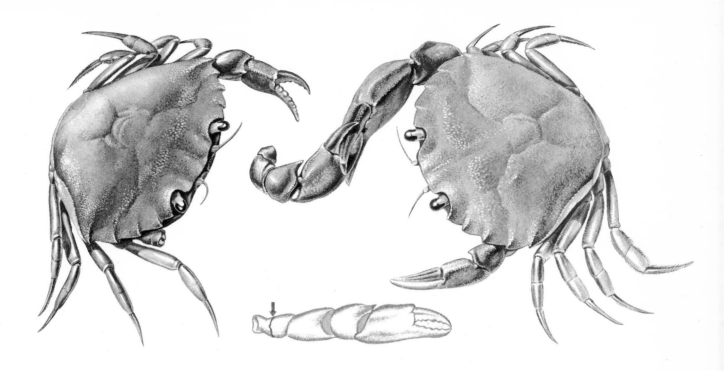

Above: If a crab's limb is broken off as sometimes happens when they fight, it can grow another when it moults. The arrow on the diagram indicates the special breaking point.

Barnacles

The little white tent-like barnacles that we see at the seaside do not look like crustaceans. No-one really knew what they were until their life cycle was studied and their young stages were discovered. These are just like the young of crabs and they show that barnacles really are crustaceans. The common barnacle of the shore closes its 'tent' when the tide goes out. It opens it again when the water returns and the animal starts to 'comb' food from the water by pushing its delicate limbs out into the water.

Crabs, Shrimps, and Lobsters

These creatures make up the most advanced group of crustaceans. They are often known as *decapods* because they have ten walking legs. Not all of them are actually used for walking. At least one pair is modified to form pincers. These are very large in some crabs and lobsters and are used for display purposes as well as for catching food. The animals are mainly carnivorous and take both living and dead material. The decapods breathe by means of gills, which are leaf-like or feathery out-growths of the body wall or the leg bases. The gills are enclosed by the carapace and they are provided with a current of water by the continuous movement of some of the limbs.

Crabs are relatively short, broad crea-

Lobster

Woodlouse

tures, in which the abdominal or tail region is small and folded tightly under the front part of the body. Most crabs walk or scuttle sideways on the sea bed, although some have broad legs and are able to swim. The shrimps, prawns, and lobsters have relatively longer bodies and their abdominal regions are clearly visible. There are several pairs of abdominal legs called *swimmerets*. By using these as paddles, the animals can swim quite well. The abdominal legs are also used to hold the eggs. If you eat shrimps or prawns you have probably seen individuals with their hind regions covered with eggs. In this condition they are said to be *in berry*.

Woodlice

The woodlice are the only group of crustaceans that have successfully invaded the land. They belong to a group called *isopods*. Woodlice are flattened, oval creatures up to about an inch long. Most of them are greyish in color. They have seven pairs of walking legs and a number of leaf-like abdominal limbs towards the back. Although they live on the land, the woodlice cannot survive in very dry conditions. They need moisture to enable them to absorb oxygen through their abdominal limbs. Some species, however, have begun to develop breathing tubes rather like the insects' tracheae. These species can survive in drier places. Some of them can roll themselves into a ball and this also helps them to survive dry periods. Most woodlice live under logs and stones. They feed on decaying material. Not all members of the group live on the land. The sea-slater lives on the sea-shore, although it is almost a land animal. Another member, known as the hog louse, lives in ponds and streams.

The Insect World

Dragonflies, butterflies, bees and beetles are just four of the many groups that make up the insect world—a world with more than a million different species and untold millions of individuals. The insects belong to the even larger group of animals called the arthropods. This name means 'jointed feet', and if you look carefully at the leg of an insect you will see that it is made up of a number of tiny joints or segments. In fact, the whole body is made up of segments. Some segments are tightly joined together, others are more loosely connected by soft membranes. The arthropods also include animals such as spiders, shrimps and centipedes. You can see the body segments very easily in a centipede. Insects can be distinguished from the other arthropods because they have three distinct regions to the body. These are the head, the thorax (in the middle), and the abdomen. The thorax bears three pairs of legs and it usually carries two pairs of wings as well. Insects also have a single pair of feelers, or antennae, on the head.

Insects live almost everywhere on the earth—on the land and in the water, in the deserts and around the poles. They have not managed to make much headway in the sea, but they are found everywhere else. Most of the insects can fly and this has certainly helped them to spread. Their small size has also helped them to invade places where larger animals could not live.

One of the features that prevents the insects from getting too large is their method of breathing. Tiny tubes carry air from the surface to all parts of the body, but the system works only over short distances and so insects can not get very fat. The fattest insect in the world is the goliath beetle from Africa. It is about the size of a fist and weighs about four ounces. No part

Below: A highly magnified photograph of the compound eyes of an insect. The numerous tiny lenses form a honeycomb-like mosaic. Each lens sends its own signal to the brain so that the insect sees an image as a pattern of dots, rather like a printed picture. The image is not very clear and insect eyes are best suited to picking out moving objects.

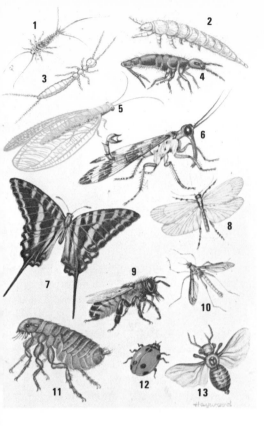

Some of the important insect groups. 1, Order Thysanura includes the bristletails and silverfish. 2 and 3, Orders Protura and Diplura contain small soil-living insects. 4, Order Collembola contains the springtails. 5, Order Neuroptera contains lace-wing flies. 6, Order Mecoptera includes scorpion flies. 7, Order Lepidoptera includes butterflies and moths. 8, Order Trichoptera includes caddis-flies. 9, Order Hymenoptera includes bees, wasps and ants. 10, Order Diptera includes true flies. 11, Order Siphonaptera includes fleas. 12, Order Coleoptera includes beetles. 13, Order Strepsiptera includes internal parasites of bees.

of the body is more than about an inch from the outside. There are many longer insects—some stick insects, for example, are more than a foot long—but their bodies are always slender. The same thing is true of butterflies and moths with large wings. At the other end of the scale, there are minute beetles only about $\frac{1}{100}$ of an inch long.

Insect Groups

There are about 25 major groups of insects, separated mainly by differences in the wings and the methods of feeding. Four groups contain rather primitive insects without wings. These include the little silverfish which is often found in dusty corners in houses. The other groups are basically winged insects. They have all had wings at some time during their history, although some of them have lost their wings again. Soil-living insects do not need wings, and nor do the parasitic insects that live among fur and feathers. Fleas and lice have all lost their wings. Some of the important insect groups are shown on this page.

Insect Life Histories

A young grasshopper shortly after it has left its egg looks very much like an adult grasshopper, except that it is much smaller and has no wings. It is called a nymph. It starts to eat grass and it soon finds that it is getting too big for its tough outer coat or skeleton. The insect rests for a short while and then the outer coat splits. The insect

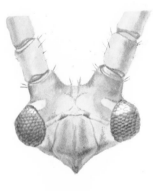

Above: The head of an aphid showing the compound eyes.

then wriggles its way out. A new and larger coat has already grown beneath the old one, but it is rather soft to start with. The insect puffs itself up with air and continues to rest until the new coat has hardened. It then expels the air, leaving room for more growth. This process of skin-changing is called moulting and it takes place four times before the grasshoppers are fully grown.

After the first moult the grasshopper nymph begins to develop its wings. They appear as tiny flaps on the thorax. After the second moult the wing flaps are larger and after the third moult they are larger still, but they do not become fully formed until after the fourth moult. The insect is then a fully grown adult grasshopper and it does not change its skin any more. This sort of life history, in which the young insect gradually takes on the appearance of the adult, is also shown by dragonflies, earwigs, bugs, and various other insect groups. The gradual change from young to adult form is called partial or incomplete metamorphosis.

Right: The small stag beetle shows the major features of insect anatomy—three body sections, three pairs of legs, and one pair of feelers.

Butterflies, bees, houseflies and beetles, have a different kind of life history. The young stages of these insects are called grubs or caterpillars, and they look nothing like the adult insects. Scientifically, they are called larvae, and they often eat quite different food from the adults. The larvae change their skins, just like the grasshopper nymphs, but there is no sign of wing flaps. Then, after two, three, four, or even more skin changes, the larvae turn into pupae or chrysalises. The pupae do not feed and they do not usually move much, but great changes are going on inside them. Instead of the young insect gradually taking on the adult form, the changes take place all at once inside the pupa. This abrupt change is called a complete metamorphosis. After a few days or a few months, depending on the type of insect, the adult comes out of the pupa. Its wings are crumpled and its body is soft, but the wings soon expand and the body hardens. The insect is then ready to fly away.

Insect Diets

Many insects feed on plants and, apart from some of the seaweeds perhaps, there is probably no plant species in the world that is not attacked by some insect or other.

Many insects, including moths and flies, have young stages which look nothing like the adults. The young stages are called larvae and they have to pass through a resting stage (the pupa) before they become adults.

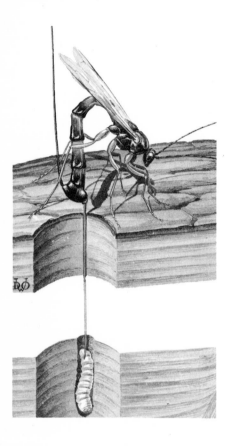

Ichneumon flies are parasitic insects whose young feed on the bodies of other insects. *Rhyssa,* shown here, feeds on wood-boring grubs. The female Rhyssa has a very long egg-laying tube with which she drills into the wood. She then lays an egg on the host grub.

Every part of the plant may be attacked, from the root to the flower and fruit. Well known plant-feeding insects include caterpillars (young butterflies and moths), greenfly, and many beetles. A whole lot more insects are carnivorous, which means that they feed on other animals. Dragonflies, bush crickets and ladybird beetles all eat other insects. Fleas, bedbugs, and mosquitos all feed on much larger animals (including human beings) by sucking blood. Then there are the omnivorous insects that eat more or less anything they can find. The cockroaches are good examples. There can be few natural materials that are not eaten by one kind of insect or another. Among the many strange things they eat we find wood, hair, dung, wax, tobacco, leather and flour.

Insect Mouth-parts

The earliest insects all had biting jaws, used for eating solid food. Jaws of this type are retained by many of today's insects, including grasshoppers, beetles and wasps. The jaws are not inside the mouth as ours

are: they are really very special legs just outside the mouth. They have sharp teeth on them and they cut the food into small pieces. Other 'legs' around the mouth help to hold the food and to push it into the mouth when it has been cut up. The jaws and the other feeding limbs are collectively known as the mouth-parts of the insect.

The mouth-parts of butterflies, flies, fleas, bugs, and some other groups have become changed during their long history and these insects now feed on liquids. The butterfly mouth-parts form a long 'tongue', called a proboscis, which is normally coiled up underneath the head. When the insect finds a nectar-filled flower it unrolls its proboscis, pushes it into the flower, and sucks up the nectar. Most moths have a proboscis, but some of them have no working mouth-parts and they do not feed when they are adult.

The mouth-parts of bugs and fleas are in the form of sharp needles which are sunk into the tissues of plants or other animals.

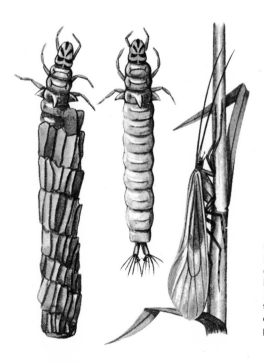

Caddis flies are moth-like insects which spend their early days in water. The larvae have soft bodies (center) but they protect themselves by building cases of sand grains or plant fragments (left).

96.

The powder post beetle is one of several species whose grubs do much damage by tunnelling into wood.

Below: The house-fly mops up liquid food with its sponge-like tongue.

The needles fit together to form narrow tubes through which the insects drink sap or blood. Many of these insects carry disease germs from one plant or animal to another. Aphids (greenfly), for example, carry many serious plant diseases.

Many flies have sharp, piercing mouth-parts which they use for sucking blood. The mosquito is a well known example, and others include the tsetse fly, the stable fly and the horse fly. Like the fleas and bugs, many of them carry diseases from one animal to another. The housefly and the bluebottle, however, have different kinds of mouths. Their mouth-parts form spongy pads and the insects mop up liquid food with them and suck it into the mouth. Solid food can be dealt with by pouring digestive juices on to the food and then sucking up the resulting solution. The digestive juices come from the food canal and often contain parts of the fly's previous meal. The flies can contaminate our food with germs picked up somewhere else.

The dragonfly is a hunter throughout its life. The adult scoops flies from the air with its spiky legs, while the young insect has a clawed lower lip which it shoots out to impale other creatures.

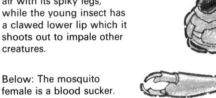

Below: The mosquito female is a blood sucker. She sinks her needle-like mouth-parts into other animals and withdraws blood.

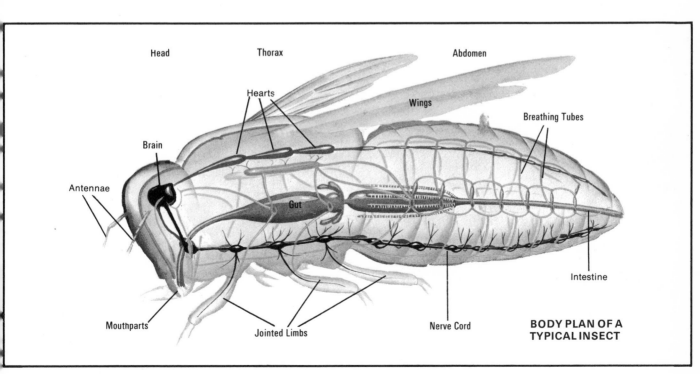

Head   Thorax   Abdomen

Hearts

Wings

Breathing Tubes

Brain

Antennae

Gut

Mouthparts   Jointed Limbs   Nerve Cord   Intestine

BODY PLAN OF A TYPICAL INSECT

Butterflies and Moths

Butterflies and moths both belong to a large group of insects called the Lepidoptera. This name means 'scale wings' and, if one looks at a butterfly wing under a microscope, one can see that the whole wing is covered with tiny overlapping scales. It is these scales that give the wings their patterns. Some scales in male butterflies also give off scent which attracts the female.

The Lepidoptera is one of the largest groups of insects and contains something like 100,000 different species. The butterflies account for relatively few species out of this total, the rest belonging to various families of moths. There is an enormous

Right: The colors of a butterfly's wing are produced by thousands of tiny overlapping scales which come off like dust if the wings are rubbed.

Below: Butterflies vary a great deal in shape and size. These species, belonging to several families, are all half life size.

size range within the moths, from insects such as the giant owlet moth of South America (30 cm from wing tip to wing tip) to minute moths no more than 3 mm across whose caterpillars live between the upper and lower surfaces of leaves. Some of the butterflies are also very large, but they have nothing comparable with the little moths.

People often want to know the difference between butterflies and moths, and they are surprised to learn that there is not a simple distinction. It is often said that butterflies are brightly colored and fly by day, whereas moths have dull colors and fly at night. This might be true for the majority of species, but certainly not for all of them. There are many dull colored butterflies and many more brightly colored, day-flying moths.

The best way to distinguish a butterfly from a moth is to look at its feelers or antennae. All the butterflies have clubbed antennae. Moths have many different kinds of antennae, but rarely have clubbed tips. Those moths that do have clubbed antennae can be distinguished from butterflies by looking at the wings. The 'shoulder' of the hind wing carries a bristle, known as the *frenulum*, which fits into a little hook on the underside of the front wing. This arrangement holds the two wings together in flight. Butterflies do not have a frenulum, There are many moths without a frenulum, but these never have clubbed antennae and they cannot be confused with butterflies.

Reproduction

The butterflies and moths belong to that division of insects which pass through four distinct stages in their life history and undergo a complete change, or *metamorphosis,* during that time. The first stage, as with all insects, is the egg. The eggs are

Ulysses' Butterfly
(Australia, New Guinea)

The Monarch
(Widespread)

The Brimstone
(Europe)

The Small Copper
(North America, Europe)

The Zebra
(North America)

The Smoky Orange Tip
(Africa)

The Fiery Acrea
(Africa)

The Zebra Swallowtail
(North America)

Rainbow Butterfly
(New Zealand)

The Camberwell Beauty
(Europe, North America)

The Regent Skipper
(Australia)

The Adonis Blue
(Europe)

The Silver Barred Charaxes
(Africa)

The Birdwing
(Australia, New Guinea)

BRIAN HARGREAVES

usually laid on the insect's food plant and they are often beautifully patterned. A caterpillar develops inside the egg and, when it is ready, it chews its way out. All caterpillars have biting jaws and, with the exception of a few 'clothes moths' which feed on woollen material, they are vegetarians. Some feed inside the stems, roots and fruits of plants, but most of the caterpillars feed on the leaves. They are generally well protected by camouflage or by hairy coats, but vast numbers are eaten by birds. Many more fall victim to ichneumon flies and other parasites but some escape and, after about four skin changes, they are ready to enter the third stage of the life history.

Many caterpillars, especially those of moths, spin silken cocoons around themselves and attach them to the branches of the food plant. Others hollow out snug chambers in the soil, while many butterfly larvae simply attach themselves to a convenient support by a pad of silk. After a short rest, the caterpillar's skin splits and is pushed off to reveal the pupa or chrysalis. Inside the pupa, the caterpillar's body is broken down and rebuilt into the body of the adult insect. This may take as little as 10 days, but it is often nine months before the adult insect emerges. When the adult first crawls out through a split in the pupal skin it is very soft and its wings are tiny crumpled objects. Its first action is to climb up some support and hang itself up to dry. It pumps blood into the veins of the wings and they gradually expand to their full size. After a few hours they are hard enough to bear the insect away on its first flight.

Sight and scent both play a part in the courtship of butterflies and moths. Among butterflies it is the male who gives out the scent, but among moths it is usually the female. Some of the male moths have extremely sensitive 'noses' on their antennae and they can smell a female's scent from more than a mile away. A freshly emerged emperor female put out in a muslin cage on a sunny afternoon will soon be surrounded by dozens of males if they are present in the area.

Feeding

A few very primitive moths have biting jaws, which they use for feeding on pollen, but the rest of the moths and the butterflies have lost their jaws. Their feeding apparatus, if they have any, consists of a slender tube which they use to suck nectar and other sweet substances. One sure way to encourage butterflies into your garden is to plant aubretia, Michaelmas daisies, and other plants with plenty of nectar.

Bright colors will not always distinguish butterflies from moths – the tiger moth at the bottom is just as bright as the fritillary butterfly. The antennae are a better guide: almost all butterflies have distinctly clubbed antennae.

The majority of butterflies and moths are quite harmless in the garden, and many are very useful because they pollinate the flowers and help them to set seed. There are some species, however, which are garden pests. The 'cabbage whites', for example, are found in most parts of the world and their caterpillars destroy large amounts of valuable greenstuff. Bufftip moth and lackey moth caterpillars do much damage to fruit trees by stripping them of their leaves. The gypsy moth, a European species, created havoc in the American forests when it found its way to North America some years ago. Another common pest is the codling moth, whose caterpillars are the 'maggots' so often found inside apples.

Insect Pests

Clothes moths eat our clothes and carpets, furniture beetles (woodworms) tunnel through our furniture, grain weevils destroy stored grain, caterpillars eat our cabbages, mosquitos spread malaria, and houseflies contaminate our food. These are just some of the insect pests which man has to put up with. It is interesting to consider where they came from and what they did before man came along to provide food and shelter for them.

Domestic Pests

One of the commonest household insects is the housefly, although it is less common today than it used to be. The housefly breeds in all sorts of rubbish but it is especially fond of horse dung. This was probably its original home, but it now finds suitable breeding sites in the garbage and rubbish heaps of human settlements. The adults lay their eggs on the rubbish and they also feed there to some extent. Then they might fly to our table and contaminate

Above: Two domestic pests, the house-fly and the cockroach.
Below: the Colorado beetle, an American insect now a widespread pest of potatoes.

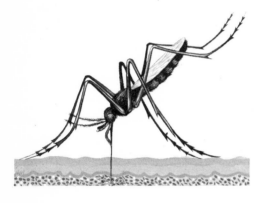

Left: A feeding mosquito plunges its 'beak' into the skin of an animal. Mosquitoes are serious pests because they carry the germs of malaria, yellow fever, and other diseases.

our food with germs picked up on the rubbish dump. The use of insecticides, together with better methods of refuse disposal, has reduced the housefly population in recent years.

Clothes moths are small insects whose caterpillars feed on woollen articles. There are several different kinds, and the adults are actually very attractive little creatures. Many of the clothes moths have been found in the nests of birds and rodents, where they feed on scraps of feather and fur. It is quite natural, therefore, that they should invade the homes of men in search of similar materials. Carpet beetles have had much the same sort of history.

The furniture beetle or woodworm and the much larger death-watch beetle have always made their homes in dead wood. Originally, their larvae tunnelled in dead

tree-trunks and helped to break them down and return the material to the soil. The beetles still do use tree-trunks out in the wild, but they are also quite happy to feed on our furniture and floorboards.

Carriers of Disease

Mosquitoes, tsetse flies, fleas and lice are all responsible for carrying disease-causing germs. Most of them are blood-sucking insects and if they take in germs when they feed on one person they are quite likely to inject some into a later victim. The germs themselves usually have to undergo certain changes while they are in the insect before they can infect another patient. For this reason most of the diseases are associated with just one kind of insect. Malaria, for example, is carried only by certain kinds of mosquitos, while sleeping sickness is normally carried only by tsetse flies.

Crop Pests

A great deal of the world's food is lost every year because insects eat it or damage it while it is growing in fields and gardens. You might wonder where the vast hordes of insect pests come from and why they have to attack our crops. The insects existed long before we started to grow crops, and it is because we grow large amounts of crops that the insects can increase and become pests. The Colorado beetle is a good example of this. It came originally from the western parts of North America, where it fed on wild plants and did no harm to anyone. Around the middle of the nineteenth century, people started to grow potatoes in the region, and from then on the Colorado beetle has been a pest. The potato plant is related to the wild plants on which the beetle fed. Consequently the Colorado beetle soon transferred its attention to the potatoes. Surrounded by abundant food it was able to increase its numbers and to spread out from its original home. Potatoes

The large white butterfly caterpillars feed on cabbages and do a great deal of harm to crops throughout Europe.

The locust is perhaps the most destructive of all insects. There are several species and they affect all of the warmer parts of the world. Huge swarms with millions of individuals build up in the semi-desert areas and then migrate. Large swarms will strip every leaf for miles when they land, and when flying they may completely blot out the sun.

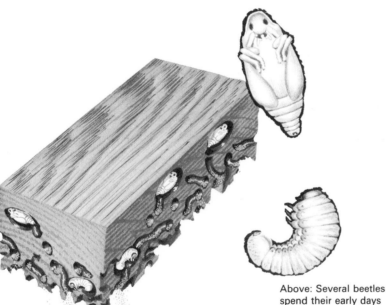

Above: Several beetles spend their early days feeding in dead wood. Well known ones include the woodworm or furniture beetle (shown here) and the death watch beetle. The damage is often undetected until the adult beetles bore their way out.

were being grown over much of the United States, and within 20 years the beetle had reached the east coast of America. It got into Europe attached to some imported plants and it is now a pest nearly all over Western Europe. Both the larva and the adult eat the potato leaves and the crop is therefore greatly reduced.

Aphids—greenfly and blackfly—are probably the most serious of the insect pests. There are many different kinds and they affect a whole range of plants. They weaken the plants by sucking sap from them, but they do even more damage by carrying virus diseases from plant to plant. These diseases weaken the plants even further and may kill them.

Controlling Pests

There are two main ways of controlling insect pests. One way is to use insecticides, which are insect-killing poisons. They usually kill the pests well enough, but there is also the danger that they will kill useful insects such as bees as well. There is a risk, too, that the insect populations will gradually become immune to the insecticides.

The other way of controlling pests is known as biological control. This means using the pests' natural enemies to keep their numbers down. There is no danger in doing this, but it is not a very practical method for large scale use because it is not easy to rear large numbers of the natural enemies. There have, however, been some notable successes with this method.

Control of domestic pests is very much a matter of keeping things clean. Houseflies can be kept down, for example, by ensuring that there is no rubbish left around in which they can breed.

Right: Grain weevils, which originally fed on grass seeds in the wild, are now serious pests in grain stores. They destroy hundreds of tons of grain every year.

The Honey Bee

Most of the insects live alone and look after none but themselves. Some of the bees and wasps, however, and all of the ants and termites live in communities of many individuals. They are social insects and each individual works for the good of the whole community.

One of the best known of the social insects is the honey bee. It came originally from somewhere in South-East Asia, but it has been taken to many parts of the world to provide men with honey. Wild honey bees normally nest in hollow trees, but they are quite happy to make their homes in the bee keeper's hives. These are not really much different from hollow trees—both are wooden boxes of a sort.

Ruled by the Queen

The honey bee colony is headed by the *queen*, who is larger than the other bees in the colony. She does nothing but lay eggs. Most of the bees in the colony are *workers*. They are females, but they are not fully developed and they cannot usually lay eggs. The workers look after the queen, feed and tend the young bees, build and clean the nest, and collect food. Although the workers do all the work, the smooth running of the colony is maintained by the queen. She produces an oily material called *queen substance* all over her body. The workers lick this off and spread it around among themselves. It unites them and maintains the 'community spirit'. A third type of bee is produced in the colony during the summer. This is the male or *drone* bee. He is rather lazy and does no work in the hive. He is important, though, because his job is to mate with a new queen when one appears.

Life in the Hive

The bees' nest consists of a number of 'combs' hanging vertically in the hive or tree. Each comb is composed of hundreds of little six-sided cells, all made of wax from the bees' bodies. The cells are used for storing pollen and honey and also for rearing the young. The queen wanders over the combs and lays an egg in each empty cell she finds. When the eggs hatch the workers begin to feed the grubs. For the first three days the grubs receive a substance called *brood food*. This is a protein-rich material made in the salivary glands of the young workers. After the first three days the grubs are fed on pollen and nectar. About nine days after the eggs were laid the grubs turn into pupae, and

The worker bees collect both pollen and nectar from flowers. The pollen clings to the bees' hairs and is then brushed into the pollen baskets on the back legs.

Below: The direction of the bee's dance tells the other bees where to find nectar in relation to the sun's position.

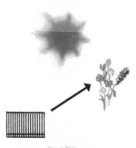

the adult bees emerge about 10 days after that.

When they first emerge from their cells the young workers spend a few days doing 'housework' and feeding their younger sisters. Then, when their wax-making glands start to work, they take over the duties of making and repairing the combs. Not until about three weeks after emerging from their cells do the young workers go out and search for food.

The bees visit flowers and collect both pollen and nectar. While doing this they pollinate the flowers. This pollinating activity is more important to us than the honey that the bees provide, because without the bees to pollinate the blossoms we would be without many of the fruits and seeds that we eat. The bees carry pollen back to the hive in the pollen baskets on their back legs. You can often see bees almost weighed down by the yellow pollen masses on their legs. The nectar is carried back in a pouch of the food canal and it is then given up to other bees in the hive who remove much of the water from it and convert it into honey.

Bumble Bees

Bumble bees are larger and more hairy insects than the honey bees. They are social insects, but they do not live in such large colonies as the honey bee. The main difference, however, is that the bumble bee colony lasts for only one year. New queens are produced in the summer and, after mating, they hibernate. In the spring they wake up and build small nests. They rear a few workers and these then take over the work of building the nest while the queen 'retires' to lay more eggs. The drones and workers all die in the autumn. The social wasps have a very similar life story, with only the queens surviving the winter. Wasps, however, make their nests with paper which they make from wood, and they feed their young on insects instead of pollen and nectar.

102

Bee Dances

The bees have ways of telling each other about good nectar sources that they have found. They come back to the hive and 'dance' excitedly on the combs. The type of dance and the speed of dancing tell the other bees where to find the nectar and off they go to look for it. They find their way about by the sun and they even make allowances for the movement of the sun across the sky during the day.

Swarming

Every now and then the population in the hive gets rather too large for comfort and the workers prepare for a swarm. They build special queen cells at the edge of the comb. These are larger than the ordinary cells and they look something like acorns. The queen lays an egg in each queen cell and the work-

Worker

Queen

Drone

ers look after these eggs just as they look after the other eggs in the comb. The grubs in the queen cells, however, are fed on brood food for the whole five or six days of their larval life. The extra protein that they receive ensures that they become fully developed females or queens. The first queen to emerge from her cell kills the other developing queens and then goes off on a marriage flight with one or more drones. She returns to the hive later to become queen. In the meantime her mother, the old queen, has left the hive with a swarm of workers to start a new colony somewhere else. A new queen is also reared when the old queen is becoming weak. The new queen returns to the hive and the old queen soon dies.

The queen honey bee normally lives for about three years and she lays over half a million eggs during this time. The workers live for only a few weeks during the summer, although those reared in the autumn will survive until the spring. Drones may live for several weeks during the summer, but they do not survive the winter. The workers refuse them food in the autumn and often throw them out of the hive to die. In winter the colony is quite small and it survives on the food stored up during the previous summer. Numbers build up again in spring when food becomes plentiful, and there may be 60,000 bees in the hive during the summer months.

Above: The bee keeper provides the bees with sheets of wax on wooden frames. The bees then build their cells on these. Some cells are used for rearing young, others are used for storing honey.

Right: The queen wanders over the combs and lays an egg in each empty cell she finds. She is always attended by workers, who lick her and touch her with their feelers to obtain queen substance.

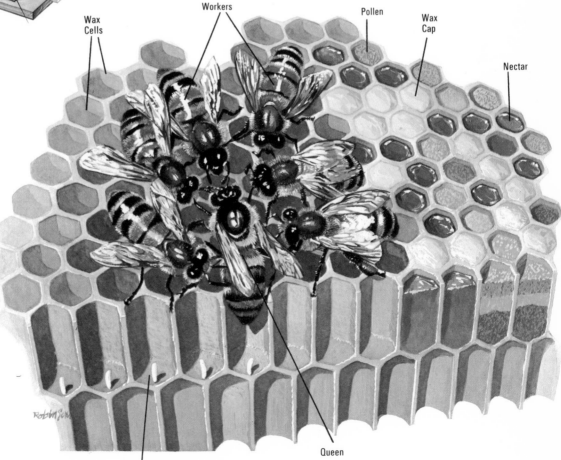

Wax Cells

Workers

Pollen

Wax Cap

Nectar

Queen

Egg

Ants and Termites

There are more than 6,000 different kinds of ants, found in all but the coldest parts of the world. All ants live in colonies and, like the honey bee, they have three main castes. The colony is headed by a mated female called the queen. Male ants are produced at certain times of the year, but most of the ants in a colony are sterile, wingless females. These are the workers and they are the most familiar of the ants—they are the ones we see crawling over the ground and the vegetation in search of food. Most of them are black or brownish in color and they are easily recognised by their narrow waists and by their 'elbowed' antennae. Many ant species, especially those living in warm countries, have more than one type of worker. Each type has a particular job to do. Many ant species also have an extra caste called a soldier. This is a large-headed ant whose job is to defend the colony with its large jaws. Some of the soldiers also use their jaws to crack open the hard seeds on which some ants feed. The ordinary workers help in the defence of the colony by using their stings and sharp jaws. Not all the species have actual stings, but some of those without them can squirt formic acid at their attackers.

The typical features of ant life are shown by the little black garden ant called *Acanthomyops niger,* although this species does not have a soldier caste. It often makes its nest under concrete paths and it frequently

Above: The wood ant's nest has many galleries and chambers. Eggs are kept in one part, larvae in another, and so on. The queen ant can be seen in her royal chamber at the center of the nest.

Above: Some ants can turn their hind ends forward and squirt acid at their attackers.

comes into houses. Winged male and female ants are produced during the summer and, when the weather is just right, they emerge for their wedding flights. All the nests in an area will release their flying ants together, so we find great swarms of them at this time. The males die after the wedding flight, but the females or queens begin their duties as nest founders and mothers. They break off their wings and find suitable holes in which to hide for a few months until their eggs are ready to be laid.

When the first eggs hatch the queen feeds the larvae on saliva. They soon turn into worker ants and set to work to build the nest. This consists of numerous chambers and passageways, but there is nothing like the cells that are found in the bees' nests. The queen remains in one chamber and devotes the rest of her life to laying eggs. She may live for several years, although the workers do not live so long.

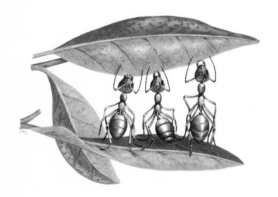

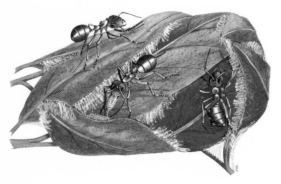

Left: Weaver ants make small nests by sewing or gluing leaves together. While some ants hold the leaves together other ants use the larvae rather like tubes of glue. They squeeze the larvae and make them secrete strands of sticky silk. Some ant species actually use their larvae like needle and thread. They pass the larvae to and fro through holes made with the jaws.

As well as building the nest, the workers collect food and look after the young ants. Plant and animal material is eaten, but many ants are particularly fond of honeydew. This is the sweet, sticky liquid given out by greenfly. You can often see ants running about on aphid-infested roses and stroking the aphid. This makes the aphid give out honeydew, and the ant then takes it back to the nest. Some ants will actually take the aphid back to their nests and install them in special chambers hollowed out around the plant roots. The aphids feed on the plant roots and the ants 'milk' them without having to go out of the nest.

Not all ant species live in quite the same way as *Acanthomyops.* Some tropical species make no permanent nests at all. They are called army ants or driver ants. They settle down for short periods now and then to rear more young, but most of the time they are on the march. A marching colony may be several yards wide and hundreds of yards long and its members will eat any animal in its path.

Many ants harvest seeds and herd aphids, but some actually grow their own crops. These farming ants live in South America and they feed entirely on fungi which they grow on beds of chewed leaves in their nests. The workers go out and bring back pieces of leaves, waving them above their heads as they go. For this reason they are often called parasol ants.

Right: Parasol ants carrying leaf fragments back to the nest.

Slave Makers

Some species of ants have no workers of their own. They are rather like cuckoos because they lay their eggs in the nests of other ants and rely on these ants to rear them. Other species have some workers of their own but rely on 'slaves' to do most of their work. Workers of the slave-making ants periodically raid the nests of other ants and bring back pupae. The ants that emerge from these pupae are the slaves.

Termites

Termites are often called white ants, but they are not related to the ants in any way except that both groups are insects. Termites live in the warm or temperate parts of the world and there are about 2,000 species. About forty species live in the United States. They are all rather soft-bodied insects and they are rather pale in color. They feed mainly on wood, and some species do great damage to trees and to buildings. Wood is not a very easy material to digest, and the termites enlist the help of tiny protozoans for this job. The protozoans live in the termite's food canal and digest the wood for it. In return, the protozoans take some of the food for themselves.

Above: The three castes of the wood ant.

Below: The queen termite becomes a huge barrel-shaped creature full of eggs. The king rests on her back, while they are protected by a ring of large-headed soldiers.

All the termites are social creatures and, unlike the ants, they have a king and a queen at the head of the colony. The workers (if any) and the soldiers are also of both sexes, whereas those of the ants are all females.

Termite colonies are started by both the king and the queen after they have had a short flight. They break off their wings and start to dig out a nest. The workers take over later and the royal couple retire to the royal chamber, where they may live for 50 years in some species. The most primitive termites make their nests in dead wood—usually in old tree trunks or logs. These termites do not have proper worker castes and the work of the colony is carried out by the young termites. They do have soldiers, however. Other termites make their nests under the ground, and the soil dug out may be heaped into huge mounds and cemented with the insects' saliva. These mounds are called termitaria and they are especially common in Australia and South Africa. They are very hard and they are full of tunnels and chambers.

Termites resemble the ants in that their workers and soldiers have no wings. The termites, however, have a much simpler life history and the young insects do not change much as they grow into adults.

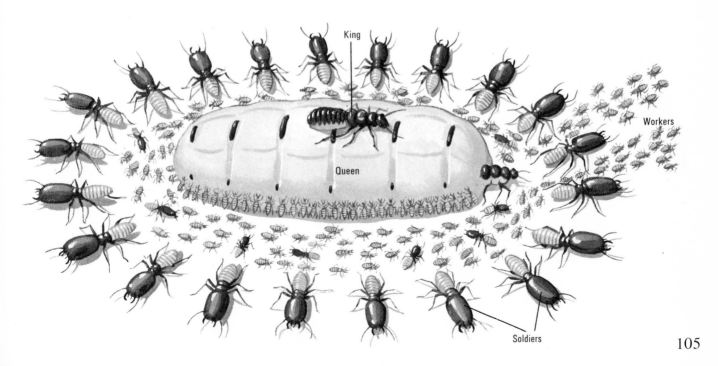

King

Queen

Workers

Soldiers

105

Spiders and Scorpions

Spiders and scorpions, together with mites, ticks, and a few other animals, form the group of arthropods called *arachnids*. They normally have four pairs of walking legs and they have no antennae. These features easily distinguish the arachnids from the insects.

Spiders

The spider's body is fairly soft and it is divided into two regions. The front part bears several pairs of eyes and it also carries the legs. Around the mouth there is a pair of poison fangs and there is also a pair of small leg-like structures called palps.

All spiders are carnivorous creatures and they eat only freshly caught animals. Most spiders eat insects, but some of the larger ones eat birds, lizards, and small mammals. The spider's venom is used to kill or paralyze the prey, and digestive juices are then poured into the victim. The spider then proceeds to suck the victim dry. Spiders have no jaws and they take only liquid food. Some spiders are harmful to man, but most of them are unable to pierce human skin with their fangs.

The Spider's Web

Some spiders chase their prey, or lie in wait for it, but many of them make traps or snares. These are in the form of silken webs. There are many different kinds of webs, but the most fascinating are the circular orb-webs spun by the garden spider and its relatives. These show up particularly well on autumn mornings when draped with dew or frost. If you look closely at an orb-web you will see how perfectly it is made. If you break a web down and then have enough patience to sit and watch you might see the spider

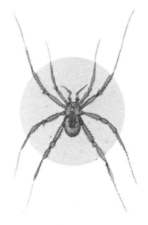

Above: A harvestman, distinguished from spiders by the absence of a 'waist'.

Below: A dew-spangled web, death trap for unwary insects.

make a new web. First of all it makes a frame and a number of radii. The latter are threads which run out from near the center, rather like the spokes of a wheel. The spider then moves round on the radii and leaves a rather open spiral of dry silk. Then, starting from the outside and using the first spiral as 'scaffolding', it lays down a tighter spiral of sticky silk. This is what traps the insects. The spider then retires to a little platform in the center of the web or else it hides under a nearby leaf, ready to rush out as soon as an insect gets caught in the web.

All spiders produce silk, whether they make webs or not. It is used for wrapping up the eggs, and also as a life-line for a spider that falls from its home. It pays out a thread as it falls, and then simply climbs back up. Young spiders also use silk to help them to reach new areas. They let a strand blow in the breeze until it whisks them up into the air. Drifting spiders have been caught thousands of feet up in the air.

The silk is produced in the abdomen. The silk glands open through tiny pores called spinnerets near the hind end. The silk is liquid to start with and it is pushed out like toothpaste from a tube, but it hardens as soon as it meets the air. Another gland adds the gum to make the silk sticky. The abdomen also contains the breathing organs. There are some tracheae like those of the insects and there are also some unusual organs called lungbooks. The lungbook consists of a tuft of thin 'leaves', rather like the pages of a book, suspended in a small chamber which is open to the air. The spider pumps air in and out of the chamber and the oxygen passes into the blood-filled leaves of the lungbook. Scorpions also have lungbooks, but most other arachnids breathe through tracheae.

Male spiders are usually somewhat smaller than the females and they have to take great care when they go courting. They have an elaborate system of signals, such as waving their enlarged palps or tugging the web in a special way. These signals prevent the female from rushing out and eating the male.

Scorpions

These animals are easily recognised because the palps are enlarged to form powerful pincers and because the abdomen is elongated into a tail with a sting at the end. Most scorpions live in dry regions and their hard external skeletons are well suited for conserving water. The majority of species are active at night and they hide under rocks and stones by day.

Scorpions eat insects and other small ani-

Above: The scorpion carries its sting above its body. The sting is mainly a defensive organ but it is not often used.

Below: A selection of spiders showing the 'waist' and the four pairs of legs. What look like antennae are actually palps. They are swollen in male spiders. The black widow (bottom left) shows the 'hour-glass' mark on the underside

mals, which they catch with their pincers. The food is then torn to pieces by the smaller pincers at the front of the head and the juices are sucked up. Scorpions, like spiders, have no jaws and they cannot eat solids. The food also provides the scorpions with their water, for very few of them ever drink. The scorpion's eyes are very poor and the bristles on the pincers seem to help in finding food. The sting is rarely used in catching food. It is mainly a defensive organ, but even then it is not often used. When a scorpion is disturbed it is more concerned with getting away than with attacking. It will sting when cornered, though, and campers in scorpion country are well advised to shake out their shoes and clothing before dressing in the morning. The effect of the poison varies with the species of scorpion. Some have very little effect on a man, but others can be fatal.

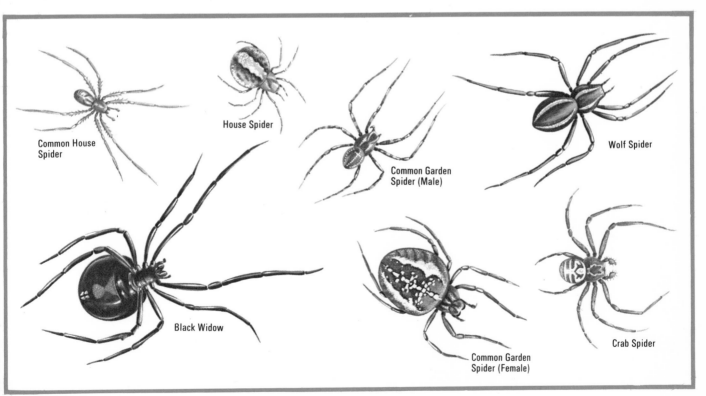

Common House Spider

House Spider

Common Garden Spider (Male)

Wolf Spider

Black Widow

Common Garden Spider (Female)

Crab Spider

107

Slugs, Snails and Bivalves

Slugs, snails, cockles, mussels, oysters, and octopuses are all *molluscs*. They are all very soft-bodied animals without backbones. Most of them are protected by a hard shell.

Snails

Snails belong to the group of molluscs called *gastropods*. They usually have a coiled shell into which they can retreat. Many also have a horny plate with which they can close the shell. Snails live in the sea, on land, and in fresh water, although we usually use the name snail only for the land and freshwater species. Examples include whelks, winkles, limpets, pond snails and garden snails.

The snail's body has three main regions. The most obvious is the flattened *foot* on which the animal glides about. It is propelled by muscular ripples. A gland near the front of the foot pours out slime which lubricates the snail's path and makes movement easier. The *head* is not clearly separated from the foot. It bears two pairs of tentacles in most land snails and one pair in the others. The eyes are at the tip of the larger tentacles in land snails, and at the base of the tentacles in other species. Most of the snail's internal organs are in the coiled *hump*, which remains inside the shell. The hump is covered by a thick 'cloak' called the *mantle*.

Above: Most snail shells are right-handed (as the lower shell) but a few are left-handed (like the upper shell).

Below: The scallop and razor shell are common bivalves. The gastropods, also shown, display a great variety of form and size.

The shell is formed by the mantle. It consists of three layers. The outer layer is thin and horny and it carries the colour pattern. The middle layer consists of large rectangular crystals of calcite, while the inner layer is formed of mother-of-pearl. Only the inner layer can be replaced if the shell is damaged, and so we often see shells with pearly patches showing through. Because the shell consists mainly of lime, there are far more snails in chalk and limestone regions than in sandy areas. The shape of the shell varies a great deal, but they nearly all coil the same way. If you hold a shell upright so that the opening faces you, the opening will nearly always be on the right. A few species have the opening on the left.

The marine snails breathe with gills. The gills lie in a cavity between the mantle and the body. The snail pumps water in and out of the cavity, and oxygen passes from the water into the blood of the gills. Some pond and river snails also breathe in this way, but the mantle cavity of land snails has been converted into a lung. The snails pump air in and out of the cavity instead of water. The gills themselves have disappeared. Many water snails breathe in this way as well: they surface every now and then to take in a lungful of fresh air.

Most of the land snails are vegetarians, feeding on dead plant material as a rule. Freshwater snails are also mainly plant-eaters, and so are many marine species. Other marine snails, however, are carnivorous. The whelk, for example, feeds on other molluscs. The snail has a peculiar 'tongue' called a *radula*. It is covered with hundreds of horny teeth, and it is used like a file to rasp away the food. The teeth are always wearing out, but the radula keeps growing and producing new ones.

Most snails are hermaphrodite animals. This means that each snail has both male and female organs. They have to pair up, however, before they can reproduce. Some snails lay eggs, and others give birth to active young. Many water snails lay their eggs in sausage-shaped masses of jelly attached to water plants. The young snail adds material to its shell from time to time, and the shell carries growth lines showing where the additions were made.

Slugs

Slugs are also gasteropods and they are closely related to snails. They are, in fact, snails which have almost or quite lost their shells. They behave in much the same way as snails, although some of them are carnivorous and eat earthworms. Many slugs do harm in gardens because they eat young

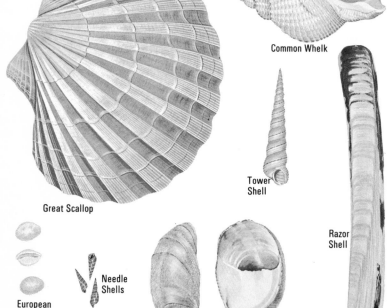

Great Scallop

Common Whelk

Tower Shell

Needle Shells

European Cowries

Slipper Limpet

Razor Shell

plants. The thick, slimy coating on a slug is its protection against drying up. It has no shell into which it can retreat and it would quickly shrivel in dry air without the slime. Even so, the slugs are confined to moist places.

Bivalves

The bivalves are those molluscs with two parts to their shells. The two valves are hinged at one edge by little teeth and by an elastic ligament. Muscles hold the shell shut and, when they relax, the hinge opens it. All bivalves live in water, and most of them are marine. They include cockles, mussels, scallops, oysters, and giant clams. The latter live in warm seas and may reach weights of 500 lbs. They are among the largest invertebrates and they are also larger than most vertebrates.

Some bivalves are cemented to rocks or attached by threads. Others live freely in the mud or sand and use the tongue-shaped foot for burrowing. Scallops can actually swim by flapping the two halves of the shell together and shooting themselves through the water.

The animals have no head, and many of those that are fixed to the bottom have no foot either. The body consists mainly of the mantle and the large gills. All the bivalves are filter feeders. This means that they filter tiny particles of food from the water. The mantle usually forms two tubes called *siphons*. Water is drawn in through one

Above: the common garden snail (top) and a slug (bottom), two garden pests.

Pearls

Very often a grain of sand gets inside a mollusc shell and irritates the animal. When this happens the mantle pours mother-of-pearl around the sand grain. This makes it nice and smooth and it stops the irritation. The smooth, round object is called a pearl and it is eagerly sought for jewelry. The best pearls come from certain types of oyster, but many bivalves make good pearls. There was once a thriving pearl industry in the north of England, based upon one of the freshwater mussels.

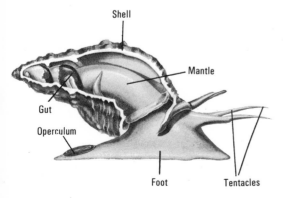

Shell
Mantle
Gut
Operculum
Foot
Tentacles

Left: Cross-section of a marine snail.

siphon and it goes to the gills. Oxygen is absorbed in the normal way, but tiny hairs on the gills waft food particles along to the mouth. The water then goes out again through the other siphon. Some bivalves have a long inlet siphon and they suck up mud by moving the siphon about on the bottom.

Most of the bivalves have separate male and female individuals, but there is no pairing in these animals. The males scatter sperms into the water. Many species release eggs into the water as well, so that fertilization takes place in the water. Other species retain their eggs until they have been fertilized by sperms drawn in with the water current. Fertilized eggs normally develop into free-swimming larvae which become widely scattered before they settle down to grow into adults.

109

Squids and Octopuses

Tales of sea monsters, furious creatures with many arms and hideous eyes, are recorded in many books and legends. Some of the tales are entirely fanciful, but some are probably based on fact, although they are much exaggerated. The descriptions of the monsters suggest that most of them were squids or octopuses.

Squids and octopuses, together with the cuttlefishes and the pearly nautilus, are known as cephalopods. They all live in the sea and they make up one of the three main groups of molluscs. They are thus related to the slugs and snails and bivalves, although they are much more active animals. Cephalopods have a more obvious head than the other molluscs and they certainly have a more efficient brain. They are among the most intelligent of the invertebrate animals. The most obvious features of the cephalopods, however, are the arms or tentacles that surround the head. Octopuses have eight arms, while most other cephalopods have ten. They are equipped with suckers and they are used mainly for catching food.

The cephalopods, like the other molluscs, are soft bodied creatures and they are surrounded by a thick sheet of skin called the mantle. This normally secretes some sort of shell or skeleton as it does in the snail. The early cephalopods had thick shells on the outsides of their bodies, but the pearly nautilus is the only living cephalopod with an external shell. The red and cream spiral shell is up to a foot across and it is composed of many chambers. The animal lives in the outer chamber, with its many tentacles exposed. The pearly nautilus lives in the warmer parts of the Pacific Ocean and spends much of its time near the sea-bed, but it can swim quite well.

The shells of other cephalopods are inside the mantle and they are much reduced in connection with the more active life of the animals. The shell of the octopus is almost non-existent, consisting only of a very thin plate or just a few chalky granules. The squid has a thin, horny plate known as a pen, while the cuttlefish has an oval chalky plate known as a cuttlebone. It is often washed up on the beach and it is frequently given to cage birds to provide them with the calcium necessary for the formation of egg shells.

Jet Propulsion

Sitting on the sea-bed, or in a marine aquarium, an octopus is continually swelling and contracting its body. This adds to its already rather sinister appearance, but the movements are only breathing movements. The mantle is very muscular and it covers all of the body except the head. It joins the body at the 'neck', where there are a few small openings. When the mantle expands, water is drawn in through these openings and into the mantle cavity. This is a large space around the body in which there hangs a pair of gills. Oxygen is removed from the water and the water is then pumped out again, not through the inlet holes but through a short funnel called the siphon.

Above: The cuttlefish uses its long arms to catch a shrimp.

Below: Close-up view of mouth of cuttlefish.

The common octopus is not an attractive animal when at rest, but it is not the dangerous animal that many people imagine. The eyes of these creatures are remarkably similar to our own eyes.

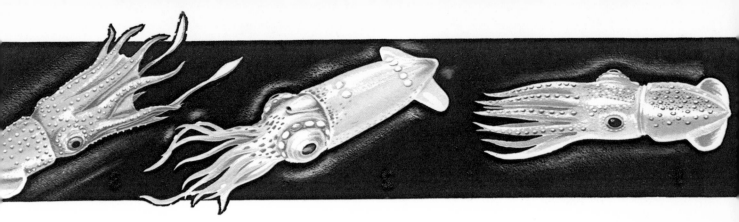

The same mechanism is also used for swimming: rapid contraction of the mantle forces a jet of water out through the siphon and the animal shoots backwards.

The squids and cuttlefishes breathe in the same way and they can also jet propel themselves about, controlling their direction by altering the position of the siphon. More usually, however, they swim by gently flapping the fin-like edges of the mantle.

Undeserved Reputation

The octopus is the best known of the cephalopods. It is often portrayed as a fearsome monster, but it does not deserve its reputation. There are about 150 different species and most of them are quite small creatures. An octopus with tentacles spanning 10 feet is a very large one, but even then the body is still quite small. Some octopuses swim freely in the open ocean and have been taken at depths of from 2,000 to 3,000 feet. Others prefer coral reefs or rocky retreats which remain unaffected by the tides. They prefer rocky coasts and they tend to hide away in crevices because they are shy creatures.

The main food of octopuses consists of crabs and bivalve molluscs. These are caught by the arms and then carried to the mouth. The mouth is right in the center of the eight arms and it has a very hard parrot-like beak which is able to crack the hard shells of the prey. The beak is also provided with poison glands.

Octopuses lay their eggs in grape-like clusters among the rocks. When they hatch the young octopuses spend some time in the plankton of the surface layers, and many of them travel a long way before settling down on the sea-bed again.

The Largest Invertebrates

Squids are much more active than the octopuses. Most of them live out in the open sea and range from the surface layers to the deepest regions. They play a very important part in the economy of the sea, for they eat

Above: These three squids live in deep water and they all have light-producing organs on their bodies (see page 143).

Below: The suckers of the octopus (bottom) are simple horny cups. Those of the squids (middle) and cuttlefishes (top) have teeth or claws.

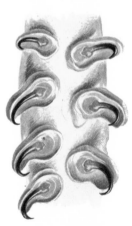

large numbers of fishes and they are themselves eaten by many other animals. Whereas the octopus is a rather shapeless bag, the squid is a slender creature with a rocket-like shape. There are some 400 species, ranging from a few inches to 60 feet in length. The giant squid, up to 60 feet including its arms, is the largest of all the invertebrates. This is undoubtedly the creature which has inspired most of the tales of sea monsters, especially when it has been seen grappling with a sperm whale. Sperm whales feed almost exclusively on these squids, but the squids are not easy to handle. Many whales bear scars showing where the squids' powerful suckers have gripped and torn their flesh. Unlike the octopuses, the squids and cuttlefishes have ten arms. Two of them are much longer and more slender than the others, but they widen out near the tip to form sucker-covered 'hands'. They can be shot out to catch prey. All the cephalopods share the horny beak of the octopus.

Apart from the nautilus, the cephalopods all possess an ink sac. This contains a thick inky fluid which can be shot out into the water when the animal is disturbed. It may form a 'smoke screen' behind which the animal can disappear, or it may simply confuse an attacker for a while.

Color Change Artists

All cephalopods can change their colours to some extent, but the best performers are the cuttlefishes. These animals are rather broader and flatter than the squids and their tentacles are relatively shorter. They generally live around the coasts and hunt for shrimps and other crustaceans on the sea-bed.

> **Thunderbolts**
> In many parts of the world, especially where the rocks consist of Jurassic clays (about 150 million years old), one can pick up smooth stones that look very much like rifle bullets. Commonly called thunderbolts, they are actually fossilised parts of the skeletons of extinct squids called belemnites. Other well known extinct cephalopods include the ammonites which are described on page 214.

Spiny-Skinned Animals

Of all the animals that can be found along the shore, the starfishes are some of the most interesting. Together with the brittle stars, sea urchins, and a few others, they make up the phylum (large group) called the *Echinodermata*. This name means 'spiny-skinned' and refers to the rough skins of the animals. The roughness is caused by numerous chalky plates embedded in the skin. The echinoderms are all marine animals of moderate size. They are easily recognised and they are very distinct from other invertebrate animals.

Like the corals and jellyfishes, the echinoderms are radially symmetrical. This means that they are more or less circular or cylindrical creatures, regularly arranged around a central point. There is no brain nor is there even any structure which we can call a head. The nervous system consists simply of a sparse network with thicker strands around the mouth. Sense organs are poorly developed and special breathing organs are generally absent. Oxygen is absorbed from the water through the skin.

The echinoderms possess a unique feature called the *water vascular system*. This is a

Right: An edible sea urchin.

Below: The starfish is a predatory animal feeding largely on shellfish. The starfish's tube feet work in unison to exert a strong pull and open the bivalve shell.

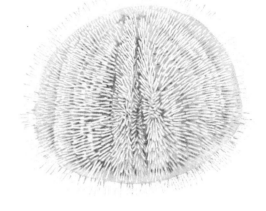

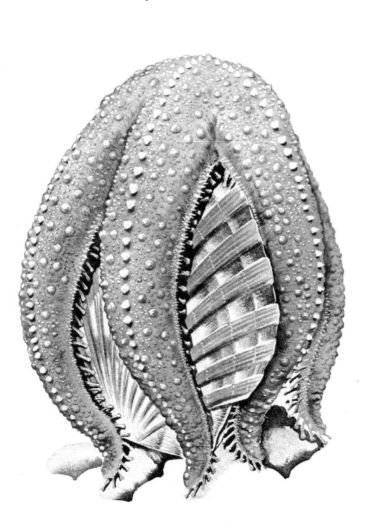

system of water-filled canals running to all parts of the body. It is in contact with the outside through a group of little pores on the upper surface of the animal. Tiny branches of the water vascular system grow out through the general body surface of most echinoderms and form *tube feet*. These are little muscular projections with disc-shaped ends. They help to absorb oxygen from the surrounding water and they also enable some of the animals to move about. Water is pumped into the tube feet and it forces their

ends out to make contact with stones or other objects. The muscles of the feet then contract and turn them into little suction cups. These grip very tightly and the animal can drag itself along.

Most of the echinoderms have separate male and female individuals, although some are hermaphrodite. Eggs and sperms are usually both released into the water, where fertilization occurs. Young echinoderms are very different from the adults and, whereas the adults all live on the sea bed, the young float freely in the plankton. This helps them to spread to new areas. Young echinoderms are remarkably similar to the young stages of the acorn worm (*Balanoglossus*). The latter is believed to be close to the ancestors of the back-boned animals. It is therefore likely that the starfishes and their relatives are more closely related to the vertebrates than to the other invertebrates. If you look at the 'family tree' on page 84 you will see that the echinoderms and the vertebrates are on the same major branch.

The Starfishes

Living echinoderms are split into five groups or classes, of which the most familiar is the starfish group. Starfishes are flattened animals and normally have five arms radiating from the central disc. The mouth is on the underside of the body, and so are the tube feet. The latter occur along the arms. The

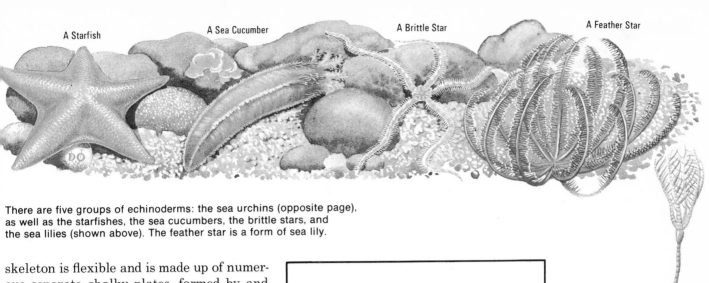

There are five groups of echinoderms: the sea urchins (opposite page), as well as the starfishes, the sea cucumbers, the brittle stars, and the sea lilies (shown above). The feather star is a form of sea lily.

skeleton is flexible and is made up of numerous separate chalky plates, formed by and embedded in the skin. Projections from these plates make the surface rough. Some projections develop into spines; others form tiny pincers. The latter are on movable stalks and they pick pieces of sand and dirt from the skin, thus keeping the animal clean.

Starfishes feed on various molluscs and other invertebrates. They are often serious pests in oyster beds. Some starfishes swallow the molluscs whole and eject the shells later, but others open the shells to get at the soft parts. The tube feet come into play again here. The starfish wraps itself around the shell and uses the suction of its tube feet to pull the two halves of the shell apart. Part of the starfish's stomach is then pushed out over the mollusc and digestive juices are poured out. The partly digested food is then sucked up and it goes into digestive glands in the starfish's arms for digestion to be completed.

Brittle Stars

These animals are often confused with the starfishes, but their arms are much more slender and they are distinctly separated from the central disc. The arms contain no digestive glands. There are tube feet on the arms, but they have no suckers and they play no part in movement. Brittle stars move by coiling their spiny arms around stones and heaving themselves along. The animals often live at great depths and they feed on small creatures. Some brittle stars catch their prey with their arms, others shovel mud into the mouth with their tube feet and digest any creatures it contains.

Sea Urchins

Sea urchins are globular animals whose skeletal plates form a rigid shell called a test. Some are quite circular, but others are oval or heart-shaped and have lost their radial symmetry. The plates bear many spines and pincers. Although there are no free arms, the tube feet grow out through

Sea Lilies

In spite of their name and appearance, sea lilies are animals. The central disc is attached to the sea bed by a stalk and it bears a number of feathery arms. The arms bear tiny hairs (cilia) which create water currents. These water currents flow towards the mouth and carry tiny food particles to it. Some members of this group of animals break away from the stalk when fully grown. They are called feather stars and basket stars.

the test in five distinct rays. The mouth of the sea urchin is on the underside and it is surrounded by a framework of bone-like plates called *Aristotle's lantern*. Attached to this are five strong teeth which project from the mouth. The urchin uses these teeth to browse on the seaweeds which form its main food.

Sea Cucumbers

These are sausage-shaped animals in which the skeleton is reduced to scattered little plates. The surface is rather leathery. There are five rows of tube feet, and those around the mouth are modified to form tentacles. The tentacles occasionally catch small creatures, but the sea cucumber obtains most of its food by scooping mud into its mouth and digesting any useful material.

Above: A sea lily.

Brittle stars differ from starfishes in that the arms are quite distinct from each other and clearly separated from the central disc. Brittle stars heave themselves along by their arms.

113

Fishes

Fishes are backboned animals (vertebrates) which, with a few exceptions, live entirely in the water. There are more than 20,000 different kinds in the world today, some living in the sea and some living in fresh water. Most of them are beautifully streamlined, and they swim easily through the water by means of their fins and tail. They get their oxygen direct from the water through their gills (see page 78). The gills lie in chambers connected to the back of the throat, and the fishes breathe by taking water in through the mouth. This is why a fish is always opening and closing its mouth. It takes a mouth-

A typical bony fish cut away to show the arrangement of the internal organs. The swim bladder, which helps the fish to stay up in the water, is shown in blue.

Below: Many tropical fishes are brightly colored and they are commonly kept in aquaria.

Neon Tetra

Yellow Dwarf Cichlid

Black Tetra

Tiger Barb

ful of water and then shuts its mouth and forces the water into the gill chambers. Oxygen is taken up by the gills and the water escapes through openings on the sides of the body.

The fishes were the earliest of the backboned animals to appear on the earth, but the

Below: Not many fishes look after their eggs or young, but Tilapia is an exception. When danger threatens it gathers its young into its mouth for safety.

first fishes were very different from those of today (see page 220). They were small, but heavily armored and very clumsy. They had no jaws and they could not bite anything. They lived mainly on the bottom, sucking up mud and filtering out small animals and decaying material as food. Fishes with jaws appeared later and most of the jawless fishes died out, but a few strange jawless fishes are still alive today. They are known as lampreys and hag-fishes.

Cartilage Skeletons

Apart from the jawless lampreys and hag. fishes, today's fishes fall into two main groups: the cartilaginous fishes, whose skeletons are made of gristly cartilage, and the bony fishes. Both groups possess two pairs of fins on the sides of the body. The cartilaginous fishes all live in the sea and they include the sharks, rays, skates and dogfishes. We often eat members of the last two groups —dogfishes are known as rock salmon or rock eel when they are cooked—and the soft, gristly nature of the skeleton is easily seen. Most of the cartilaginous fishes can be distinguished from bony fishes by looking for the gill openings. The sharks and their relatives have five gill slits on each side, but the

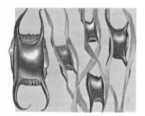

Above: Many shark-like fishes, including the skate and the dogfish, lay their eggs in horny cases known as mermaids' purses.

114

Thick-lipped Mullet

Cod

Rudd

Dace

Wrasse

Mackerel

Salmon

Norway Haddock

Perch

Bass

Barbel

Sailfish

Blue Shark

European Plaice

Sting Ray

Male Enlarged

Illustrations not in proportion to size.

Deep Sea Angler Fish with Parasitic Male

115

gill slits of the bony fishes are hidden under a flap called the operculum. Cartilaginous fishes can also be distinguished by their asymmetrical tails. The upper part of the tail fin is larger than the lower part, whereas most bony fishes have symmetrical tail fins.

Despite their soft skeletons, some of the sharks grow to a great size. The whale shark is the largest of all living fishes. It reaches lengths of more than 50 feet. The cartilaginous fishes have very rough skins, which can be used as sand-paper. The roughness is due to the tooth-like scales that cover the skin. Some shark's teeth, carried on the skin of the lips, are simply enlarged and razor-sharp versions of the normal body scales. It seems likely that all vertebrate teeth have been derived from this type of scale. Most sharks feed on other fishes, but the skates and rays are bottom feeders and prefer shellfish. Instead of having sharp cutting teeth, the skates and rays have blunt teeth well suited to cracking and crushing mollusc shells.

Bony Skeletons

Nearly all of today's fishes are bony fishes, with skeletons of real bone. Examples include herrings, cod, perch, stickleback, salmon and plaice. A few primitive species retain the bony armor of their ancestors, but the armor is reduced in most species to a covering of flat circular scales. Some fishes have no scales at all. The body is generally flat-

Above: The true flat-fishes, such as the flounder, lie on one side. During the fish's growth one eye moves round so that both are on the top side. The head becomes distorted.

The seahorse, a very unusual fish.

Below: The bichir is a primitive bony fish with lungs, although not related to the true lung fishes.

tened from side to side and much more slender than that of the sharks.

Apart from the various bottom-living species, the bony fishes all have what is known as a swim bladder, or air bladder. This is a white or silvery pouch lying just above the food canal. In some of the more primitive fishes, the swim bladder has a duct connecting it with the throat. The swim bladder of most bony fishes, however, is completely closed. It contains oxygen and it acts as a sort of buoyancy tank, enabling the fish to float effortlessly at any level in the water. As the fish swims deeper, the swim bladder produces more oxygen to compensate for the greater pressure. The gas is absorbed again as the fish rises.

Those fishes whose swim bladders are connected to their throats can gulp air into them. Some of them can stay alive for a while out of water because they can breathe in this way.

Reproduction in Fishes

The majority of fishes lay eggs, and some of them lay enormous numbers. A female cod may contain up to nine million eggs. Very few of them ever reach maturity, of course. Fish eggs and young fishes form a very important part of the diet of other sea creatures. Not all fishes lay vast numbers of eggs. In general, the number of eggs depends upon the degree of care that the eggs and young fishes receive from their parents. The cod simply scatters her eggs in the water and leaves them. She has to produce large numbers to ensure that at least some escape being eaten. At the other end of the scale, the little freshwater bullhead lays only a few dozen eggs. These are deposited in a hollow scooped out under a stone on the stream bed, and the male keeps a watch over them until the young fishes are several days old.

The stickleback is another fish that cares for its eggs and young. The male adopts a territory and defends it against other males by displaying his bright breeding colours. He then makes a nest with water plants and twigs and he encourages one or more females to lay their eggs in it. After the eggs are laid the male stands guard over them and ensures that they get plenty of oxygen by fanning a current of water over them with his tail. He does the same for the young fishes until they are able to look after themselves.

The seahorses—fishes, in spite of their strange appearance—have an even more unusual breeding behavior. The male has a pouch on his belly and the female lays her eggs in it. The eggs develop in this pouch and the young seahorses emerge a few at a time.

FISHES WITH LUNGS

Early on in the history of the bony fishes two distinct groups developed. One group possessed fins supported by numerous fine bones or rays. The other group had muscular fins, with sturdy bones in the center. Both groups had air bladders connected to the throat. The ray-finned fishes could move about only by swimming through the water. They could breathe with their gills and their air bladders gradually developed into buoyancy tanks, often losing their connection with the throat. Almost all of today's bony fishes have descended from this group. The fishes with muscular fins, however, could crawl about on the bottom. They could also crawl out of the water and breathe by gulping air into their air bladders. These fishes gradually evolved into land animals (see page 226), but a few remained fish-like and are still with us today. They are known as lung fishes.

There are only six species of lung fishes—one in South America, one in Australia, and four closely related ones in Africa.

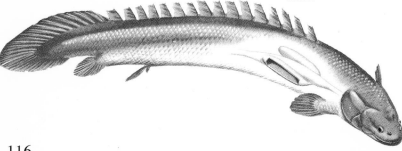

Amphibians

Amphibians, represented by frogs, toads, newts, salamanders, caecilians, were the first backboned animals to leave the water and live on land. That happened some 400 million years ago. Even today the amphibians have not completely broken away from the water. They have moist skin and they thrive only in damp places. Nearly all of them have to go back to the water to breed. The name amphibian means 'double life' and indicates that the animals live partly on the land and partly in the water. The life history of the common frog, illustrated on this page, is typical of the whole group. The larval stages, known as tadpoles, are fish-like and they breathe with gills. They gradually change into the air-breathing adults.

The modern amphibians are very different from the early ones that first crawled out of the water. There are two main groups of amphibians living today. These are the frogs and toads on the one hand and the tailed amphibians (newts and salamanders) on the other. A third group, (caecilians), is found in the warmer parts of the world, but its members are not very common. They look like large earthworms and they spend their lives burrowing in the soil.

Frogs and Toads

These are the jumping amphibians. They use their long, webbed hind legs for both jumping and swimming. We tend to call the smooth skinned members of the group frogs and the rougher skinned ones toads, but this is not a very scientific division. Many so-called toads are actually more closely related to the common frog than to the common toad.

Some frogs and toads live permanently in the water, especially in swiftly flowing streams. They are flattened animals and they have no lungs. They breathe entirely through their skin. Some other frogs and toads live entirely on the land. They lay eggs in damp soil and the tadpole stage is passed inside the egg. Each tadpole has, in effect, its own private pond. When the animals finally come out of their eggs they are already small frogs. The majority of frogs and toads, however, lead lives like that of the common frog. The tadpoles live in water, but the adults spend most of their time on the land and return to the water only for a short time during the breeding season. Not all of them actually use ponds or streams for breeding. Some tree frogs make use of the small 'ponds' formed in the forks of trees or between the leaves of other plants.

The frogs and toads normally leave their eggs to look after themselves, but a few species show some concern for the welfare of their young. The European midwife toad mates and lays its eggs on land. The eggs are in a string of jelly and the male toad carries them about wrapped around his back legs. The tadpoles begin to develop inside the eggs and then, when they are ready to hatch, the male enters the water and releases the eggs. From then on the life history is just the same as that of normal frogs and toads. A few frogs and toads build nests for their eggs, but one of the strangest habits is shown by the Surinam toad of South America. This is one of the very flat toads that spends almost all of its time in the water. The female has a long egg duct which carries the eggs on to her back when they are laid. The eggs then sink into small hollows in her skin and the skin grows over each hollow to form a lid. The eggs hatch and the tadpoles complete their development in these little pouches. After about four months the lids open and out come tiny toads.

Toads usually nave a thicker, rougher skin than frogs and their back legs are shorter. Toads cannot jump as well as frogs and often crawl about.

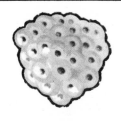

The Frog's Life History

Some adult frogs hibernate during the winter, hidden under logs or stones or buried in mud. They awake in spring and the males return to the water. Their loud croaking attracts the females, and the animals pair up in the water. The female may carry the male about on her back for a while and he then fertilizes the eggs as soon as she lays them. The eggs are covered with jelly which swells up as soon as they get into the water and protects them.

After fertilization the little black egg starts to grow and, within a few days, it develops a distinct head and tail. About 10 days after the egg is laid the tadpole emerges from the jelly and clings to a water weed by means of a sucker. The mouth is not yet fully formed. By this time it has developed three pairs of feathery external gills. During the next few weeks the tadpole grows rapidly and important changes take place. The external gills gradually disappear and they are replaced by internal gills rather like those of fishes. The hind legs start to form when the tadpole is about six weeks old, and the front legs appear a few weeks later. The diet has now changed from plant material to small water creatures. Lungs have formed inside the body by this time and the tadpole is starting to visit the surface to gulp air. The legs continue to grow and the tail begins to shrink. The animal takes on the typical frog shape and soon leaves the water to feed on small insects among the vegetation. The whole process takes about 12 weeks, depending on temperatures, but the frog takes a further three or four years before it is fully mature.

The frogs and toads that live permanently in the water have lost their tongues and they grab their food with their jaws. They eat all kinds of animal food, both dead and alive. The land living species normally have tongues which are fixed at the front. They are sticky at the tip and they can be flicked out remarkably quickly to pick up insects or some other small creature.

Newts and Salamanders

Most of the tailed amphibians are called salamanders. They vary from a few inches to about 5 feet in length. Newts are really only a particular group of salamanders. Apart from their tails and slender, lizard-like bodies, the newts are quite similar to the frogs and toads. The eggs are fertilized in-

A frog showing the moist glistening skin which is characteristic of these animals.

while still in the mother's body, but the young are not born until they have passed through the tadpole stage and turned into tiny adults. These salamanders are completely terrestrial creatures, although they still have to live in damp places because of their moist, thin skins.

Salamanders that do not grow up

Many salamanders spend all their lives in the water and quite a number of them have no lungs. Some of them actually retain their feathery external gills all their lives. This makes them look like overgrown tadpoles, although they grow up in other ways and can reproduce themselves. One of the best known animals of this type is the axolotl. It is the tadpole of a Mexican salamander. It *can* grow up into a proper salamander, although it does not do so very often. The mud-puppy and some of the other salamanders never grow up properly.

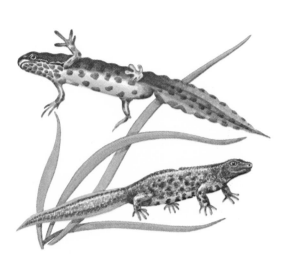

Above: Male newts develop bright colors and large crests on their tails in the breeding season
Below: the fire salamander and an apodan, one of the worm-like amphibians.

side the female's body and she usually lays them separately on the water weeds, but otherwise the life history is very similar to that of the frog. One small difference is that the newt tadpole keeps its external gills until it becomes an adult.

The European spotted salamander is one of many viviparous species. This means that it brings forth active young. The eggs hatch while still in the mother's body. She then enters the water and the tadpoles are born. Some other salamanders never go into the water. Some of them lay eggs in damp soil and the tadpole stage is passed while still in the egg. Miniature adults emerge from the eggs. The black salamander of the Alps takes the process even further. The eggs hatch

Below: The midwife toad, a European species in which the male looks after the eggs.

118

Reptiles

Reptiles are cold-blooded vertebrates with dry, scaly skin. They include snakes, lizards, crocodiles, and turtles. Most reptiles live in the warmer parts of the world, where their cold-bloodedness is not much of a handicap. There are some in the temperate regions, however, and the adder reaches as far north as Finland. The animals are active throughout the year in warm regions, but those of the cooler areas hibernate during the winter.

The reptiles first appeared some 300 million years ago. They evolved from some kind of amphibian. The two most important features which separated the early reptiles from their amphibian ancestors were their waterproof skins and the tough shells around their eggs. These meant that the reptiles were independent of the water. They could colonise the land much more effectively than the amphibians had done and for more than 200 million years they ruled the Earth (see page 224).

Living reptiles fall into three main groups. These are the tortoises and turtles, the crocodiles and alligators, and the lizards and snakes. A fourth group is represented

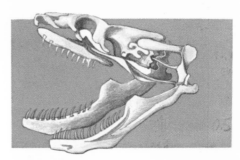

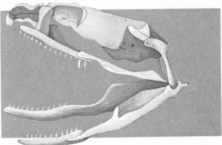

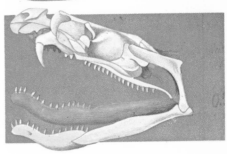

Non-poisonous snakes, such as garter snakes and boas, have no fangs (top right). Some snakes have fangs at the back of the mouth (center) but these are not usually very dangerous. The most dangerous snakes are the front fanged species, such as the cobra (right). Cobra fangs are rigid, but viper fangs (above) fold away when the snake shuts its mouth.

Left: The thorny devil, an Australian lizard.

only by the New Zealand tuatara (see page 236). The lizards and snakes are very successful animals, with nearly 5,000 species between them. The other groups, however, are much smaller. There are just over 200 different kinds of tortoises and turtles, and about 21 species of crocodiles and alligators. These animals are merely remnants of much larger groups that existed in the past.

Reptiles have decreased in size as well as in numbers since the days of the giant dinosaurs. A few snakes and crocodiles may reach 30 ft in length, but they have nothing like the bulk of the larger dinosaurs.

Tortoises and Turtles

These animals make up a very ancient group and they have changed very little during the 200 million years that they have

been on the Earth, although some of them have returned to life in the water. There is rather a muddle over the names of some of these animals. An individual may be called a tortoise in one country, a terrapin in another, and a turtle in a third. But there is a tendency to use the name tortoise for those living on land, terrapin for the edible species, and turtle for the marine and freshwater species.

The animals are easily recognised by the box-like shells in which they live. These shells are made of bone and they are normally covered by horny plates. There are gaps for the limbs and the neck, and the head can usually be withdrawn into the shell by bending the neck. Most species bend the neck vertically, but some species in the southern hemisphere bend their necks from

Many animals trick their enemies by pretending to be something else (see page 138). This frilled lizard raises a collar of skin around its neck, making itself look larger and fiercer than it really is.

side to side when they withdraw their heads.

The land-dwelling tortoises feed mainly on plants, but the water-dwelling species are mainly carnivorous. None of them has any teeth, and the jaws are covered with a sharp, horny beak.

All the members of this group lay hard-shelled eggs. The marine turtles come ashore to lay their eggs on some sandy beach. The eggs are very nutritious and men collect large numbers for food. This egg-collecting has helped to make the larger turtles quite rare.

Crocodiles and Alligators

These creatures are the closest living relatives of the great dinosaurs. They are large animals, with thick skin and bold scale patterns. The skins have long been popular for ladies' handbags, and the crocodile populations have suffered a great deal as a result. The animals live mainly in tropical rivers, although some species venture out into the sea. They spend most of their time basking on the river bank or else floating just under the surface of the water. The eyes and nostrils are on the top of the snout and they just break the surface when the animal is floating. The animals swim well by moving their powerful tails from side to side. They can also walk on land, although they are rather ungainly because their hind legs are longer than the front ones. In this respect, they resemble the dinosaurs. The long back legs also indicate that the crocodiles' ancestors walked on their back legs only.

Crocodiles and alligators feed on a variety of other animals. Fishes are their principal food, especially when they are young, but they also catch birds and mammals which come down to the water to drink. The jaws are armed with wicked-looking teeth.

Left: The egg-eating snake feeds on birds' eggs. It takes the whole egg into its throat and then breaks it with special 'teeth'. The pieces of shell are then spat out.
Right: the cobra is a very poisonous snake and it increases its fearsome appearance by raising a hood of skin on its neck.

Crocodiles and alligators look very much alike, but they can be distinguished by their teeth. The fourth tooth in the lower jaw is visible on the outside when the crocodile's mouth is closed, but it is hidden in the alligator. The alligator's snout is also somewhat blunter than that of the crocodile. There are actually only two kinds of alligator, one in the southern part of the United States and one in China. There are many different kinds of crocodiles. Other members of the group include the caimans, closely related to the alligators, and the gavial. The latter animal has a very long, narrow snout and it feeds on fishes. It lives in various Asian rivers.

Some crocodiles bury their eggs in sand, but others, together with the alligators, lay them in heaps of rotting vegetation. The heat of fermentation helps to incubate the eggs. The mother always stays around the nest and, when she hears the young animals breaking out of their eggs, she removes the covering material and helps them out.

Lizards

The lizards are a very varied group of reptiles. Most of them are only a few inches long, but some are much larger. The komodo dragon, for example, reaches a length of 12 feet. This is the largest lizard and it lives in Indonesia. The lizard's body, like that of the snake, is covered with small scales. The head is often armored with bony plates. Most lizards have four limbs, but some species are legless and look very much like snakes. The slow worm is a good example.

Lizards nearly all feed on animal material, such as insects, but some are vegetarians. Among the most famous of the insect-eaters are the chameleons, which catch flies by

Above: Turtles can withdraw their heads into their shells by bending the neck sideways (top) or vertically. The shell has played a large part in the survival of these slow-moving primitive creatures.

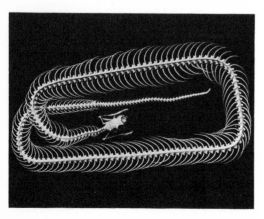

Snakes have very many vertebrae and joints in their backbones. This is what enables them to coil up so tightly.

Skeleton of a harmless snake.

shooting out their sticky tongues with remarkable speed and accuracy. Only two species of lizard are poisonous. The best known of these is the gila monster (see page 165).

Most lizards lay eggs, with rather thin papery shells, while others bring forth active young. The western blue spiny lizard is one of the live bearing species.

Snakes

Snakes are all legless reptiles, although the boa constrictor and some of its relatives have small claws near the base of the tail. These claws are the remains of legs, and it is clear that the snakes have descended from lizard-like ancestors. Many people are frightened of all snakes, but not all snakes are harmful. In fact, only about 150 out of over 2,000 different kinds are at all dangerous to man.

Snakes live in all kinds of places, from the sea to the driest desert. Even though they have no legs, they manage to get about quite easily as long as the surface is rough. Many of them have broad scales along the belly and these scales can be lifted and dug into the ground. By digging in all the scales in turn the snakes can pull themselves smoothly forward. Other snakes throw their bodies into loops from side to side and they glide forward by pushing backwards against small lumps on the ground.

All snakes feed on animals. The constricting snakes catch their food by wrapping their bodies around it and squeezing it until it suffocates. These snakes, which include the anaconda and the python, are not poisonous but they are often very large. The other snakes catch their prey with their mouths. They strike very rapidly and sink their teeth into the prey. The poisonous snakes have special hollow teeth called fangs. These inject poison into the wound and the poison kills the prey. The non-poisonous snakes, such as the garter snake, overcome their prey by strength.

The prey is swallowed whole, and the snake's body is specially built so that it can swallow animals wider than its own body. The lower jaw is separate from the skull and the mouth can open very wide. The snake gradually works itself over its food by using its backward-pointing teeth. The ribs of the snake are not joined on the lower side of the body and they can open out to let the meal move down the food canal.

Most of the snakes lay eggs with leathery shells, but some species bring forth active young.

The crocodiles can be distinguished from alligators by the fourth tooth of the lower jaw. It is visible in the crocodile when the mouth is closed.

Birds

It is impossible to mistake a bird for any other member of the animal kingdom because of its feathers. No other kind of animal has feathers. We do not know exactly when the first birds came into being, but they were probably the last great group of animals to appear on the Earth. It is believed that they arose from reptiles. Birds' legs are still covered with scales like those of the reptiles, but the scales on the rest of the body have developed into feathers.

A bird is a warm-blooded creature and its feathers undoubtedly help to keep it warm. The feathers also enable a bird to fly. Birds probably descended from some kind of tree-living reptile which used to glide from branch to branch. The front limbs became broader and eventually evolved into wings (see page 224).

Their wings enabled the birds to spread far and wide and to reach all parts of the Earth. There are nearly 9,000 species living

Above: The masked love bird (top) and the great black cockatoo.

Below: A painted stork

today, with a very wide range of sizes, shapes, and habits. Many birds have now lost the ability to fly. They include ostriches, emus, kiwis, and penguins. The penguins' wings have been transformed into flippers which are used for swimming, but the wings of the other flightless birds are weak and useless. There are many flightless birds in New Zealand and on other isolated islands. The birds arriving on these islands found few enemies and there was never any need for them to fly away. They gradually lost the ability to fly. Many of New Zealand's flightless birds are now being killed by cats and other mammals introduced by man.

Ostriches and emus have evolved into large running birds, well suited to the grasslands on which they live. The ostrich is the largest living bird. A male may reach a height of about nine feet and a weight of about 300 lb. But even the ostrich is dwarfed by some of the extinct species. The elephant bird, which lived in Madagascar until a few thousand years ago, was a large relative of the ostrich. It weighed nearly half a ton, although it was hardly any taller than to-

Lined Tiger Heron

Hummingbird

Agami Heron

Osprey

Teal

Winter Wren

Great Bustard

Emperor Penguin

Ostrich

Golden Pheasant

Macaw

Shoveller Duck

Bullfinch

Kingfisher

Rook

Falcon

day's ostrich. The tallest birds of which we have any record were the moas of New Zealand. Some of them lived until a few hundred years ago and they reached 13 feet in height. The moas and elephant birds were unable to fly. The heaviest flying bird alive today is the mute swan. It reaches 50 lb in weight, although several birds have greater wingspans. At the other end of the scale, the smallest bird is the bee hummingbird. Found in Cuba, this bird has a body less than an inch long, although its beak and tail make it up to about two inches. It weighs less than one tenth of an ounce.

How Birds Fly

Any flying object, whether a bird or an aeroplane, must have two forces acting on it. It must have *thrust* to push it along and it must have *lift* to keep it up in the air. Some birds glide and soar almost effortlessly on air currents, but most birds fly by flapping their wings. In flapping flight the wings provide both thrust and lift. Thrust is provided by the twisting movement of the feathers at the wing tips. This movement forces the air backwards and therefore drives the bird forwards. The feathers form a stiff front edge to the wing, allowing it to cut cleanly through the air. The feathers also produce the curved, or cambered, upper surface which is responsible for the lift. Air passing over the curved upper surface of the wing is 'stretched out' a little. Its pressure is lower than that of the air pressing up on the underside of the wing. As long as the pressure difference is large enough, the extra pressure on the underside will keep the bird up in the air.

The greater the area of a wing, the more lift it can provide. Large birds, such as swans, therefore have large wings to lift them. Narrow wings are best for fast flight and for birds such as albatrosses that glide through fast-moving air. Small birds, with little weight, have short wings, but albatrosses have much longer wings. Vultures, buzzards, and other birds that fly slowly or soar round and round have very broad wings. These provide plenty of lift at slow speeds.

Small birds usually take off merely by jumping into the air, but larger birds often need to run in order to get the air moving over their wings. On landing, the wings are held back and the body is tilted so that it acts as a brake.

Feeding

Birds use up a great deal of energy when they flap their wings in flight. They also need a lot of energy to keep their bodies at a temperature of about 106°F. This means that

There is a great variety of bird beaks. They include the hooked beak of the macaw which can be used as a third leg, the sifting beak of the shoveller duck, the seed-cracking beak of the bullfinch, the dagger-like beak of the kingfisher, the stout 'all-purpose' beak of the rook, and the sharp-edged hooked beak of the flesh-eating falcon.

birds need a lot of food. Many birds, especially the smaller ones, spend nearly all their time looking for food. No living bird has any teeth. The horny beak is used for catching and cutting up food. A beak suitable for crushing hard seeds would obviously be unsuitable for catching fish or for picking up small insects. The birds therefore have a wide variety of beaks, each type suited to the diet. Most kinds of food are eaten by birds, although very few species feed on leaves. Most birds feed on seeds or insects. Many also eat fruit, and small animals such as worms and snails. Smaller numbers of birds eat fish, mammals, and other birds.

Sight and Hearing

The sense of smell is poorly developed in birds. They rely on the senses of sight and hearing to find food. A blackbird can actually hear a worm tunnelling just under the ground and it can dig down in just the right place to catch it. Woodpeckers also seem to hear insects under the bark of trees. Sight is extremely good in birds. Hawks hovering high in the sky can spot a tiny mouse in the grass below and swoop down to catch it. Kingfishers on their perches can see the outlines of fishes in fast moving water and dive down to get them. The kingfishers often miss, but it is difficult for us even to see the fishes, let alone catch them. The skimmers also feed on fishes, but they have a different way of catching them. They fly low over the water and use their large bottom jaws as 'fishing nets' to scoop things from the water.

Nest Building

Many animals build nests in which to rear their young, but none of them builds such elaborate nests as the birds. The nests are used only during the breeding season and they are not permanent homes, although some birds return year after year to the same nest. At the start of the breeding season the males of many species adopt territories. They stake their claims by singing all around the boundary of the chosen territory. They defend the territory against other males, but they attract females by singing and displaying their plumage. The feathers are often very bright at this time of year. When the birds have paired up they begin to build the nest. Some species make do simply with a rocky ledge or with a hollow in the ground, but others make very elaborate nests. The birds collect twigs, grass, moss, hair, mud, and other materials and weave it all together into a cup-shaped nest. Many species add a roof as well, especially in tropical countries where monkeys and other animals might take the eggs.

Above: Diagrams to show the various stages in wing movement during the flight of a pigeon
Far right: Nests of weaver finches—masses of interwoven grasses providing safe shelter for the chicks.

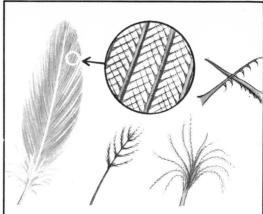

Feathers
There are several different kinds of feathers. The most obvious are the *contour* feathers which cover the outside of the bird. There is a stiff, horny shaft in the center. The bottom part of the shaft, where it joins the body, is called the *quill*. The flat part of the feather, called the *vane*, consists of hundreds of tiny branches called *barbs*. These are all hooked together to give a smooth, streamlined surface.

When the nest is complete the eggs are laid. Some species lay only one egg but most species lay between four and ten. Some species lay as many as twenty eggs. The eggs have to be incubated—that is, they have to be kept warm. The parent birds keep them warm by sitting on them. Sometimes both parents take it in turns to sit on the eggs, sometimes only one parent sits on the eggs. The other one brings food. When the eggs hatch the little birds need almost continuous feeding. The parents are very busy at this time, often bringing insects to the nest at the rate of four every minute.

Some young birds, especially those nesting on the ground or on the water, are fully feathered when they hatch and they are able to run about right away. Most tree-nesting birds, however, are blind and naked when they hatch. The parents have to continue to keep them warm until their feathers have grown. Small birds probably do not live more than four or five years in the wild. Larger birds live longer. In captivity, parrots and ravens have been known to live for 70 years.

Mammals

The mammals are the dominant animals in the world today. They are not the most numerous animals, but many of them are large creatures and they live in a wide variety of habitats. They have a marked effect on other animals and on the land in general. One of the main factors which have helped the mammals to dominate the earth has been their intelligence. Mammals have large brains and they are able to learn quite quickly. Another thing which has helped them has been their ability to maintain their bodies at constant high temperatures regardless of their surroundings. This has enabled them to invade the coldest regions of the earth.

Mammals are easy to recognise by their furry coats. All mammals have some hair, although there is not much of it on an elephant or a whale. Other features that distinguish the mammals are the warm blood, although this is also shared by birds, and the feeding of the young on milk from the mother's body. There are also various internal features which separate the mammals from the other vertebrates. For example, the lower jaw of a mammal consists of just a single bone on each side. Experts can use this fact to decide whether fossil bones come from mammals or not.

The mammals came to prominence quite recently in the earth's history—only about 70 million years ago in fact, when the dinosaurs became extinct. Thousands of different kinds of mammals have roamed the earth since then—many of them weird 'experimental' types—but there are less than 5,000 species alive today. Although they still dominate the earth, several of the orders of mammals are on the decline. The aardvark is the only member of its order, there are only two elephants, and several more orders contain only a few species. The only really large groups are the bats, rodents, carnivores, and hoofed mammals. In fact, these four groups contain more than 80 per cent of all living mammals.

Living mammals are arranged in three major groups. These are the *monotremes*, the *marsupials*, and the *placental mammals*. The first two groups are almost confined to the Australian region, and the world is ruled by the placental mammals—the group to which we ourselves belong.

The Monotremes—mammals that lay eggs

The mammals originally descended from egg-laying reptiles. Most of the mammals gave up the egg-laying habit, but one small group broke away from the others very early in their history. They did not give up laying eggs, and they also kept many other reptile features. A few of these primitive mammals are still alive today. They are called monotremes and they include the duck-billed platypus and the spiny ant-eaters. The latter are also called echidnas. Although they lay eggs, they still look after their young and feed them on milk, so they are true mammals. They also have hair and warm blood, al-

Below: The order of mammals. (Not drawn to scale)

Pipistrelle Bat

Black Rhino

Aardvark

African Elephant

Hyrax

Flying Lemur

Brown Rat

Platypus

Lion

Gorilla

Rabbit

Common Shrew

Black Buck

Nine Banded Armadillo

Pangolin

Common Seal

Kangaroo

Dugong

Sperm Whale

still returns to its mother's pouch from time to time until it is quite able to look after itself. The wallabies, koala bears, opossums and all the other marsupials start life in the same way, although some of the opossums have no real pouch.

Most of the marsupials live in the Australian region today, although some live in America. At one time they were found all over the world, but the placental mammals gradually replaced them in most places. Australia and South America had become separated from the rest of the world before many of the placentals reached them. The marsupials thrived there for a time. South America was later re-connected to North America and placentals from the north wiped out most of the marsupials. A few opossums survived, however, and the Virginia opossum even managed to invade North America. It still lives there quite happily.

Marsupials are still quite numerous in Australia, but their numbers have dropped a great deal since Europeans introduced cats, dogs, foxes, and other placental mammals. Several species, especially the smaller ones, have become extinct.

Above: The koala is one of Australia's most famous animals. It is a marsupial and it feeds almost entirely on eucalyptus leaves.

Left: The echidna or spiny anteater, one of the monotremes or egg-laying mammals.

though they are not as good at keeping their temperature steady as the other mammals. The monotremes are found only in Australia and New Guinea.

The Marsupials—mammals with pouches

The marsupials include such well known animals as the kangaroo and the koala bear. In some respects, they are half way between monotremes and placentals. They do not lay eggs, but the young are born at a very early stage of development. The red kangaroo may be over six feet tall, but it gives birth to a tiny baby only an inch long. This tiny baby is quite unable to look after itself, but it does manage to crawl into its mother's pouch. This is a large fold of skin on the mother's belly and her milk glands open into it. The baby stays in the pouch for some months and feeds on the milk. Eventually it is large enough and sufficiently well developed to get out of the pouch and hop about, but it

Right: Prairie dogs are American members of the squirrel family, part of the largest of all mammal orders – the rodents.

Right: The American dormouse is one of the smaller members of the rodent order.

The black bear is a common animal in North America. Although it belongs to the flesh-eating group of mammals (the carnivores), the black bear eats a lot of berries and other plant material.

Below: Walruses are relatives of the seals. They use their tusks to rake up shellfish from the sea-bed.

Bottom: Two zebras, close relatives of the horse.

The Placental Mammals

This is the most advanced and most intelligent group of mammals. It is the group to which man belongs. The female keeps her babies inside her body for a relatively long period and she feeds them through a special organ called the *placenta*. This is attached to the mother's womb and the baby is connected to it by the *umbilical cord*. This cord carries blood vessels from the baby to the placenta. In the placenta the blood vessels from the baby lie very close to those of the mother. Food and oxygen can then pass from the mother's blood into that of her baby. The umbilical cord breaks when the baby is born and it must feed and breathe for itself.

Because they are fed for a while inside their mother's body, placental babies are much more highly developed when they are born than the babies of marsupials. But many of them are still quite helpless. The young of dogs, cats, rabbits, and many others are blind and naked when they are born.

Below: An alert lioness, another member of the flesh-eating group.

Bush-babies (above) and vervet monkeys (above right) are two representatives of the highest order of mammals – the primates. Man is classified in this order.

Their mothers have to do everything for them. The young of horses, antelopes, and other animals that normally live on the open plains are able to get up and run about as soon as they are born. This is essential for them because they have to keep up with the herd. A helpless young animal out in the open would soon be found and eaten. But even the babies that can run about when they are born need milk from their mothers. It is some time before they go off on their own. During the time they are with their mother they learn a great deal about hunting or finding food. This period of learning while in the care of the mother (and father as well with some animals) has surely helped the placental mammals to rise above the other animals.

The placentals have managed to invade all habitats—the land, the air, the sea, and the fresh waters. They have also adapted themselves to a wide variety of foods.

Rhinoceroses are hoofed mammals belonging to the same group as the horse and tapir. There are five species. Only the two African species—the black rhino and the white rhino—are at all numerous today. Even these number only a few thousand.

Instinct and Learning

The thrush does not have to be taught how to build its nest, and nor does the spider have to be shown how to make its web. The animals know automatically how to perform these tasks. Behavior patterns of this kind which are inborn, are called *instincts*.

Instincts are very common in the animal world, particularly among the lower orders of life. They account for a great deal of animal behavior, including reaction to danger, courtship displays, rearing and feeding of the young, and migration. The behavior pattern needs something to trigger it off. The trigger is called a *releaser*. It might be a noise, it might be a visual signal, or it might be something that happens inside the animal.

Each kind of animal has its own types of instinctive behavior, and all members of a species will normally behave in the same way. The instincts are inherited and they are just as much part of an animal as its shape and color. Behavior patterns have evolved in just the same way as colors and shapes, and they are just as important in adapting the animal to its surroundings and helping it to survive. The wasp beetle (page 138) is an interesting example. It has evolved the colors of some of the smaller wasps, but the colors would not have been much good if the insect had not also evolved the scuttling behavior of the wasps. The two features together confer a great deal of protection on the harmless beetle.

Some instincts are very simple actions. Woodlice, for example, scuttle about in an agitated fashion if exposed to bright light or dry air. They keep this up until they reach a dark or damp place and then they settle down. This is a simple action, but it ensures that the animals keep themselves in the right conditions. Other instincts are much

Although many animals do fight, they spend more time threatening their opponents than actually fighting. A show of strength, such as is being given by this elephant, is usually sufficient to scare off a weaker opponent.

Below left: Blue tits soon learn to peck through milk tops to get at the cream. Although this is a new habit for the bird, it is not too different from the natural habit of pecking at insects.

Below: The courtship of the whooping crane.

more complicated. Nest-building, for example, involves collecting various kinds of materials and joining them together to make the layers of the nest. Even more complex are the engineering feats of the beaver. This clever animal knows instinctively how to fell trees and cut them into logs. He also knows how to dig little canals and how to float his logs along them. He then uses the logs to build dams which raise the water level and allow him to build safe homes in the middle of the ponds.

Courtship behavior is also a very complicated business for many animals. Pheasants and peacocks display their fine feathers to the females. Male redstarts hold singing contests to impress the hens. Male fishes often perform elaborate 'dances' to engage the attention of females, while deer and many

other animals fight among themselves. There are usually several stages to this sort of behavior, and each stage normally requires a separate trigger. In courtship behavior, for example, the female's reaction to one stage will trigger off the next stage in the male's display.

Some instincts are perfect right from the start. For example, a silk moth only makes one cocoon in its life, but it is perfectly made. Other instincts have to be perfected by practice. Birds know how to fly by instinct, but they need a bit of practice before they master it. The same is true of walking in other animals.

Above: The male lyre bird displays his fine plumage to the female when courting This display triggers off the female's mating behaviour.

Because instinctive behavior is triggered off automatically by the releaser, animals can often be misled. A male robin will attack a bundle of red feathers from another robin, but it will ignore a stuffed robin whose breast feathers have been painted brown. The attacking behavior is triggered off simply by sight of the red breast feathers. Baby herring gulls instinctively peck at the red spot on the parent's beak. This makes the parent regurgitate food for the chicks. The chicks will peck at a model of the beak as long as it has a red spot. The spot is the releaser.

Left: These baby shrikes have just left the nest. They know how to fly, but they will have to have several practice flights before they master it.

Learning

The lives of the lower animals are ruled almost entirely by instinct. The members of

Right: The woodpecker finch of the Galapagos Islands has developed the habit of using a thorn to dig insects out of crevices.

Below: Thresher sharks co-operate to round up and catch the herrings on which they feed.

a species all react in the same way and the animals lack any kind of personality. Mammals, birds, and some other animals are able to learn to some extent. They remember things that have happened before and they can sometimes modify their instinctive reactions as a result. Much of the behavior of adult birds and mammals is probably a combination of instinct and learning. Birds reared by themselves can sing quite well, but their songs are often quite different from the songs of wild birds. It seems that the ability to sing is instinctive, but the ability to sing the right song has to be learned.

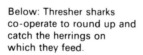

Instincts and Hormones

Some instincts are always present in an animal, but others appear only at certain times. Instincts which develop only at certain times are often due to the action of hormones. These are chemical substances made in the body and released into the blood stream. They have marked effects on behavior although they do not actually cause behavior patterns to begin. A releaser is still necessary, and the effect of the hormone is to make the animal sensitive to a signal which might not affect it at other times. Hormone activity is high during the breeding season and a cock bird will display vigorously when he sees a female. Another cock bird will trigger off some form of threatening or fighting behavior. Out of the breeding season, when the relevant hormones are not active, neither the hen nor the cock makes any impression.

Animal Parasites

On pages 47 and 48 of this book there are some pictures of parasitic plants. These are plants which 'steal' some or all of their food from other organisms without giving anything in return. But not only plants live in this way. Many animals are parasites too. They live in close association with other creatures, called the hosts, and they take food from them. They do not usually kill the hosts, but they weaken them.

Internal Parasites

Internal parasites are the parasites that live inside the bodies of their hosts. The most important members of this group are tapeworms, roundworms, and liver flukes. Some live in the blood system of the host animal, but the majority live in the food canal. Some worms merely attach themselves to the wall of the host's stomach or intestine

Right: These strange sputnik-like objects growing on a rose leaf are galls caused by the grubs of a small ant-like creature.

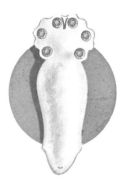

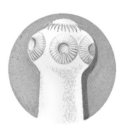

Above: Parasites all have some way of attaching themselves firmly to the host. The louse (top) has claws; the fluke (middle) has suckers; and the tapeworm (bottom) has both hooks and suckers.

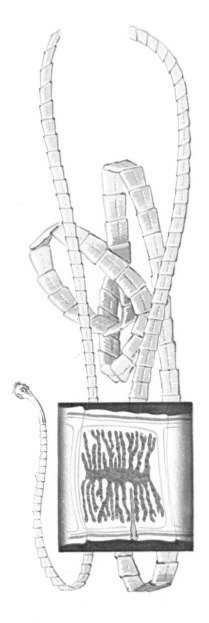

Left: The tapeworm's body consists of a small head and a string of segments containing little more than eggs. One segment with developing eggs is shown enlarged.

Oak Apples and Other Plant Galls

A soft pinkish outgrowth is a common sight on oak twigs in the summer. This is the oak apple. It is not the fruit of the oak tree, but an abnormal growth caused by an insect. If you cut it open you will see a number of small grubs inside. In the spring a small brown ant-like creature laid its eggs in the oak twig, and when the grubs hatched out they caused the twig to swell up around them. They then started to feed on the soft tissues, and they are in effect parasites of the oak tree although they do not seem to do it any harm. Growths of this kind are quite common on many kinds of plant. They are called galls. Most of them are caused by insects, but some are caused by microscopic creatures called mites. The little red pimples seen on sycamore and maple leaves are galls caused by mites.

and absorb the digested food. Worm-infested animals may therefore be thin and hungry. Other kinds of worms actually eat the host's tissues, rasping them away with sharp, horny teeth.

Nearly all animals have worms—even humans get them from time to time—and the worms do not do much harm in small numbers. It is only when their numbers build up that they start to weaken their hosts. Farm animals are normally dosed with worm-killing medicine every so often to prevent the worm population from getting too large.

A magnified photograph of a flea. Fleas are parasitic only in adult life.

Some parasitic worms, including the roundworms, have a fairly simple life history. Their eggs pass out with the host's droppings and they wait on the ground until another animal takes them in with its food. They can then develop into new worms. Tapeworms and liver flukes, however, usually have a much more complicated life cycle. The common tapeworm of dogs lives in the food canal. Its eggs pass out with the droppings and many of them end up in and around the dog's sleeping quarters. Here, there are usually a number of young fleas feeding on the rubbish. They take in some of the tapeworm eggs and the eggs hatch inside them. The young tapeworms grow in the flea and then they get back into the dog when it licks its coat and swallows the flea. The liver fluke, which lives in the liver of sheep, spends part of its life growing in a water snail.

The internal parasites do not have much work to do: everything is provided for them and they do not have to move about very much. These parasites do not therefore have much in the way of muscles or sense organs. Like the parasitic plants, they concentrate all their efforts into reproducing themselves. Huge numbers of eggs are produced, but only a few manage to reach the right kind of host. Most of the eggs perish.

It is not only the vertebrate animals that have parasites. Many caterpillars and other insects are attacked by ichneumon flies. These are rather special parasites because

Above: A blood-sucking louse is an external parasite on warm-blooded animals.

they do kill their hosts, but not until they have finished with them. The female ichneumon fly, which is something like a slender wasp, seeks out a caterpillar of the right kind and lays her eggs inside it by piercing it with her egg laying tube. The eggs hatch and the grubs begin to eat their host. They eat only the non-vital parts, such as muscles, to start with and the infected caterpillar becomes very 'lazy'. When the ichneumon grubs are fully grown they finish off their host and then turn into pupae themselves. These pupae, in their yellow silk cocoons, can often be seen surrounding dead caterpillars in the autumn.

External Parasites

The external parasites live on the outside of their hosts and most of them suck blood. Examples include ticks, lice, and fleas. The latter are parasitic only in adult life, for the young fleas live by eating rubbish in the nests of their hosts. External parasites can usually move about a bit more than internal parasites, but most of them have strong claws or hooks with which they hang on to the fur or feathers of the host.

Apart from weakening their hosts by sucking blood, many of the external parasites carry disease-causing germs. Fleas can carry plague, a disease of rats which can also affect humans. Lice can carry typhus fever, and ticks can carry many serious diseases of cattle. The ticks are related to the spiders. They attach themselves to cattle and other warm-blooded animals and bury their mouths in the host's flesh. They then suck blood and gradually swell up. Cattle in tropical countries may carry thousands of ticks and they have to be sprayed regularly with insecticide in order to keep them free of these parasites.

Social Parasites

These creatures display a special form of parasitism. They do not parasitise an individual animal, but they take advantage of a family or a community. One of the best known social parasites is the cowbird. This bird does not bother about making its own nest and it simply lays its eggs in the nests of other birds such as sparrows and starlings. The cowbird is much bigger than these other birds, and when the young cowbird hatches from its egg its first action is to throw out the other eggs or nestlings. The young cowbird then does nothing but take food from its 'foster parents' until it is big enough to fly away.

Similar examples are found in the insect world. Bumble bees are industrious creatures that make nests and bring up their young. There are, however, some closely related insects called cuckoo bees. These do not make their own nests and they do not have workers. They lay their eggs in the nests of bumble bees and leave the bumble bee workers to look after them and to feed the grubs that hatch.

Animal Partnerships

The animals described on the previous two pages are parasites. They live in close association with other creatures, but the associations are very one-sided, with the parasites getting all the benefit. There are many other associations, however, in which both partners derive benefit from living together. This type of association is called *symbiosis,* which simply means 'living together'.

Animal Partners

One of the best known examples of symbiosis involves a hermit crab and a sea anemone. Hermit crabs have soft hind parts and they protect themselves by living in old sea shells, such as whelk shells or winkle shells. The crabs carry the shells about with them, but they cannot add to the shells at all. They have to find larger ones from time to time to accommodate their growing bodies. A crab sometimes chooses a shell on which a sea anemone is growing, or else a sea anemone might start to grow on an occupied shell. This could be the beginning of a partnership. The base of the anemone may spread as it grows and cover part of the crab's body as well. The growth of the crab and the anemone then keep pace with each other and the crab has no need to change its home any more. As the hermit crab moves about in search of food the anemone is carried into new feeding grounds as well, and it certainly benefits from the

The cattle egrets on these buffaloes feed on the insects that the animals disturb. In return, they warn the buffaloes of approaching danger.

association. The hermit crab also benefits because it is protected by the stinging cells of the anemone.

Bird Partners

There are many symbiotic partnerships between birds and large game animals. It is quite common to see small birds running about on the backs of animals such as buffaloes, hippopotamuses and rhinoceroses. These birds are known as oxpeckers and, far from annoying the animals, they perform a useful service by eating the ticks and other blood-sucking parasites attached to them. Both partners benefit from this arrangement, and so it is a clear example of symbiosis. It is also said that the birds warn the animals of approaching danger, although these large animals do not have any serious enemies other than man.

Cattle egrets also associate with buffaloes and antelopes. The birds feed on the insects disturbed by the animals as they walk through the grass. In return, the birds warn the animals of approaching danger. This is a somewhat looser association than that between the oxpeckers and their mammals, but it is still a mutually helpful association.

Perhaps the strangest partnership is that between the Nile crocodile and the Egyptian plover. This little black-and-white bird feeds on the leeches that infest the crocodile. It is even allowed inside the crocodile's mouth, where it removes leeches and scraps of food adhering to the teeth.

The hermit crab can get along quite nicely without an anemone sitting on top of it, and the game animals do not really

need their feathered friends. But there are animals that cannot live without their symbiotic partners. This is especially true of the termites and various other wood-eating creatures. Wood is a difficult material to digest, and very few animals can produce the necessary digestive juices. Termites rely on hordes of tiny protozoan animals to do the job for them. The protozoans live in the stomachs of the termites and, in return for breaking down the wood, they get shelter and also take a certain proportion of the food themselves. Young termites get their protozoans when they are given partly digested food by older termites in the nest.

Plant Partners

The most obvious partnerships between plants are those found in the lichens. These plants, described on page 32, consist of algae and fungi living in a symbiotic association.

The roots of clovers, peas, and all other members of the pea family bear little swellings called root nodules. These nodules contain certain kinds of bacteria which are able to convert atmospheric nitrogen into more complex substances called nitrates. Nitrates are very important plant foods, and so the plants possessing root nodules derive great benefit from them. The bacteria get shelter and certain other food materials from the plant. Farmers try to include peas or beans or clover in their crop rotation because the decaying roots and nodules add nitrates to the soil. The farmer then does not have to add so much fertiliser for his next crop.

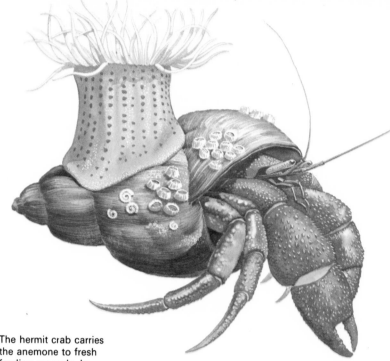

The hermit crab carries the anemone to fresh feeding grounds. In return, the anemone protects the crab with its stinging cells. The barnacles living on the shell have no effect upon the crab and do not enter into any partnership.

The crocodile, normally a sinister creature, is quite happy to let the plovers feed on the leeches on its skin. It even lets them into its mouth to clean its gums.

Many fungi enter into symbiotic associations with the roots of other plants. Such associations are called *mycorrhizas*. Most forest trees have networks of fungal threads around their roots, and the threads also invade the outer layers of the roots. The fungi help the roots to absorb water and minerals from the soil, and in return they get certain carbohydrate materials from the trees. Some fungi are very particular about the trees with which they associate. The red-and-white fly agaric toadstool is almost always associated with the roots of pine or birch trees. Some trees will not grow if the right sort of fungus is not present.

Mixed Partners

Partnerships between plants and animals are quite common. Many corals and other aquatic animals contain green algae in their bodies. These tiny plants were once believed to provide oxygen for the animals through their photosynthetic activity, but they actually produce very little oxygen. Their main role is probably in using up various waste materials produced by the animals.

Termites employ tiny animals to digest their food for them, but many other animals employ bacteria. Most grazing animals have special pouches somewhere in their food canals where they keep large populations of bacteria. The bacteria break down the tough cellulose of the grass and other plants, producing sugars and other materials that the grazing animals can absorb.

Survival

The animal world is engaged in a non-stop struggle to find food and to avoid being eaten. The animals that survive best are those with the best forms of attack and defence.

Camouflage

During their long history the animals have developed many different ways of protecting themselves. Large numbers of animals rely on camouflage and they blend in very well with their surroundings. Camouflage helps animals to hide from their enemies and it also helps the hunting animals, such as the leopards, to stalk up on their prey without being seen. One of the simplest forms of camouflage is known as *countershading* and nearly all animals use it. The underside of the body is generally somewhat lighter in color than the upperside, thus tending to cancel out the shadows. This helps the animal to merge with the background, even when it is out in the open. Irregular stripes and patterns are very effective at camouflaging animals because they break up its outline. The tiger is very difficult to see as it stalks through the long grass, and many fishes also use this method of camouflage. Many moths that sit on tree trunks during the daytime have irregular stripes. When they settle on the bark they

Above: The coloring of many animals blends well with their natural habitats.

The pale wing tips and thorax of the buff-tip moth (below) look just like broken twigs when the insect is at rest (left).

Below: many young deer are marked with white spots. These help to conceal the animals in the woodlands.

Above: Prawns become lighter or darker, according to the background, by decreasing or increasing the size of the red pigment cells.

Some animals with stings or poisonous bites like the coral snake (above) are marked with bold colors. These are called *warning colors*.

The porcupine is adequately protected by its barbed quills.

The red underwing caterpillar is almost invisible as it lies along a twig. Shadows are obliterated by a fringe of hairs along the underside as shown in the enlargement.

line up their patterns with the cracks and they 'disappear'.

Some animals are able to change their colors to match the backgrounds on which they are sitting. The true chameleon is the best known of these creatures, although it is not so good at changing its color as certain other animals. One of the best color-change artists is the cuttlefish. It can change color more quickly than anything else. The skin of the cuttlefish (and of its cousins the squids) has three layers of pigmented cells. One layer is black, one is red, and the outer one is yellow. Each pigment cell is controlled by a nerve and it can contract or expand very quickly. The animal can produce a whole range of colors from black, through reds and browns, to yellows acccording to which cells are expanded. When the black cells are expanded the animal looks dark; when they are contracted it looks much paler. Because each cell has its own nerve the animal can expand the cells in one part of the body and not in another. It can therefore produce a variety of patterns.

Prawns can also change their colors to match different backgrounds: on a sandy background they will be yellowish, whereas in a weedy pool they will be much darker. Their skins contain red, white, and yellow pigment cells. In a weedy pool, where it is fairly dark, the brain causes the release of hormones (chemical messengers) which ex-

pand the red pigment cells. This makes the animal darker, but the change is much slower than in the cuttlefish. Many fishes, such as the plaice, can also change color to match either sandy or stony backgrounds. In experimental work some of these fishes have even adopted a checkered pattern matching a chess board, but animals can produce only those colors associated with the pigments in their skins. They cannot match every background.

Animal Twigs

A number of animals, especially among the insects, have taken camouflage even further. They have developed or evolved remarkable similarities to objects in their surroundings. The majority resemble leaves and twigs, some twig-like caterpillars even bearing 'warts' which look like buds. One group of bugs look just like prickles on a stem, and a number of moths even resemble bird droppings. This type of camouflage is called *protective resemblance* and it certainly does protect the insects. Few birds would bother to peck at something that looks like a twig or a prickle.

Spines and Stings

It is usually the defenceless animals that employ camouflage to hide from their enemies. Those with some form of defence do not need to hide. One common form of defence is the possession of spines. Hedgehogs can roll up into very prickly balls when attacked and they are then safe from most of their enemies, although badgers seem to know how to deal with them. Several spiny fishes, such as the porcupine fish and the pufferfish, inflate themselves with water when attacked and present a formidable problem to their attackers. Other animals may have a more active form of defence in the possession of stings. Bees, wasps, and scorpions are good examples. The stings of bees and wasps are actually the modified egg-laying organs and it is only the females (queens and workers) that can sting. A further group of animals, including such diverse creatures as the skunk and the ladybug, protect themselves by having an unpleasant taste or smell.

Warning Coloration

Far from hiding themselves, the animals with unpleasant tastes or stings are often marked with bold colors and they are very conspicuous. For example, many wasps are clearly marked with black and yellow. Ladybugs are also black and yellow or black and red. These bold color combinations are known as *warning colors*, because they warn birds and other animals not to touch the insects. Young birds will try these unpleasant insects, but they will not eat them. They soon learn that the bold pattern is associated with something nasty and they leave the insects quite alone after that.

Mimicry

Many of us have drawn back hastily from a harmless hoverfly at some time or other in the belief that it was a wasp. It certainly looks like a wasp at first glance and even the birds are fooled as well. They

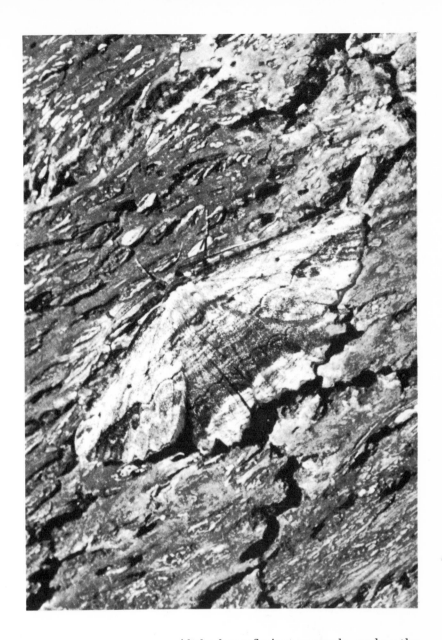

Above: The waved umber moth settles with its wing pattern lined up with the bark crevices and it almost disappears.

avoid the hoverfly just as we do, and so the insect survives because it resembles a wasp. This is an example of *mimicry*, and we say that the hoverfly mimics the wasp. There are many examples of mimicry among insects. The boldy marked bees and wasps, which are unpleasant because of their stings and their nasty taste, are mimicked by a variety of harmless insects. These include several different kinds of hoverflies, beetles, and moths. The bees and wasps are the *models*, and the other insects are the *mimics*. Young birds try to eat the models, but they soon learn to leave them alone. They also leave alone any other insects that look like these models. The mimics are therefore protected by their appearance, although they are harmless and perhaps good to eat. This rather clever arrangement will work, however, only if the unpleasant models are much more common than the mimics. If the mimics were common the young birds would associate the bold pattern with pleasant food and would attack all insects with that pattern.

The type of mimicry described above was first discovered by a naturalist called Henry Bates in the middle of the 19th Century. It is called Batesian mimicry and it involves an unpleasant model and one or more harmless mimics. There is, however, another form of mimicry known as Müllerian mimicry. It was first described nearly 100 years ago by a German zoologist named Fritz Müller. In this form of mimicry, best shown by various tropical butterflies, two or more inedible or otherwise unpleasant species share the same warning pattern. An insect-eating bird therefore has only one pattern to learn before it avoids all the insects sharing it. Imagine that a bird has to try 50 unpleasant insects before it learns the warning pattern. If each species had a different pattern, each young bird would have to try 50 of each species before learning to avoid them. But if five species all share the same pattern, only 10 of each species need die before each bird learns to avoid them.

In their behavior and appearance the hoverfly (right) and the wasp beetle (below right) mimic certain wasps. They are left alone by birds and other predators although they are actually harmless.

Below: The eyed hawkmoth is quite inconspicuous as it rests on a tree trunk. When disturbed, however, it raises its front wings and displays two large eye spots. These are enough to frighten many predators away.

Eye Spots for Protection

A number of small animals, notably fishes and insects are decorated with eye spots of various sizes. Small eye spots near the tail of a fish or on the wings of a butterfly might deceive any enemy and cause it to strike at these relatively unimportant parts of the body. The animal then escapes without much harm. Larger eye spots are often used to frighten enemies because they look like the eyes of much larger creatures. The eyed hawk moth rests on tree trunks by day and is not particularly conspicuous. But if it is disturbed it will raise its front wings and expose the eye spots on the hind wings. This is quite enough to deter some of the moth's enemies.

The cinnabar moth and its caterpillar have a very unpleasant taste and they advertise their unpleasantness with bright warning colors.

The Canada goose migrates between Alaska and Mexico.

The white stork, a summer visitor to Europe.

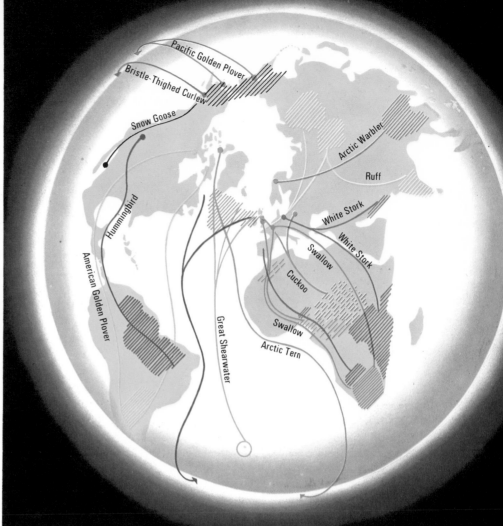

Migration

'The cuckoo comes in April . . . .' This is the beginning of a well known English rhyme and it indicates that the cuckoo is one of those birds that spend the winter in the tropical and sub-tropical regions and return to the temperate regions in the spring. The birds breed in the temperate regions and then, when autumn approaches, they return to the warmth of the tropics. This regular movement between two regions is called migration. It is of great importance to the birds because it enables them to find food throughout the year. Nearly all of the birds that migrate between the tropics and the temperate regions are insect-eating species. Other well-known migrating species are various warblers, the swifts, the martins and the swallows. The common swallow, called the barn swallow in America, is one of the most widely distributed birds. Apart from the most northerly regions, it breeds over nearly all the Northern Hemisphere. When autumn arrives the birds move south and, depending where they started off, they reach winter homes in Argentina, South Africa, and even Australia.

Not all birds migrate between the tropics and the temperate regions. Many species

The migration routes of many birds have been discovered by putting little rings on their legs and recording where they have been found. As the map shows, some species travel thousands of miles.

The purple martin travels between Canada and Brazil.

move between the temperate regions and the polar regions, especially the Arctic. A good number of these species are ducks and wading birds. The lakes and marshes all freeze over during the long northern winter, and so the birds fly south in search of open water and feeding grounds.

Land Migrations

Although birds are perhaps the most obvious of the migrating animals, they are by no means the only ones. Land-living mammals do not cover such large distances as the birds, but many of them make important migrations. Animals living in harsh climates avoid the worst of the winter by moving into more sheltered places. The caribou (reindeer), for example, spends the summer wandering over the open tundra in the far north, but it moves southwards into the coniferous forests for the winter. Mountain animals, such as the chamois and various species of sheep and goats, spend the summer on the upper slopes, but move down into more sheltered regions for the winter.

The Monarch Butterfly

Many butterfly species migrate. One of the best known of the migrating species is the monarch or milkweed. It is an American butterfly and during the summer it spreads

139

far into Canada. In early autumn the insects begin to move southwards, gradually joining up into huge swarms which fly down to their winter quarters in Florida and Mexico.

Migration in the Sea

Marine creatures have almost as much freedom to migrate as the birds. Many whales move from their polar feeding grounds into the warmer seas before bringing forth their young. Some fishes also cover large distances to and from their breeding grounds. The eel starts its life in the Sargasso Sea in the Western Atlantic. The young eels take about three years to reach the coasts of Europe and the United States, and are still less than three inches long. They then move up the rivers and stay there for several years while they grow up. Each year the fully grown eels move downstream to the sea and make their way across the Atlantic. These adult eels do not return to the rivers again.

The life history of the Atlantic salmon is just the reverse of that of the eel. The fishes spawn in the upper reaches of the rivers and the young salmon spend about two years there. Then they make their way down to the sea, where they spend a further year or two before returning to the rivers to breed. Just as a swallow or a house martin may return year after year to the same nest, the salmon normally returns to the river in which it was born. It goes back to the sea after breeding and may survive long enough to make several journeys to the breeding grounds.

All the migrations mentioned so far have been seasonal ones, taking place at certain times of the year. Some creatures, however, migrate every day. These animals are the little crustaceans and other members of the plankton that float in the sea. They spend about 12 hours in the surface layers and then they sink down to lower layers for about 12 hours. When they come up again the surface currents have brought fresh food supplies and so the little creatures benefit from their short journey each day.

Finding the Way

The urge to migrate is quite instinctive in animals: they do not have to think about it. Internal changes, together with changes in temperature and length of day trigger off the movements. How the migrating animals find their way has always been a puzzle to naturalists. Memory must certainly play a part in the return of birds to the same nests year after year, but it cannot explain how young birds find their way to Africa several weeks after their parents have gone. The birds use the sun to guide them, and they know instinctively that they must keep at a certain angle to the sun during their journey. Butterflies and many other migrating animals use the sun to guide them. Fishes may use other methods. There is evidence that salmon generate a weak electric field around them, and they may use this to detect the direction of the Earth's magnetic field. In other words, the fishes may be using their own bodies as compass needles.

Above: The lemming is famous for its *emigrations*. These are mass movements without any later return. Every few years their numbers build up to such levels that the animals move away in large hordes. Most of the emigrating animals die, but those that are left behind can find enough food for themselves and they settle down to normal life again. If some of the animals did not move away they would all die. Other animals that emigrate include locusts and honey bees.

Below: Caribou on the move: huge herds of these animals move south into the forests in the autumn and then go back to the tundra in the spring.

Hibernation

Many animals living in the cold and temperate regions disappear in the autumn because they cannot withstand the cold, or because they are unable to find food during the cold season. Birds usually migrate to warmer lands, but many other animals go into *hibernation*. This is a very deep sleep, during which the body processes slow down almost to a stop. Very little energy is consumed and the body temperature, even in mammals, falls to within a degree or two of that of their immediate surroundings.

A large number of invertebrate animals pass the cold season in the egg stage. Most of the insects do this, and their eggs are extremely resistant to the cold. Many butterflies and moths pass the winter in the pupa stage or chrysalis, often safely buried in the soil. Some species hibernate as caterpillars, and these can sometimes be found among the dead leaves in ditches and hedgerows. A few butterflies and moths hibernate as adults. When the days grow cold they find some dry corner—perhaps in a hollow tree or a garden shed—and go to sleep until the first warm days of spring. Several species of snail hibernate under stones and logs. They seal up their shells with mucus which hardens into a glassy material. The common garden snail often hibernates in colonies, with many shells stuck together by this hardened slime.

Frogs, tortoises, snakes and lizards are well known hibernators. They all bury themselves away from the effect of frost. They often huddle together and this habit undoubtedly helps to conserve moisture.

Fishes do not really hibernate, although some species become lethargic and partly bury themselves in the mud. The temperature at the bottom of a lake does not fall below 40°F as a rule and most fishes are able to remain active throughout the winter.

Hibernation in Warm-blooded Animals

Some birds become drowsy in cold weather, but true hibernation is not known among the birds. The nearest approach to hibernation is shown by the whip-poor-will, an American night-jar or goatsucker. It sleeps for much of the winter, but its temperature does not fall as much as in the true hibernators. It stays at about 68°F, compared with the 102°F of the active bird.

Many mammals, such as bears, badgers and tree squirrels, sleep for several days at a time during the winter but these are not really hibernating. True hibernation, where

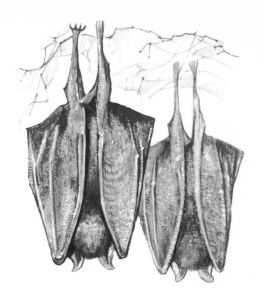

Right: The bats of cool climates sleep throughout the winter in caves and other secluded places. They hang upside down and their temperature falls to a very low level. Moisture from the air often condenses on their fur and gives them a glistening appearance.

the body temperature falls almost to that of the surroundings, is found in only a few groups of mammals. Bats and other insect-eating mammals such as hedgehogs hibernate, and so do various groups of rodents. These include dormice, ground squirrels, and hamsters. But even though they go into a deep sleep, they wake up periodically during the winter and they may feed on stored food.

Before they go into their winter sleep, many animals put on weight in the form of fatty deposits. These are drawn on during the winter. Some hibernators also lay in a store of food.

Colder weather, shortage of food, and the shorter days of autumn probably all play a part in bringing on the winter sleep. The end of hibernation is probably brought about by the warmth of the spring, and the animals then gradually speed up their body processes until their temperatures return to normal and they can start to move about. The waking process uses up a great deal of food and energy, and it is likely that many hibernating animals never wake up because they have not enough food reserves to provide the necessary energy. This is especially true if the animals have been disturbed and woken too much during the winter.

The lungfish aestivates in the mud of its dried up river.

Aestivation—the Summer Sleep

In some of the warmer parts of the world, especially where there is a well marked dry season, a number of animals hide away and go to sleep to avoid the heat and drought. This 'summer sleep' is called *aestivation*. Many desert-dwelling animals aestivate. They include various frogs and toads, as well as some of the desert rodents. The African lungfish also aestivates. It lives in sluggish rivers which often dry up during the summer months. When this happens the fish burrows into the mud and surrounds itself with mucus. It can breathe air, which it obtains through a narrow 'chimney' from the burrow, and it remains moist until the rains come and the river starts to flow again.

Nocturnal Life

When night falls a great change-over takes place in the animal world. Birds, butterflies, and most of the other creatures that are active by day seek their resting places. Deer, foxes, badgers, hedgehogs, field mice, and many smaller animals wake up and venture out into the open.

The air temperature drops when the sun goes down, and the air also becomes damp because it cannot hold so much water vapour when it is cool. If it cools down very much, some of the moisture may condense and form dew. This is very important for small creatures such as slugs, snails, and woodlice which have no waterproof coats. They can come out at night without risk of drying up, but they have to spend the daytime hidden away under stones and leaves. Earthworms may also leave their burrows at night and search for food on the surface. They make quite a rustle as they move among the leaves, but they are not easy to find because they

Above: The fox hunts mainly by night, using its keen eyes and its sense of smell.

Left: The bat has weak eyes and it finds its way about by sending out high-pitched sounds and listening for the echoes. The directions of the echoes tell the bat whether anything is in its way or not.

retreat rapidly into their burrows if disturbed. Most desert-dwelling creatures also come out at night and get much of their water supplies by eating dew-laden food.

The light is poor at night and many of the nocturnal animals have large eyes which gather in as much light as possible. The owl is a good example. Our own eyes are also quite good for seeing at night, although the cells which detect color do not work at low light levels and we see everything in shades of grey. Cats, foxes, and some other nocturnal animals increase the sensitivity of their eyes by having a reflecting layer at the back. You can see this layer shining brightly if the animal passes through the beam of a torch or a car headlight. The light passing back through the sensitive part of the eye helps to produce a brighter image.

Not all nocturnal animals have good eyesight. Many of them concentrate on the other

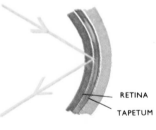

RETINA
TAPETUM

Cats have a 'mirror' behind the eye and this makes the eye much more sensitive. During the daytime the pupil is reduced to a narrow slit, preventing too much light from entering.

senses for finding their way about and for finding food. The sense of touch is very well developed in many nocturnal animals, whether they have good eyes or not. The large hairs or whiskers on the faces of cats and mice are sense organs and the animals react rapidly if these whiskers are touched. The sense of smell is also very important for nocturnal animals such as badgers, hedgehogs, and field mice. The moist night air holds scent much better than dry air does.

The sense of hearing is another very important one for nocturnal animals, as well as for many creatures that are active by day. The animals hear their food and their enemies and they also hear other members of the same species. This helps them to meet and mate. Many of the sounds are beyond the range of human hearing, but the animals' ears are specially tuned to pick up the sounds that are important to them.

The Bat's Radar

Most of the insect-eating bats fly by night. They have very weak eyes, but they hardly ever bump into anything. In one experiment, some thin wires were stretched across a darkened room and some blind-folded bats were released in the room. They flew about quite easily without bumping into the wires. This showed that the bats were not using their eyes at all to find their way about. They actually used a system known as echolocation, which is rather similar to the radar system used for locating ships and aircraft. The bats give out a very high-pitched sound —far too high for us to hear it—and listen for the echoes bouncing back from nearby objects. The echoes are picked up by the bats' large ears and the bat can actually detect the direction of the echo. If the echo comes back from directly in front, the bat can change course so as not to collide with the object.

142

This all happens at very high speed, but even more amazing is the fact that the bat can pick up echoes from flying insects and then change course to catch them.

Animal Lights

Many animals produce their own light. It helps them to find their way about and it also helps them to find mates. One of the most famous of the light-producing animals is the glow-worm. This is really a kind of firefly, but the female has no wings and she does perhaps look a little like a worm. Her light-producing organ is on the underside of her hind end, and the greenish light is produced by a chemical reaction. Man-made lights all give off a fair amount of heat as well, but the glow-worm's light-producing reaction is very efficient and gives out almost

Above: many cave-dwelling animals, which never meet the light, are blind and very pale in color. The olm, which lives in south east Europe, is a relative of the newts and salamanders.

no heat at all. The female shines her light as she sits in the grass, and the winged, lampless male flies to her. Fireflies are also beetles. In some species, both male and female insects produce light. Their courtship involves an exchange of flashing signals. The New Zealand glow-worm is actually the larva of a fly that lives in caves. The luminous larvae spin sticky threads which hang from the cave roof. Mosquitos and other insects are attracted to the lights and they get entangled in the threads. They are then eaten by the glow-worms.

Light-producing animals are common in the sea, especially down in the depths where there is no daylight. Jellyfishes, prawns, squids, and fishes all have luminous species. Eyes are well developed in some of them and the lights help them to find mates and food. Some deep sea angler fishes have luminous 'bait' on their rods, and this lures smaller fishes to their doom. Several squids and prawns baffle their enemies by giving off clouds of luminous slime. These distract their enemies and allow the squids and prawns to escape.

Above: the firefly (left) and a click beetle, two insects that produce light.

More than about 2,000 feet down in the ocean it is completely dark, but many species of animals carry their own lights around with them. Some have luminous patterns which enable them to recognise members of their own species. Some of the angler fishes (center) dangle luminous 'bait' above their mouths and lure their prey with it.

Animal Electricity

Certain fishes have the remarkable ability to generate high voltage electrical energy in their bodies. This is not the simple sort of static charge produced when we comb our hair or pull off a nylon jumper. It is a powerful discharge, sufficient in some instances to stun a horse. The fishes use these electric shocks for paralysing their prey and for defending themselves. They can also use them for navigational purposes.

All animals produce some electrical energy, for an electric impulse is involved every time a signal travels along a nerve and into a muscle. The electric fishes have simply modified this system. Their electric organs, or 'batteries', are composed of special muscles which have lost the ability to contract. Each muscle fiber becomes flattened and its nerve supply becomes concentrated at one side. The modified muscle fibers are known as *electroplates*, and large numbers of them are stacked up to form the

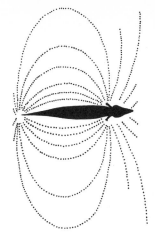

Above: The pattern of the electric field surrounding the Nile fish. The pattern is disturbed when the fish approaches an object, so the fish can take avoiding action.

Below: Some well known electric fishes.

battery. Each electroplate is insulated from its neighbors by a layer of jelly and connective tissue.

A normal muscle contracts when it receives an electrical signal from a nerve ending. The muscles of the electric organs cannot contract, however, and the energy is not converted into movement. It moves on as an electric discharge. If all the electroplates are discharged together, a powerful shock can be sent through the surrounding water. The fishes can also produce less powerful discharges by firing off only part of the battery at one time.

Electric Fishes

Electric organs are found in a variety of fishes, in the sea and in fresh water. One of the best known of the electric fishes is the electric eel, which lives in the rivers of the Amazon basin in South America. It is a cylindrical fish, reaching six feet or more in length, but it is not really an eel. The electric organs run about three-quarters of the way along the body and they can generate up to 550 volts—more than enough to kill the frogs and fishes on which the eel feeds. Larger animals coming too close to the eel are likely to be stunned by the shock. The fish can also produce smaller discharges which help it to navigate in the murky streams which are its home.

The eel-like Nile fish also finds its way about in muddy water by means of its own electricity. It surrounds itself with an electric field generated in its tail, and it is very sensitive to disturbances in this field. If it swims too near a rock it will detect a change in the field and take avoiding action. The fish can also use this system to seek out crevices for safe hiding.

The electric organs of the electric catfish form a sheath around most of the body. They can generate about 350 volts and are used for stunning the catfish's prey. The electric catfish lives in the rivers of tropical Africa.

Marine electric fishes normally generate lower voltages than freshwater fishes. This is because salt water is a much better conductor of electricity than fresh water. It requires a lower voltage to push a current through it. The skates and rays are the most important electric fishes in the sea. Most of the species have batteries in their tails and produce low voltages of unknown function. The various species of electric rays, however, have large batteries on each side of the head. They can produce a large current at up to 200 volts. The discharge is used for both defence and feeding. The fishes wrap themselves around their prey and electrocute it.

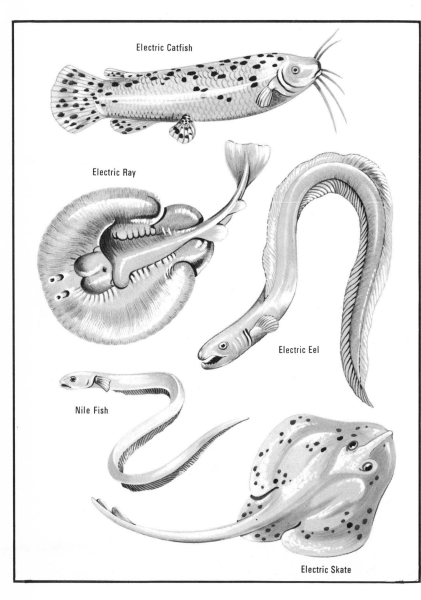

Electric Catfish

Electric Ray

Nile Fish

Electric Eel

Electric Skate

Dictionary of Animals

Aardvark A strange African mammal with no close relatives. Its bulky grey body is about 6 ft long, including the 2 ft tail, and about 2 ft high. It has a long narrow head and large ears. It feeds mainly on termites, which its digs out of the ground with its very powerful legs.

Aardwolf An African member of the hyaena family, somewhat larger than a fox. The coat is yellowish grey with black stripes, and there is a bushy tail. It is nocturnal and feeds mainly on termites, which it picks up with its long tongue.

Abalone A single-shelled marine mollusc related to limpets and also known as an ormer or earshell. The body is snail-like and fringed with tentacles. The shell has a line of holes across it, through which water is exhaled. Some abalones are among the largest of shellfish. They live in many parts of the world but are commonest in warmer waters.

Accentor A member of a small family of birds found in Europe and Asia. The commonest species is the dunnock, also called the hedge sparrow, although it is not a sparrow at all. Accentors have slender beaks and feed on insects and small seeds.

Acorn Worm A strange animal that lives in the mud of the sea-bed, often at great depths. The 'acorn' is the head end of the animal, which sticks out of the mud and is bright orange or red. There are several species, ranging from about 2 inches to 6 ft in length. The body is rather like that of an earthworm at first glance, but the animal breathes by means of gills. It is something of a link between the invertebrates and the vertebrates.

Addax A rare whitish antelope, related to the oryx. It lives in the deserts of North Africa and can live almost indefinitely without drinking. It gets enough water from the plants that it eats and from the dew on them.

Puff adder—an African relative of the European adder.

Adder A snake of the viper family, often simply called the viper. It has a fairly stout body up to about 2 ft long, and has a characteristic black zig-zag mark along the back. It lives in Asia and Europe and it reaches further north than other snakes. It is the only poisonous snake in Britain. It feeds mainly on lizards, mice and voles.

Agouti A South American rodent looking rather like a long-legged guinea pig. There are several different kinds, all brownish in color and about 20 inches long. Agoutis live mainly in woods and feed on leaves, roots and fallen fruit.

Albatross A bird of the petrel family, characterised by tube-like nostrils on top of the beak. Albatrosses are large birds, with goose-sized bodies and very long, narrow wings. The wandering albatross has a wing span of 12 ft or more—greater than that of any other bird. Most of the 13 species live in the Southern Hemisphere and they spend nearly all their lives at sea, gliding gracefully over the waves.

Alder Fly An insect related to the lacewings. The young alder fly lives in water and the adults are never found far from ponds or streams. They have rather smoky wings with thick black veins. Fishermen often use artificial alder flies as bait for trout.

Alligator A large reptile closely related to the crocodiles, but distinguished by its blunter snout and by the fact that the fourth tooth in the lower jaw is not visible when the mouth is shut. Alligators live in warm rivers and spend much of their time basking on the banks. Adults feed on fish, birds, and small mammals that come to the river to drink or swim. There are two species of alligator, one in North America reaching 15 ft or more in length, and a much smaller one in China.

Ammonite An extinct animal whose coiled shell is often found in rocks of the Mesozoic Age. There were many different kinds and they were related to octopuses and the nautilus.

Amoeba A member of the Protozoa, the simplest group of animals. It is very small—the largest species can only just be seen with the unaided eye—and usually lives in water. The animal is continuously changing its shape as it moves about. It feeds by engulfing tiny bacteria and other particles.

Anaconda The largest snake in the world, with lengths often exceeding 25 ft. The anaconda lives in South America and spends a lot of time in the water. It is often called the water boa. It feeds mainly on birds, deer and rodents. It is not a poisonous snake, but kills its prey by twining tightly around it until it cannot breathe.

Anchovy A small herring-like fish, of which there are several species. Anchovies are most abundant in warmer seas, but they are also found in the shallower regions and bays of cooler seas. Like the herring, the anchovy lives in huge shoals and feeds on the tiny plankton that floats near the sea surface.

Anemone Sea anemones are simple marine animals related to jellyfishes. The body is soft and jelly-like and bears numerous tentacles. The tentacles are covered with stinging cells which are used to catch fishes and other small animals. Sea anemones live attached to the rocks and shrink into small round 'blobs' when the tide goes out.

Anglerfish Anglerfishes are rather rough-skinned, bottom-dwelling fishes which catch their food with a sort of 'fishing rod'. The 'rod' is a modified fin spine, and it hangs near the mouth with a worm-like flap of skin acting as the bait. Small fishes investigating the bait are quickly snapped up. The anglerfishes have very large heads and mouths. They live in warm and temperate seas and some species go down to great depths.

Ani The anis are among the world's most unusual birds—almost everything they do is out of the ordinary. The four species live in North and South America and are related to the cuckoos, although they do not lay their eggs in other birds' nests. Their nesting behavior is unusual, however, because they live in colonies and several pairs usually combine to build a communal nest. The birds are about 15 inches long and shiny black in color. They feed on insects, which they usually collect on the ground.

Anole Anoles are small, slender lizards with long tails and triangular heads. They are related to the iguanas and are found only in the Americas. Most of them live in the trees and climb by means of sharp claws and adhesive pads on the toes. Anoles can change color with their mood, from brown to green.

Ant Ants are insects belonging to the order Hymenoptera, which also contains the bees and wasps. There are several thousand kinds of ants and they can usually be recognised by their slender 'waists' and by their sharply bent antennae. The ants all live in communities, and most of the work is done by wingless worker ants. Winged ants occur at certain times of the year and fly off to mate. The males then die and the females or queens return to earth to form their own nests or to swell the size of existing ones.

Giant anteater

Antbird The antbirds are a family of small birds living in South and Central America. The beak is usually noticeably hooked and the birds feed mainly on insects and other invertebrates, not necessarily ants. Antbirds live mainly in the forests, but they occupy many different habitats within the forest and there are many different kinds. Some look like thrushes, and others like shrikes, so we have names like ant-thrush, ant-shrike, and so on.

Anteater The three kinds of anteater live in South and Central America and they feed on ants and termites. The giant anteater is one of the world's most unusual mammals. It has a long cylindrical snout at one end and a huge bushy tail at the other. The whole animal is six or seven feet long and it lives on the ground. Its black, grey and white colors help to conceal it. The silky anteater and the two-toed anteater (or tamandua) are smaller and they live in the trees. Both have prehensile tails, with which they can grip the branches. All of the anteaters have long, slender tongues which they use to lick up ants and termites. They have no teeth.

Antelope Antelopes are hoofed mammals belonging to the same family as the cows and

145

A Uganda Kob, an antelope related to the waterbuck.

sheep and goats. They have horns which are unbranched and which do not fall off each year. Both males and females may have horns. Antelopes live in Africa and Asia, often forming huge herds. Gazelles, oryxes, and eland are all antelopes.

Ant lion An insect resembling a dragonfly but more closely related to the lacewing fly. The adults have long slender bodies and four narrow netted wings. They fly very lazily in the sunshine. The name 'ant lion' refers more to the young insects. They live where the soil is sandy and they make little pits for themselves. They lie more or less buried at the bottom of the pit and wait for an ant or some other insect to tumble down the loose sandy sides. They then pounce on it and eat it. Ant lions live in many parts of the world.

Aphid Aphids are better known as 'greenfly' or 'black fly', for these are two of the common-

est of the many types of aphid. Also called plant lice, the aphids are sap-sucking bugs and they do an immense amount of damage to crops by sucking sap and by spreading germs that cause plant disease. Aphids are all small insects, sometimes with wings and sometimes without.

Arapaima One of the largest freshwater fishes, the arapaima is rather like a pike. It has a long cylindrical body and a small flattened head with a jutting lower jaw. It is usually seven or eight feet long, but some specimens grow to 15 feet and may weigh over 400 lb. It lives in South America and is one of the more primitive of today's fishes.

Archerfish The archerfishes, of which there are five species, get their name because they obtain insect food by 'shooting' the insects down with streams of water drops. The jets shoot up from the surface and bring the insect down from the surrounding plants. The fishes then snap them up. They have been known to hit insects from a distance of six feet. Archerfishes, which grow up to about 12 inches in

length, live in the coastal waters and swamps of South East Asia and extend as far as northern Australia.

Arctic fox A small fox with short ears and muzzle, giving it a somewhat cat-like appearance. It lives in the far north and its greyish yellow summer coat turns white in winter. Like the polar bear, the arctic fox has hairy soles on its feet. This helps it to get a grip on the snow and ice, as well as helping to keep its feet warm.

Argonaut Also called the paper nautilus, this little marine animal is distantly related to the octopus and the cuttlefish. There are six species, all living in the warmer seas of the world. They move about in the surface layers by jet propulsion, squirting jets of water out through the breathing tube. They appear to feed on small fishes, and they themselves are eaten by larger fishes.

Armadillo There are 20 species of armadillo, distributed from the southern part of the U.S. to Argentina. The best known species is the 9-banded armadillo, which reaches as far north as the United States. The largest armadillo is the giant armadillo of the South American forests. It reaches a length of three feet. Armadillos are known for their ability to roll up into a tight armored ball when threatened. The 3-banded armadillo, or apara, is particularly adept in this way. Armadillos feed mainly on ants and termites, ripping open the nests with their strong claws and licking up the insects with their long tongues. They also eat plants, insects and snakes. Some species damage buildings by tunnelling underneath them.

Army Ant Also called driver ants or legionary ants, these insects are often more than an inch long. They are confined mainly to the warmer parts of the world and they have no permanent nests. The colonies move about in long columns, with large-jawed workers on the outside to defend them. It is very difficult to divert a marching column and the best thing to do is to get out of the way until they have all gone. Any animal in their way will be eaten. Insects, rats and snakes are all readily consumed, and even large animals such as goats and horses may be reduced to skeletons by a colony of these ants.

Arrow poison frog This name is given to several frogs living in South and Central America. They are generally very brightly colored and they secrete a very powerful poison which the native Indians use on their arrows. The most poisonous frog is the Kokoi, which lives in Colombia. It is only an inch long, but it can provide enough poison from its skin to kill 50 men. The male arrow poison frogs carry the tadpoles around attached to their backs.

Arrow worm A little animal, up to 4 inches long, that floats in the plankton of the sea. The body is transparent except for a pair of tiny black eyes, and it is very difficult to see this creature unless it has just had a meal. Arrow worms have spiny jaws and they catch small animals by darting after them and spearing them.

Asp A snake closely related to the adder and distributed over the warmer parts of Europe. Its colour ranges from greyish brown to orange or red, and it rarely exceeds 2 ft in length. It can be distinguished from the adder by its turned-up snout and by its yellow eyes. The adder's eyes are coppery-red.

146

Ass The ass is a small kind of horse. Two species are known—one in North Africa and one in Central Asia. A race of the African species has been domesticated for a long time and we know it as the donkey. There may well be no truly wild asses left in Africa today, but the Asian species survives in several countries.

Assassin Bug There are more than 3,000 kinds of assassin bug, so called because they grab and kill their insect prey rather violently. Assassin bugs have a powerful curved beak which is thrust into the prey and which then sucks up their juices. Many assassin bugs mimic their prey and can mix with them freely without causing alarm. They are truly wolves in sheep's clothing. Most assassin bugs live in the tropics.

Atlas moth One of the largest moths in the world, the atlas moth is found in India and other parts of South East Asia. It is mainly brown, but there is a transparent spot on each wing. The moth may reach a wing span of 10 inches. The atlas moth belongs to the same family (Saturnidae) as the European emperor moth and the Indian tussore silk moths, but it does not provide useful silk.

Avocet A graceful black-and-white wading bird with a long upturned beak. There are four kinds, one in Eurasia, one in North America, one in South America, and one in Australia. The upturned beak allows the avocet to feed on floating animals (and seeds) around the shores of the seas and rivers. The beak is held just under the surface, so that the end is more or less flat. The bird then walks along, sweeping its head from side to side and scooping material from the water.

Armadillo

Axis Deer The two species of axis deer live in India, Burma and neighboring countries. One species, known as the chital or spotted deer, is one of the prettiest of all deer. It is a small deer, rarely more than 3 ft high, and it has only small antlers, but its beauty lies in its white-spotted red coat. The other axis deer is quite different. Its squat pig-like appearance has given it the name hog deer. It has a speckly coat and small antlers. These deer are preyed upon by wild dogs and cats, as well as by large snakes such as pythons.

Axolotl The axolotl is a sort of Peter Pan—it never grows up. It is a brownish newt-like creature about 6 inches long and it breathes by means of feathery gills, just like the young newt or salamander. But, whereas the young newt eventually develops lungs and leaves the water, the axolotl retains its juvenile appearance. Although it does not have a grown-up look, the axolotl is able to reproduce itself, and it therefore does not need to grow up. For a long time, scientists were puzzled to know just what sort of animal the axolotl was. The puzzle was solved about 100 years ago in Paris, when some axolotls turned into salamanders. The axolotl is therefore a form

The axolotl—an animal Peter Pan.

of salamander that does not normally change into the adult form. Axolotls live in Mexico.

Aye-Aye The aye-aye is one of the many strange animals that live in Madagascar. It is closely related to the lemurs. It is about the size of a cat, dark brown in color, and it has a long bushy tail. Fruit and insects make up the bulk of the aye-aye's diet. It feeds at night and uses its very long narrow middle finger to dig insects from holes and crevices. This long finger is also used for combing the fur and for picking the teeth. The aye-aye is now very rare because its forest home is being cut down to make way for agriculture.

Babbler Babblers are song birds related to the warblers and flycatchers. They live mainly in South East Asia, but extend into Africa and Australia. One species, the wren-tit, lives in America. The babblers are short-winged birds and they do not fly well. They live near the ground in the forests, searching for berries and insects in the undergrowth. They get their name from their almost continuous but varied song.

Baboon Baboons are Old World monkeys that have given up life in the trees and now live mainly on the ground, although they return to the trees to sleep at night. There are several species, living mainly in Africa, but they also extend into Arabia. Baboons live in tribes, or troops, of up to 50 individuals, headed by one or more old males. They eat virtually anything. Lions and other large cats are the main enemies of the baboons, but three or four male baboons, armed with vicious teeth, are often more than enough to fight off a lion or leopard.

Backswimmer Backswimmers are water bugs that get their name because they swim on their backs, using their long hind legs as oars. Like most other water bugs and water beetles, the backswimmer has to come up for air every now and then. It hangs upside-down from the water surface and renews the bubble of air that it carries on its underside. The backswimmer is often called the water boatman, but this is really a different insect. It looks like a backswimmer, but it does not swim on its back. Backswimmers have sharp beaks with which they pierce other insects and suck their juices.

Badger The badger is a short stocky animal, about 3 ft long. Its coat looks grey from a distance, but the hairs are actually black and white. The most striking feature is the black-and-white striped head. Badgers are related to the stoats and weasels. They are carnivorous creatures, but they also eat a variety of plant food. Insects, mice, berries, eggs and worms all appear on the badger's menu. The European badger is found all over Eurasia. The American badger is quite similar, but it lacks the black-and-white striped head of its European cousin.

Bagworm The bagworms are moths whose caterpillars make little cases from silk and plant fragments. Most of them feed on trees, and some are considerable pests. When mature, the bagworm larva fixes its case to a twig and

turns into a pupa. The adult males can fly well, and so can some of the females, but some species have wingless females that cannot even crawl out of their cases. They mate where they are and lay their eggs in the case. Most of the bagworms are tropical insects.

Bald Eagle This striking bird used to be found all over North America and it was adopted as the emblem of the United States. Today it is extremely rare. It is not really bald, and it gets its name from the pure white feathers on its head and neck, which contrast vividly with the brown feathers on the rest of the body. The bald eagle feeds mainly on fish.

Bandicoot Bandicoots are rat-like marsupials with long pointed snouts and naked ears. There are 19 species, ranging in size from that of a rat to that of a rabbit. They live in Australia and New Guinea. Their food is very varied and consists of roots, leaves, seeds, insects, worms, slugs and mice. Some species are more carnivorous than others. Some bandicoots dig burrows, others make untidy nests among the undergrowth.

Barbary Ape This is not really an ape at all, but a monkey of the macaque family. It is about 2 ft long and has an extremely short tail. Barbary apes live in North Africa and a small colony lives in Gibraltar. These are the only monkeys living in Europe, although nobody knows whether they are native or introduced.

Barber fish These fishes which come from a wide range of families, are also called cleaner fishes. They provide a service for other fishes by cleaning them of parasites, patches of dead skin, and dirt. They also clean wounds and help them to heal. While this is going on, the fish being cleaned co-operates with the barber, even allowing it to enter the mouth and gill cavities. This is a remarkable form of behavior, in which both fishes benefit. The barber eats the parasites.

Barbet Barbets are colorful birds related to woodpeckers. They live in the tropical forests in both Old and New Worlds and they use their strong beaks to bite out nest holes in dead trees. They feed mainly on fruit.

Bark Beetle These insects include a variety of small species whose larvae live by tunnelling just under the bark of trees. The tunnelling activity eventually separates the bark completely from the wood and the bark falls. Each species has its own tunnelling pattern and this is easily seen when the bark falls. Bark beetles normally attack only weak trees; but they do some damage by carrying fungus diseases.

Barnacle Barnacles are those very common sea-shore creatures whose white tent-like shells coat rocks and pier legs. When the tide is out, the plates of the shell close up and the animal is hidden, but the plates open when the tide returns and the animal starts to feed. Feathery arms come out and rhythmically comb food particles from the water. Despite their unusual appearance, the barnacles are crustaceans and are therefore related to crabs and shrimps. The young barnacles swim freely before settling down and making a shell.

Barnacle Goose A small black-and-white goose, with a conspicuous white face. Flocks appear in Europe in autumn, and they disappear again in the spring. Not until 1907 did anyone discover where they spent the summer, and it was once believed that they developed from barnacles on the seashore. We know now that the geese breed in the far north—in Greenland and Spitzbergen.

Barn Owl The most widely distributed of all land birds, the barn owl is found in all continents except Antarctica. It is very pale in colour and has probably been responsible for many 'ghost' stories. It has a very white, round face. Barn owls live in hollow trees and old buildings, such as barns, and they eat mainly rats and mice. They are very useful birds, but are less common in many places than they used to be.

Barn Owl

Barracuda Barracudas, of which there are about 20 species, are slender pike-like fishes with vicious-looking teeth. The larger species, which may reach 8 ft in length, can be quite dangerous and some fishermen fear them more than sharks, but there are not very many reports of them attacking men. They feed mainly on herrings and other surface-feeding fishes. They are found in the Atlantic and Pacific Oceans and the Black Sea.

Barracudina About 50 species of barracudina live in the world's oceans. They are pale slender fishes, usually living at great depths. Very little is known about them, but they probably play a large part in the diet of other deep-sea fishes. Some barracudina species reach at least 2 ft in length when mature.

Basilisk Basilisks are lizards living in tropical America. They are up to 2 ft long, and the males have a crest on the head and back. The animals always live near water and they can actually run across the surface, using only their back legs. They will also run on land in this semi-erect position. Basilisks eat small animals and insects and are quite harmless.

Basket Star A marine animal related to starfishes. The central disc has five major arms, but these fork many times and they tend to curl up when they are dead. This gives the animal a basket-like appearance. Basket stars live on the sea-bed and capture small creatures with their tentacles.

Basking Shark One of the largest fishes, the basking shark may reach 40 ft or more in length. It is most common in the North Atlantic, but it is found in all the temperate seas of the world. Despite its great size, the basking shark has tiny teeth and rarely attacks man. It feeds on plankton, which it strains from the water with its very large gills.

Bass There are about 400 species and related forms living in both salt and fresh waters. Some of the larger salt water species may weigh well over 100 lbs. The true bass lives in the Mediterranean and the Eastern Atlantic. The stone bass, or wreckfish, likes to live in and around ship-wrecks. It is an aggressive fish, reaching a length of 6 ft.

Bearded Lizard This stout-bodied lizard, also called the bearded dragon, lives in Australia. The body is up to 2 ft long and it is covered with small spines. There is a pouch under the jaw and this can be inflated so that the spines look like a beard. The animal feeds mainly on insects. Like the basilisk, it can run on its two back legs, balancing its body with its long tail.

Beaver The second largest rodent in the world, the beaver is about 3½ ft long, including the tail. It weighs up to 75 lb. There are two species, very much alike. One lives in northern Europe and the other lives in North America. Beavers always live near water and they make their homes in river banks or in 'lodges' built of branches in the middle of a pond. Even the pond is made by the beavers. They use their chisel-like front teeth to cut down small trees which they then use to build a dam across a stream. The dam creates a pond. The beavers feed on bark, which they strip from trees. The front feet are clawed, for digging, and the hind feet are webbed. Beavers are excellent swimmers and they use their broad tails as rudders.

Bed Bug One of the many insect pests that man has acquired during his history. It is a small flat brownish insect that feeds by sucking blood. It has no wings and it hides in crevices or among bedclothes during the daytime. It comes out to feed when we are asleep at night. Bed bugs are much less common now than they used to be.

Bee-Eater The bee-eaters are very spectacular birds distributed over most of the warmer parts of the Old World. They are brightly colored and they have curved beaks. They nest in sandy banks and their nesting colonies may contain thousands of birds. They feed on insects, which they catch in flight. They destroy large numbers of locusts.

Bell Bird A very well named bird, of which several species live in the Southern Hemisphere. They are not all closely related and are named

A beaver about to bring down a tree.

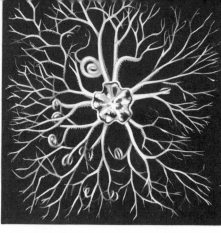

A basket star

from the metallic ringing sound of their voices. The crested bell bird of Australia collects caterpillars in the breeding season and, after paralysing them with a nip, it stores them in its nest to provide a store of fresh food for its young.

Beluga The belugas or white whales are related to the dolphins and porpoises. The young animals are grey, but the skin gradually becomes white by reduction of the black pigment. Belugas usually grow to about 14 ft in length and they are fairly slow swimmers in comparison with some of their relatives. These whales live in and around the Arctic Ocean.

Bichir A primitive freshwater fish of tropical Africa. It has a small, but broad head and its 2 ft body is covered with an armor of scales. A series of small fins runs nearly all along the back. The bichir has an air bladder connected with the throat, and it is able to breathe air in addition to obtaining oxygen through its gills. The front fins can be used as props, almost like legs, and the bichir may well represent one of the stages through which land animals passed during their evolution from fishes.

Bighorn One of six species of wild sheep, the bighorn lives in the Rocky Mountains. The males have enormous horns which curve around the back of the head and come forward again to the eyes. The horns are marked off into segments, each of which

148

represents one year of the animal's life. The domestic sheep must have arisen from one of the bighorn's Eurasian relatives, such as the mouflon or the argali. The woolly coat of the domestic sheep, however, is not shared by the wild species. They have coarse hair, rather like that of goats.

Binturong The binturong or bear-cat is a tree-dwelling carnivore found in South East Asia. It is about 4 ft long and has black fur. Its tail, which accounts for nearly half the animal's length, has very long fur and it is partially prehensile. The animal cannot swing by its tail, but it wraps it around branches as a brake or a fifth limb. The binturong is clumsy on the ground, but it is an excellent climber. It feeds on birds and small mammals, together with a fair amount of fruit.

A male Bird of Paradise (left) performs acrobatic displays to impress the female.

The bee-eater's beak is well suited for picking up insects.

Bird-Eating Spider This large spider measures over 7 inches across the legs. It lives in the jungles of the Amazon basin. It does not spin a web, but catches birds and small mammals by dashing after them. The bird-eating spider has few enemies because it is covered with poisonous hairs. It is often called the tarantula a name loosely applied to any of the large hairy hunting spiders found throughout the world.

Bird of Paradise The 43 species of birds of paradise live in the forests of New Guinea and surrounding islands. Four species reach the northern parts of Australia. The female birds are rather dull in color, but the males of most species have beautiful plumes which they use in the most elaborate courtship displays.

Birdwing The birdwings are magnificent butterflies living in the forests of South East Asia and the neighboring islands. Some of them have wingspans exceeding 10 inches. The females are a little larger than the males, but they are less colorful. The males have velvety black wings, marked with beautiful shining greens and blues. Many of the species, especially the males, are attracted to decaying meat and to damp places where water seeps out of the ground.

Bison, a magnificent species that was once nearly exterminated.

Bison Bison are massive, ox-like creatures weighing up to 3,000 lb. The largest ones stand up to 6 ft at the shoulder. The hair on the head, neck, shoulders, and forelegs is long and shaggy. Two species are alive today: the European bison or wisent, and the American bison. The latter is often wrongly called the buffalo and it has longer hair than the wisent. Both species were almost exterminated during the last 100 years, but numbers have now increased again due to proper game management.

Bitterling A minnow-like fish, about 3 inches long, living in the streams of central and eastern Europe. During the breeding season the male becomes brilliantly colored and the female develops a long pink egg-laying tube. The male guides the female to a freshwater mussel and she then lays her eggs in the mussel's breathing tube. The male releases his sperm, or milt, and this is drawn into the mussel with the breathing current. The bitterling eggs are then fertilised and they stay there until they hatch. The young fishes also stay in their safe 'hotel' until they have finished up their supply of yolk and are

ready to venture out into the water.

Bivalve One of the large group of molluscs that have two parts or valves to their shell. Examples include cockles, mussels, oysters and razor shells.

Black Bear There are five species of black bear, living in the Americas and southern Asia. They are much smaller than the grizzly and the other brown bears, weighing only about 500 lb at the most. The American black bear is very common in North America, especially in the national parks. Like its relatives, it is a good tree climber. The black bears eat a wide variety of food, including fruits, berries, rodents and insects.

Blackbuck Also known as the Indian antelope, the blackbuck is found over much of India and Pakistan. It is about 4 ft long and the male has remarkably twisted horns. It is one of the fastest land animals and has been credited with speeds of 50 mph. It can also leap well. The blackbuck is one of the few antelopes in which males and females have different colors. The males are dark brown above, while the does and young animals are yellowish brown.

Black Molly This little fish, a favorite with aquarists, is a black variety of the sailfin or some related fish (genus *Mollienisia*) which

Black Molly

comes from the rivers of Central America. The black varieties crop up from time to time in nature and they are bred artificially by the aquarists. The fishes can be recognised by the large sail-like dorsal fin of the male.

Black Widow This infamous spider has a number of sub-species distributed over the warmer parts of the world. The North American sub-species is noted for its powerful venom which produces severe pain and paralysis in a man. It can sometimes cause death, but fatal bites are rare. The female spider—the one that does the damage—is about half-an-inch long and shiny black in color. There is a red 'hour-glass' mark on the underside of her spherical abdomen. The male is much smaller. The name of the spider comes from the belief that the female always eats her husband after mating, but this does not always happen.

Blesbok These are small antelopes, very similar to the bontebok. They stand a little over 4 ft at the shoulder, but they are sturdy creatures and they weigh over 150 lb. There is a white blaze on the face and a white patch at the base of the tail. The short horns diverge from each other near the base and then become more or less parallel. Blesbok were once in danger of extinction. Many now

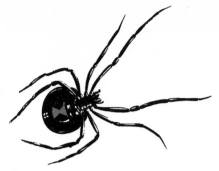

Black Widow spider

live a semi-domesticated life on South American farms.

Blind Snake This name applies to several hundred small non-poisonous snakes belonging to two families. They live throughout the warmer parts of the world. They are worm-like creatures, usually only about 6 inches long, although some species reach 3 ft. They are not completely blind, but their eyes are very small and they can do no more than distinguish light from dark.

Blister Beetle Blister beetles, of which there are many species, are found throughout the world although they are most common in warm dry climates. They get their name because their bodies contain a substance called cantharidin, which produces blisters on the skin of people touching the insects. The only British species has a shiny green body and it is known as the Spanish fly.

Bluefish A voracious fish living in most of the warm seas of the world. It may weigh as much as 25 lbs, swims in large shoals, and attacks all kinds of fishes with astonishing ferocity. The bluefish occurs in exceptionally large numbers in the western Atlantic and is caught commercially for its tasty flesh.

Blue Whale This is the largest animal that has ever lived. It can reach 100 ft in length and nearly 150 tons in weight—as much as 30 elephants. It is a whalebone whale, that is it has no teeth. Inside its mouth there are plates of horny material called baleen, or whalebone. These strain the little planktonic animals from the water and the whale then swallows them, for this largest of all animals feeds on some of the smallest. During the summer, blue whales live in the polar seas, but they move to warmer seas in the winter, and their calves are born in these warmer regions. The calves are 24 ft long when they are born. The species is in great danger of dying out because whalers have killed so many of them.

Boa Boas are found in the warmer regions of the Americas and Madagascar. They vary in size from one foot to more than thirty feet. They kill their prey by suffocation. Mammals and birds form the main diet, although some smaller species feed on lizards.

Bobcat The bobcat is the wildcat of America. It is actually a small lynx, about 3 ft long including its short tail. Its tufted ears indicate its relationship with the lynx, and the tufts are believed to increase the efficiency of the ears. The bobcat hunts mainly at night and it is an excellent climber. It feeds mainly on rabbits

(cottontails) and small rodents. It ranges from Mexico to southern Canada.

Bollworm This name is applied to the caterpillar of several moths which attack the bolls, or seed pods, of cotton. Bollworms are widespread in the cotton-growing regions of the world and do a great deal of damage to cotton and also to many other crop plants.

Bonito This name is applied to several species of mackeral-like fish. The common bonito, found on both sides of the Atlantic in the tropical and sub-tropical zones, is a steel-blue fish about 2 ft long. It is beautifully streamlined and has a number of small finlets on the hinder part of the body. Similar species occur in other parts of the oceans. They are predatory fishes, hunting smaller species by sight. Their white flesh is good to eat.

Bontebok An antelope closely related to the blesbok, although a little larger than the latter and with a much more conspicuous white rump. The white face, white legs and white rump combine with the dark brown back to give the animal a very attractive appearance, and the name bontebok actually means 'painted buck'. The bontebok was once common in South Africa, but it is now a rare animal and most of the survivors live in reserves.

A blister beetle

Booby This name is applied to six species of the gannet family that are confined to the tropical regions. They are goose-sized birds with thick necks and large heads. They are clumsy on the land, but they are powerful and agile fliers. Five of the six species have white bodies, but the brown booby (the commonest species) has a brown back. The birds breed on coasts and islands and, like the gannets, they feed by diving for fishes out at sea.

Boomslang One of the rear-fanged snakes, in which some of the back teeth carry venom. Back-fanged snakes are not generally dangerous to man, but the boomslang is an exception. Its venom is extremely poisonous, though only a small amount is produced and the snake does not bite unless handled. Boomslangs are African snakes and they live in the trees, feeding on lizards, birds and small mammals. They average 4 ft in length and are usually green or brown in colour.

Bower Bird There are several species of bower birds living in Australia and New Guinea. Some of them are beautifully coloured, but their most fascinating feature is the bower or courtship house which is built by the male. The bower pattern varies with the species, but usually consists of a platform of twigs and grasses partly or completely roofed over. The male normally decorates it with attractive pebbles, flowers, and other bright objects, and then he entices the female into it. Courtship and mating take place in the bower, but the female then makes a nest and lays her eggs elsewhere. The male's bowers normally get better and better each year, as he improves his technique and artistry.

Bowfin This fish, which lives in some of the Great Lakes and many North American rivers, is a living fossil with many primitive features. The head is covered with bony plates and the rest of the body bears thin scales. A soft dorsal fin runs most of the length of the body, which may be up to 3 ft, and the body ends in a rounded tail. The bowfin lives in sluggish waters and can breathe air with its lung-like swim bladder. It can live out of water for up to 24 hours. The male looks after the eggs and young in a nest, which may hold the eggs of several females at one time.

Bream A silvery fish living in slow moving streams and lakes throughout most of Europe. It belongs to the carp family and reaches a length of about 2 ft. Its body is greatly compressed from side to side and it is very deep. Adult bream feed on the bottom of the rivers or lakes by stirring up the mud and eating the small animals they disturb. The white bream

is a related fish, but the sea breams belong to quite different groups.

Brine shrimp A primitive relative of the crabs and lobsters. It likes very salty water and is found in many parts of the world from Greenland to Australia and from the West Indies to Central Asia. Brine shrimps are less than half an inch long with two pairs of antennae, two compound eyes on stalks and a third small eye in the middle of the head. It swims upside down with its underside to the light.

Bristlemouth One of the commonest fishes in the ocean, the bristlemouth lives in deep water where it preys upon small planktonic animals. They have small eyes and rows of light-organs (*photophores*) on their flanks. Some species are darker than others, depending on how deep in the water they live.

Bristletail These wingless insects are found all over the world, the best known being the silverfish, often found in kitchens or pantries, or among books. They are up to one inch long and get their name from the three slender bristle-like tails. The silverfish will eat anything but rarely becomes a pest.

Bristleworm This is the name given to a large group of marine worms in which the body carries a number of prominent bristles. Examples include the ragworms, lugworms and

fan worms. The latter animals use a fan of bristles to trap food particles in the water.

Brittlestar The brittlestar, a distant relative of the starfish, is found in seas all over the world. It usually has five arms (sometimes more) joined to a central button-shaped body. These arms break off easily as the name implies, but can be regrown. Many brittlestars are delicately colored, others light grey. They range in size from one-fiftieth of an inch to 4 ins across their bodies with a maximum spread of up to 2 ft across the arms. The arms are covered with hard plates and spines and along the underside are rows of tube feet which are used for feeding.

Brown bear The brown bear is found over a large area of the Northern hemisphere and there are several sub-species. These include the Kodiak, Syrian, and grizzly bears. They vary in size, but all are large omnivorous animals, up to 9 ft or more in length and weighing up to 1,650 lb. They have poor sight and rely on their senses of smell and hearing which are acute.

Brush-tail opossum Also known as the vulpine or fox-like opossum, the brush-tail opossum is found in all parts of Australia, Tasmania and New Zealand. It is about 2 ft

A brown bear showing the small eyes and alert ears.

long and its fur is thick and woolly, ranging in color from silver grey to dark brown or black. Its tail is prehensile at the tip. The brush-tail opossum is mainly vegetarian.

Budgerigar The commonest member of the parrot family in Australia, the budgerigar travels in huge flocks and, being a seed-eating bird, it is something of a pest in grain-growing areas. The wild bird is bright green with black, yellow and blue markings but many different color varieties have been produced by bird fanciers since the 'budgie' was brought to Europe as a cage bird in the nineteenth century. The name budgerigar is derived from an aboriginal word said to mean 'good food', so the bird has presumably been eaten by the Aborigines for a long time.

Brush-tail opossum

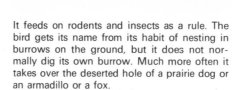

Bushbuck

Bullfrog The bullfrog is a large North American frog, in which the adults reach a length of 8 ins. It is usually greenish black on top and the underside is pale. Females are generally larger than males. Unlike the majority of frogs, the bullfrog is rarely found away from water. It feeds on a variety of insects caught at the surface of the pond, but will also eat other animals, including small snakes and mammals.

Bullhead Several fishes are known by this name, but the true bullhead is a small spiny fish found in clear streams over much of Europe. It is about 4 ins long, sometimes a little more, and has a very broad head. The basic color is brownish, but it is usually mottled and it can alter its color to suit different backgrounds. The bullhead is also called the miller's thumb, while its relatives in North America are called catfish.

Burrowing owl A small owl, only 9 ins high, living over a wide area in the Americas. It is a plains dweller and hunts by both day and night.

Bushpig

It feeds on rodents and insects as a rule. The bird gets its name from its habit of nesting in burrows on the ground, but it does not normally dig its own burrow. Much more often it takes over the deserted hole of a prairie dog or an armadillo or a fox.

Burying beetle This is the name given to various beetles, mostly of the family Silphidae, which feed on the corpses of small animals. The beetles are attracted by the smell of the corpses and they usually work in pairs—one male and one female—to bury the carcases by dragging out soil from under them. The beetles feed on the carcases and also lay their eggs on them. The young beetles feed either on the decaying flesh or on other creatures, such as fly maggots, which they find there. Burying beetles, also called sexton beetles, play a very important role in nature. Thanks to their activities we do not very often find rotting corpses lying about.

Bushbaby The four species of bushbaby, also called galagos, are relatives of the lemurs and are distant cousins of the monkeys. They all live in the African bush or scrub regions and they move around in the trees at night, feeding on insects, fruit, and even birds' eggs. They are very agile and make great leaps from branch to branch. Bushbabies are very furry creatures with large eyes and ears. They also have long furry tails. The largest bushbaby is about 30 ins long including the tail, and the smallest is only 13 ins long, including a tail 8 ins long.

Bushbuck This is a smaller relative of the kudu and nyala antelopes. It is about 30 ins high at the shoulder and weighs up to 170 lb. It is basically brown, but it has a number of

white patches and stripes. Males are distinctly larger than females and have a bushy mane along the back. The sharp horns may reach 22 ins in length, but females only rarely bear them. Bushbuck are mainly browsing animals, living in the forest and bush throughout much of Africa.

Bush dog A South American dog only distantly related to domestic dog but behaving very like it when tamed. It stands about 15 ins high at the shoulder, but its legs are relatively short. It is brown in color and, unlike most animals, it is paler on top than below. It ranges from Panama to Paraguay and hunts in small packs. Very little is known of its habits and, being nocturnal, it is seldom seen.

Bushpig The bushpig, also known as the red river hog, is a stout-bodied pig found over most of Africa. The head is large in proportion to the body and it bears tufted ears. The adults are reddish brown or black and weigh up to 300 lb. They are about 4 ft long. Their main food is roots and fruits, and they will push against small trees to bring the fruit down. Quite often they bring the whole tree down as well.

Bustard The bustards are large birds with stout legs and strong toes. There are 22 species, found in many parts of the Old World although most species live in Africa. They live mainly on the ground in open country, and they take a variety of plant and animal food. The birds run well, but they can also fly in spite of their size. The great bustard is as large as a turkey, weighing up to 32 lb and having a wingspan of 8 ft

Bustard

Butcher bird

Turkey vulture

Butcher bird The true butcher birds comprise six species of birds living in Australia and New Guinea, although the name is often applied to shrikes in other parts of the world. The name refers to the habit of impaling their prey on thorns, to form a sort of larder. The true butcher birds are stocky birds with mainly black-and-white plumage. Like the shrikes, they feed on insects, lizards and small mammals.

Butterfish An eel-like fish living between the tidemarks on both sides of the North Atlantic. Also known as the gunnel, the butterfish is about 6 ins long and it is brownish green with darker markings. The name refers to the slippery nature of the fish. It is actually a kind of blenny.

Buzzard A name loosely applied to the New World vultures, particularly the turkey vulture of the United States. All feed largely on carrion. The term is also used in Europe to describe many true birds of prey and the common "buzzard." This large hawk is distributed over much of Europe, Asia and Africa. It feeds mainly on small mammals which its seizes with its powerful talons.

Cacomistle An omnivorous relative of the raccoon, found in southwest United States and Mexico. The cacomistle has been given a variety of names including ringtail, raccoon fox and cat-squirrel. It is a small furry animal with a fox-like face and a long squirrel-like bushy tail ringed with black. Its fur is valuable and is sold as 'civet cat' and 'California mink'.

Caddis fly The name given to the insect order Trichoptera, of which there are between 4,000 and 5,000 species throughout the world. The adult insects look rather like moths and fly mainly at night. Most of the larvae live in fresh water and breathe by external gills. Many of the larvae build themselves tubular 'houses' of stones, small shells, or leaf fragments.

Caecilian A limbless amphibian with a long cylindrical body. The skin is smooth and slimy and has small scales embedded in it. The animals vary in size from 7 ins to $4\frac{1}{2}$ ft and are usually a blackish color. The eyes are small and generally useless, but they have a sensory tentacle on each side of the head. Caecilians live under-ground in warm regions, from Mexico to Argentina, and in South and Southeastern Asia, and parts of Africa. There are about 75 species.

Caiman A close relative of the alligator, differing mainly in having the skin of the underside covered in bony plates. There are five species, the smallest being the dwarf caiman up to 4 ft long, the largest the black caiman up to 15 ft long. They are found in the north of South America and southern Mexico.

Camel There are two species of camel: the Arabian or one-humped and the bactrian or two-humped. They are desert animals, although only the Bactrian camel still survives in the wild. A dromedary is a special breed of the one-humped camel used for riding. Both species are remarkably well adapted for desert life. Their humps consist largely of fat, which can be used to provide both energy and water when food is scarce.

Canary A finch coming originally from the Canary Islands and neighboring islands in the Atlantic. The wild bird is brownish yellow, but many yellow varieties have been produced in captivity. The canary is a very popular cage bird, with a sweet song. Like all finches, it is a seed-eater. It will produce hybrids with green-finches and other members of its family.

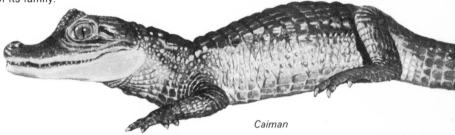
Caiman

Cape buffalo Reputed to be the most dangerous of the African big game animals, this bulky ox-like creature stands some 5 ft high at the shoulder and adult males may weigh up to 1 ton. It has large horns which sweep first downwards and then upwards. The animal lives throughout Africa south of the Sahara and there are many local races. Cape buffaloes are not normally dangerous unless they are unexpectedly disturbed. They then panic and charge at their opponents.

Cape hunting dog A ferocious carnivore, only distantly related to the domestic dog, that ranges over most of Africa. It stands about 2 ft

Caecilian

at the shoulder and has a mottled black, yellow and white coat. It also has abnormally large round ears. The animals hunt in packs and run down the antelopes on which they feed. They have a weird call that has been described as being like an oboe or the whinny of a horse.

Capercaillie The largest member of the grouse family, the capercaillie is a bird of the northern coniferous forests. It is found over much of northern Europe and Asia. During the winter the birds feed mainly on the leaves of pines and other conifers, but they move into more open woodlands in the summer and eat heather, grass and various berries. The male is about 3 ft long and his plumage is predominantly black or dark grey. He has a fine tail, which he fans out and displays to the brown hen in the breeding season.

Capuchin The capuchins, of which there are several species, are small South American monkeys. The body is about 1 ft long, but the prehensile tail may be nearly twice as long as this. The monkeys live in troops of up to 40 individuals and they feed mainly on fruit and insects. Capuchins are the traditional organ grinders' monkeys and they are exceptionally intelligent.

Capybara Also called the water pig or water cavy, the capybara is the world's largest rodent. It may reach over 4 ft in length and it looks like an enormous guinea pig. The species is found over much of South America and it lives in marshland or around the banks of rivers and lakes. It swims very well and feeds mainly on aquatic plants.

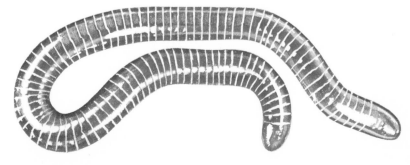

153

Cape Buffalo, a dangerous adversary because of its tendency to charge.

Caracal Also called the desert lynx, the caracal is a medium sized cat. It has a short reddish-brown coat, long legs, and long black ears topped by very long tufts of hair. The caracal ranges over much of Africa and extends across the Middle East to India, although it is now rare in Asia. It keeps mainly to open country and feeds on a variety of small antelopes and other mammals. It will also catch birds in flight by leaping up at them.

Common Carp

Caracara Caracaras are birds related to falcons, although they are unlike them in form and habits. There are several species distributed throughout the Americas. They are predatory birds, taking a variety of animals for food, but they will also eat carrion. In this respect they resemble some of the vultures. The caracaras will also chase and bully other birds until they release the food they are carrying. The caracaras then drop down and collect it.

Cardinal A familiar songbird of North America, also known as the red bird. The male has a black 'bib', but otherwise he is bright scarlet. The female is more somber, owing to a greater amount of brown in her feathers. Both sexes have a conspicuous crest and a stout, seed-cracking beak. Cardinals, which belong to the finch family, are distributed over much of Mexico and the United States.

Caribou These animals are the reindeer of North America and Siberia. They belong to the same species as the European reindeer, but the reindeer is usually regarded as a semi-domesticated race of the species. Caribou are up to 5 ft high at the shoulder and they weigh up to 700 lb. They are generally brownish grey, although some individuals may be white and some may be black. They are unusual among deer in that both sexes have antlers. Caribou spend the summer on the open tundra, feeding on lichens, grasses and other plants. In the autumn most of them move back to the forests and spend the winter there feeding on lichens and browsing on the soft twigs of the aspens.

Carp The common carp is the most widely distributed fish of its family, and one of the most widely distributed freshwater fishes in the world. It is a native of Japan, China and Central Asia, but it has been introduced to many other regions. It can survive for quite long periods out of water as long as it is kept moist. The carp prefers warm, shallow lakes and rivers with muddy bottoms, for it feeds on the mud-dwelling insects and worms. It is a favorite fish for ornamental pools and it can live for 50 years or more. The wild fish is olive green, but many color varieties have been produced by breeders.

Cassowary This large impressive and flightless bird lives in the dense rain forests of New Guinea and northern Australia. The black plumage consists of coarse bristle-like feathers, but the head and neck are naked and brightly colored. The wings are very small and are hidden from sight. The birds are very shy and are rarely seen. They can run remarkably rapidly through the undergrowth and usually have well-worn tunnels. Fruits and insects are the main foods. There are three species.

Cat-bear The cat-bear, also called the red panda or lesser panda, is the nearest living relative of the giant panda. It lives in the Himalayas. Its fur is a rich chestnut brown, with white marks on the face and dark rings on the long bushy tail. Although classified as a carnivore, the cat-bear feeds almost entirely on fruit and leaves. It spends most of its time in the trees,

154

sleeping by day and coming out to feed by night.

Catfish There are 11 families of catfish, found mainly in fresh water. They frequent slow-moving streams and muddy ponds in North and South America, Asia and Africa. At least 2 families are marine and most are scavengers. One of the largest is the blue catfish of the southern United States which may weigh as much as 150 lbs. Catfish have a broad head, 3 pairs of whiskers or barbels, small eyes and may be covered with large scales or be naked and slimy.

Catshark Catsharks are small sharks distinguished by their picturesque patterns of stripes and mottlings. Strangely enough, some of the dogfishes belong to this group. One of the less well known members is the shy-eye, a South African species which curls its tail over its head when caught, as if trying to hide its eyes.

The cat-bear, or red panda

Cave fishes There are 32 known species of fishes that spend their whole lives in underground caves and lakes. They belong to several families but, because they share a way of life, they have many features in common. They have little or no pigment, and appear pinkish because the blood shows through the more or less scaleless skin. Most of the species are blind. The largest is only about 8 ins long. Cave fishes probably feed mainly on insects that fall into the water, but one species is reputed to feed on bat droppings.

Centipede Centipedes are arthropods, distantly related to spiders, crabs, insects and millipedes. They differ from all these, however, in having one pair of walking legs on almost every segment. Millipedes have two pairs of legs on each segment. Although the word centipede means '100 feet', the creatures have varying numbers of feet—from 15 to 177 pairs. They do not have waterproof skins and all have to remain in damp places. Many live under logs and stones. They are active hunters, killing insects and other small creatures with their poisonous claws. They are good things to have in the garden, although some of the large

tropical species (up to 1 ft long) can give man a painful bite.

Chameleon True chameleons are lizards which are famed for their ability to change color, which they do by altering the distribution of pigment in their skins. Other unusual features include eyes which can be swivelled around independently, and the long tongue which is catapulted out to catch insect food. Chameleons are rather flattened from side to side and they are often decorated with brilliant colors, 'warts', and horns. There are about 80 species, most of them living in Africa. They range from about 2 ins to 2 ft in length.

Chamois The chamois is a goat-like mountain animal found in various parts of Europe and Asia Minor. It is a sturdy creature about 4 ft long and weighing up to 90 lb. Both sexes have sharply pointed horns. The long coat is tawny in summer and darker in winter. The animals live mainly in the alpine forests and feed on a variety of plants. They are somewhat intermediate between goats and antelopes, and their relatives include the gorals and serows of Asia.

Cheetah The cheetah is a long-legged cat with a beautifully spotted coat and a characteristic black stripe running from the eye to the corner of the mouth. It is a very fast animal—probably the fastest in the world—and it catches antelopes by running them down. This contrasts with the habits of most cats, which tend to stalk slowly up on their prey and then pounce. The cheetah once ranged from India to South Africa, but it is now very rare in many places. Its main stronghold is now in East Africa.

Chevrotain Also known as mouse deer, the chevrotains are the smallest of the ruminant animals. They are rarely more than 1 ft high at the shoulder. They are not true deer, and they lack antlers. Their heads are relatively small. There are four species, living in the forests of West Africa, India, and the Malayan region. They eat grass and leaves mainly, but also take insects, fish, and small mammals.

Chickaree This is the red squirrel of North America. It is about two-thirds the size of the grey squirrel, and has a tawny brown coat. The eyes are ringed with white. The chickaree is found throughout the coniferous and deciduous forests, and a related species is found along the west coast of the United States. The squirrels are active throughout the year and actually tunnel through the snow to find food.

Chimaera The chimaera or king herring is one of a small group of fishes somewhat intermediate between the cartilaginous fishes (sharks and rays) and the bony fishes. They have cartilaginous skeletons, but have gill covers like those of bony fishes. The king herring has a large head, and the body then tapers back to the tail. There is a poison spine on the back. It feeds mainly on bottom-dwelling molluscs and crustaceans, and reaches a length of 5 ft.

Chimpanzee One of the great apes and the nearest in intelligence to man, the chimpanzee is one of the most popular animals. The animals live in the rain forests of Africa and spend most of their time in the trees. Most of them live in small bands, but there is little social structure and members come and go as they please. They eat mainly fruit and other vegetable food, but they also eat meat, sometimes hunting quite large animals such as monkeys and young pigs.

Chinchilla Best known for its remarkably soft grey fur, the chinchilla is a South American rodent. It is about 10 ins long and looks rather like a squirrel, although its long back legs are more like those of a rabbit. Chinchillas live in burrows in the drier mountain areas, but they have been hunted for a long time and wild ones now live only in parts of Chile. They feed on coarse grasses and herbs.

Chipmunk

Chipmunk The chipmunks are a group of ground squirrels with about 18 species. They are found in North America and Asia, where they live in open country and feed on berries, grasses, and a wide variety of small animals. They are great hoarders of food for the winter, although they do not actually hibernate. Chipmunks range from 4 to 10 inches in length, excluding the bushy tail.

Chough The two species of chough are members of the crow family. They are a little larger than jackdaws and have glossy black coats. The common chough, which ranges in a fairly narrow belt from Ireland to China, has a long, curved, red beak and red legs. The alpine chough has a shorter, yellowish beak and lives in the mountainous regions of Europe and Asia. The common chough is found mainly by the sea in Europe. Both species are scavenging birds and they pick up a variety of animal food.

Chuckwalla The chuckwalla is a plump lizard usually about 1 ft long and weighing between 1 and 2 lbs when mature. It is a

Common chough

Cockatoo

mottled, brownish-black creature and it lives in the hot desert regions of North America. Unlike most lizards, it eats plant material and shows a particular preference for flowers. When disturbed, the chuckwalla scuttles into a rock crevice and wedges itself there by digging its toes in and also by inflating its body with air. It is then very difficult to dislodge.

Cicada Cicadas are bugs related to the froghoppers or spittle bugs, although they are usually very much larger. The empress cicada of Malaya has a wingspan of 8 ins. There are about 1,500 species, mainly found in the tropical regions. The males (and in some species the females too) 'sing' by rapidly vibrating little membranes on the sides of the body. This makes a very shrill whistling noise which can be very unpleasant to the human ear. The insects feed by sucking sap from trees. The young cicadas feed on the roots of the trees and some species take 17 years to reach the adult state.

Climbing perch

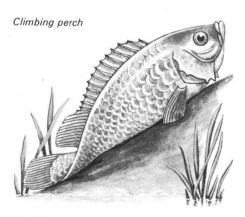

Cichlid The 600 species of cichlid fishes, favorites with tropical fish keepers, live in rivers and lakes all over Africa and throughout most of South and Central America. There are also two species in India. The body is flattened from side to side and many species are brightly coloured. At breeding time the male adopts a territory and defends it against other males. When a female joins the male both may help to defend the territory while they clean up a patch of gravel to receive the eggs. The eggs are continuously fanned by the parents and they are carried at frequent intervals to new quarters. The young fishes are often taken into the mouths of their parents when danger threatens.

Civet The 15 species of civet are related to the mongoose and are somewhat intermediate between cats and weasels. They have long narrow bodies and pointed muzzles, but their coats are cat-like. They live in Africa and Asia and keep mainly to the forests. They eat small animals mainly, but fruit is often taken as well.

Clawed frog Although sometimes called the clawed toad, this animal belongs to the group of tongueless frogs. It is about 4 ins long, and the front legs are rather short and weak. The back legs, however, are very strong and carry large webbed feet. The three inner toes of each back foot have sharp black claws. The animals spend most of their time in the water and, like many amphibians, can change their colour to match various backgrounds. Having no tongue, they cannot catch food in the normal way for frogs: they use their front teeth to catch small fish and other aquatic creatures.

Click beetle Click beetles, of which there are many species, are rather narrow insects up to about one inch long. Most of them are rather dull brown in color, although some species have bright wing cases. The insects get their name from their ability to flick themselves up into the air if they fall on their backs. This enables them to right themselves, and in doing so they make a loud click. Click beetle larvae are known as wireworms, and several species are serious pests on farms and in gardens.

Climbing perch A greyish-green or silvery fish, about 9 ins long, found across southern Asia from India to the Phillipines. Part of the gill cavity is separated off and functions as a lung. The fish can therefore live out of water for quite long periods. It can walk about on the land by propping itself up on its fins and spiny gill covers.

Coalfish A member of the cod family about 3½ ft long. It gets its name from its dark back, but it is also known as the pollack. It lives in the North Atlantic and the Mediterranean and is fished commercially.

Coati A small carnivorous mammal related to the raccoons. The body is nearly 2 ft long and the banded tail adds another 2 ft or more. The forehead is flat and runs down to a long mobile snout. The general color is reddish brown. The three species are found in the forests of South and Central America, but they are extending their range northwards at the moment. Coatis are most attractive and industrious little animals, foraging nearly all the time for invertebrates and small vertebrate animals. They will also eat fruit and they often climb trees to reach it.

Cobra Cobras are poisonous snakes found in Africa and Asia. They are medium sized snakes, some of them averaging about 7 ft in

length. There is a characteristic hood behind the head, formed by long movable ribs which swing out and stretch the skin. The cobra rears up and expands the hood when it is disturbed or excited. The snakes feed mainly on rodents, which they kill with a paralysing venom. The African spitting cobra can spit venom up to 8 ft. The largest species and most deadly snake in the world is the African king cobra—up to 18 ft.

Cockatoo The 16 species of cockatoo differ from the other parrots by having prominent crests on their heads. Most species are white, but a few are black or grey. Cockatoos live in the Australasian region, often forming large bands. Some species are pests because they take a lot of grain and often attack fruit too.

Cock-of-the-rock This name is given to two uncommon birds of South America. Both belong to the cotinga family and are recognizable by their fan-like crests. Females are dull brown, but the males are brilliantly colored. One species is orange and one is bright red. About a foot long, the birds live in the rain forests and feed mainly on fruit.

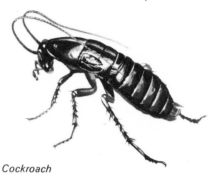

Cockroach

Cockroach Cockroaches are generally medium, or fairly large insects, flattened, and generally having rather leathery front wings. Females often have reduced wings and are therefore unable to fly. Most of the 3,500 species of cockroaches live in the tropics, but some of them have become established as pests in buildings in the cooler parts of the world. The cockroaches are scavenging insects, feeding on a variety of dead plant and animal material. Our food is therefore ideal for them and, as well as eating it, they foul it with oily secretions and odors.

Cod The second most valuable food fish after the herring, the Atlantic cod lives on the continental shelves around the North Atlantic. It can reach 6 ft in length and weigh more than 200 lb, although it is usually a good deal smaller. The fish is olive green or brown. During the summer the cod feeds mainly on other fishes, but it is more of a bottom-feeder during the winter.

Colorado beetle

Coelacanth A very primitive fish belonging to an order whose other members have been extinct for more than 70 million years. It is a living fossil, and was discovered only in 1938, when a specimen was caught by a South African fishing boat. Several other specimens have been caught since. The coelacanth has fleshy lobes in its fins and very heavy scales. It represents a very early stage in the evolution of the bony fishes.

Colobus monkey The three species of colobus monkeys live in the dense forests of Central Africa. They are uncommon and rarely descend from the trees. They have relatively long fur and reach up to 5 ft in length, including about 3 ft of tail. Colobus monkeys feed mainly on leaves.

Colorado beetle A native of North America, the Colorado beetle was quite harmless until

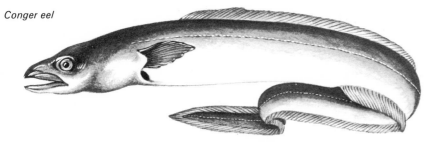
Conger eel

potato growing reached its home. It then transferred its attention to potatoes and has since spread over large areas of America and Europe. By eating the potato leaves, it drastically reduces the yield of tubers. The beetle is nearly half an inch long and has black and yellow stripes, while its larva is dark pink with black spots.

Comb jelly Small marine animals related to jellyfishes, the comb jellies are found in all the oceans. Most species are globular and, because they bear rows of little hairs, they are often called sea gooseberries. They are usually colorless and quite transparent, and they swim about by waving their hairs or cilia. Most species live in coastal waters.

Condor Condors are large American 'vultures'. They are not related to the true vultures of the Old World, although they look very much like them. They are ugly birds, with naked heads and necks and wicked-looking hooked beaks. Like the true vultures, they are carrion feeders and they also attack young and wounded animals. There are two species, the Californian condor and the Andean condor. The former is very rare and lives only near Los Angeles, but the Andean condor is found in the Andes for nearly the whole length of South America. Both birds have wingspans of about 10 ft and are among the largest of all flying birds.

Conger This is a stout-bodied marine eel, normally about 5 ft long, but sometimes reaching 9 ft. Weights of up to 100 lb are sometimes recorded. The body is brown or silvery and it is scaleless. The conger lives near the coast in many parts of the world, although it prefers rocky coasts. It is a voracious animal, eating more or less any animal that it can catch. It often hides among the rocks and darts out to catch food with its powerful jaws and formidable teeth.

Coot Coots are water birds related to the

moorhens and also to the cranes. They are dark in color, but can be recognised by the brilliant white beak and the shield that extends above it. The coot does not have webbed feet, but its toes are provided with flaps which act as paddles in the water. They make floating nests from rushes and other plant material. There are 10 species, seven of which are found in South America.

Copperhead There are 3 species of copperheads in the U.S. Although venomous, their bite rarely kills. They inhabit upland forests, southern swamps, and the semi-arid regions of the southwest. Most individuals are less than 3 ft long and bear live young. Copperheads hibernate during the winter in dens.

Coral Corals are small animals related to sea anemones and differing from them mainly in possessing a hard chalky skeleton. Some corals live by themselves, but the majority are colonial creatures. Their skeletons are joined together in great masses and they often form coral reefs. All corals live in the sea, and reef-forming kinds live only in warm waters.

Coral snake This name is given to many strikingly colored snakes, with patterns of rings running around the body. The snakes are found in the Americas and in Africa, Asia and Australia. They are related to the cobras and possess a highly potent venom. Although extremely dangerous, they are secretive by nature and rarely bite unless touched. Two species are native to the United States.

Cormorant There are about 30 species of cormorant in the world. They are large sea birds, generally with glossy black plumage although some species have white underparts. The beaks are slightly hooked. The birds live around the coasts and most species fly and swim well. They feed almost entirely on fish, which they catch by diving after them. On land, the cormorant often stands with its wings outstretched as if it is hanging them out to dry. The shag is a species of cormorant also known as the green cormorant.

Cottontail The cottontail is a small rabbit ranging from southern Canada to South America. There are about 13 species, varying in size from 10 to 18 ins long. The ears are short and rounded and the tail is white underneath, giving rise to the name cottontail from its resemblance to a cotton boll. The young are born in shallow depressions in the ground rather like those of the hare, but young cottontails are blind and helpless at first.

Cormorant

Coyote

Coucal Coucals are members of the cuckoo family, with about 27 species in Africa, Asia and Australia. They are large birds compared with cuckoos, the pheasant coucal of Australia being about 23 ins long with a long pheasant-like tail. They do not lay their eggs in other birds nests, but make their own.

Courser A long-legged, plover-like shore-bird, living in the Old World from Africa to Australia. There are about 9 species and the cream-colored courser, that occasionally ventures north into Europe, is a small bird about the size of a starling with pale sandy coloring, black primary wing feathers and a broad, black and white eye stripe.

Cowbird A small American bird of which there are several species, with dark glossy plumage and a finch-like bill. Cowbirds get their name from the way they follow cattle to feed. Some of them have the cuckoo's habit of laying their eggs in other birds' nests.

Cowrie A carnivorous mollusc related to the sea snail but with the shell whorls enveloped during growth until only the final whorl is visible. There is a long slit on the underside opening with its sides rolling inwards like a scroll. In life, the shell is covered by the cowrie's mantle, thus keeping it from being scratched or damaged. Most cowries live in the Indian and Pacific oceans.

A Crowned crane (left) and a Sarus crane (right).

Coyote A relative of the wolf, the coyote is slightly smaller, measuring about 4 ft from nose to tail. The fur is tawny colored and it has a bushy black-tipped tail which it carries low behind the hind legs. Coyotes used to live over the western part of North America but their range has decreased considerably over the last century. They eat a variety of food including rabbits, small rodents, insects and vegetable stuffs.

Coypu A large rodent related to the porcupines. Its native home is among the swamps and watercourses of South America, but it has been reared on farms for its fur and is now established in the wild in North America and in several European countries. The coypu is well adapted for life in the water with webbed hind feet and a soft waterproof underfur—sold as 'nutria' by the furriers. It eats the aquatic and waterside vegetation.

Crab spider So called because of the curvature of their legs and the way they scuttle sideways like sea-shore crabs, crab spiders are found almost everywhere. They range from very tiny to not much more than one-quarter inch. They do not make webs but lie in wait for their prey. They are sometimes found in flowers, the colors of which they often match perfectly.

Crane Cranes are elegant long-legged birds, the largest of which stands about 5 ft. high. Most of the 14 species are now rare. They breed in lonely marshes and most species migrate long distances from summer to winter quarters and vice versa. They are found in all continents except South America. Cranes feed on leaves and fruit, together with a fair proportion of insects. They have remarkably loud voices, which carry for a mile or more.

Crayfish The crayfish is a freshwater crustacean, rather like a small lobster. There are several species, distributed widely over the world although absent from Africa and most of Asia. The largest species lives in Tasmania and may reach a weight of 9 lb. Crayfishes eat a variety of small water creatures, and some actually leave the water at night to feed on nearby vegetation.

Cricket Crickets are insects related to grasshoppers. There are about 900 species, distributed all over the world. They have long hind legs, used for jumping, and they also have very long slender antennae. The males produce a shrill 'song' by rubbing the bases of their wings together. The wings are held flat over the back, with the edges bent sharply down at the sides. Bush crickets, often wrongly called long-horned grasshoppers, are quite similar although less flattened. The egg-layer of the bush cricket is broad, whereas that of the true cricket is slender and needle-like. Crickets eat more or less anything and are very common on rubbish dumps.

Crocodile Crocodiles are large scaly reptiles, the nearest living relatives of the dinosaurs. They are found in the warmer parts of the world, inhabiting rivers and estuaries. There are about 15 species. Some specimens exceed 20 ft in length, but large ones are now rare as a result of hunting. They do not usually go far from the water and spend a lot of time basking in the sun. Young crocodiles feed mainly on insects, but they turn to fishes as they get larger. Adults more often catch birds and mammals that come to drink at the water's edge.

Crossbill Crossbills are parrot-like finches in which the tips of the beak are crossed. Females are olive green and, apart from the beak, they resemble greenfinches. Male crossbills, however, are bright red, marked with black. The three species breed in coniferous forests over most of the Northern Hemisphere and feed almost entirely on conifer seeds. The unusual beak is well suited to wrenching off the cone scales to get at the seeds.

Crow The crow family includes about 100 species of bird, including the raven, rook, magpie, and jay. Of the crows themselves, the carrion crow is the best known. It is about 19 ins long and appears black all over, although close inspection reveals blue and purple in the plumage. It eats almost anything, from grass and fruit to small mammals or shellfish on the seashore.

Cuckoo The European cuckoo is a greyish hawk-like bird about 13 ins long. It covers thousands of miles each year on its migrations between Europe and tropical Africa. It is a social parasite, laying its eggs in the nests of other birds. It eats insects, including the hairy caterpillars avoided by most other birds. There are many other birds in the cuckoo family, but not all of them lay in other birds' nests. Only the European cuckoo has the characteristic 'cuc-coo' call, which has given the name to the whole family. Cuckoos include the road-runner of the Western United States.

Curassow The 12 species of curassow are related to chickens and other game birds. They range in size from a pheasant to a small turkey, and weigh up to about 10 lb. The male is generally dark brown or black, while the females tend to be paler. Curassows have long tail and legs, and are readily identified by their horny 'helmets'. They live in Central and South America and spend most of their time in the trees. They feed mainly on fruits and leaves.

Curlew The common curlew is a large wading bird, with streaky brown plumage and a downcurved 5 ins beak. It breeds all the way across Europe and Asia, frequenting moorlands and marshes for most of the time. It gets its name from the two-syllable call that it utters. The curved beak is used for digging small animals from the soil or the mud. Several other species breed in the Northern Hemisphere, and some cover huge distances in their annual migrations to the Southern Hemisphere.

The spotted cuscus, an Australian marsupial.

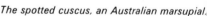

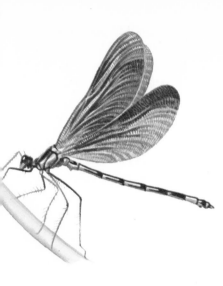

A damselfly, distinguished from the dragonfly by the more slender body and the similarity of the two pairs of wings.

Cuscus The cuscus is an Australasian tree-dwelling marsupial about 3 ft 6 ins long including its long prehensile tail. There are over 6 species, and they fill the niches occupied by monkeys in other tropical regions. They are nocturnal and they feed on fruit, leaves and insects.

Cuttlefish Molluscs related to the squids and octopuses, the cuttlefishes live in all the seas of the world. There are about 80 species, varying from about 1½ ins to 5 ft in length. The body is shield-shaped and its margins form a continuous fin. The head bears eight arms and two long tentacles, all bearing horny suckers. Inside the body there is a chalky 'shell', often washed up on the beach as the 'cuttlebone'. Cuttlefishes feed on shrimps and small fishes, which are caught by the tentacles. The animals are famed for their ability to change colour.

Dab A small flatfish living in the shallow coastal waters of Western Europe. It reaches a length of about 17 ins and is brown in color. It can be recognised by the sharp bend in the lateral line just behind the gill cover. The dab is a rather 'lazy' fish and it feeds on various bottom-dwelling invertebrates.

Daddy-long-legs This is the name given to several long-legged crane-flies that are common in gardens and grasslands. The larvae are known as leatherjackets, and they feed on plant roots. Some of the long-legged harvestmen (relatives of the spiders) are also called daddy-long-legs.

Damselfly Damselflies are slender-bodied relatives of the dragonflies, with rather weak flight. The front and hind wings are both the same size and shape. The body is often brightly marked with blue. Young damselflies, like young dragonflies, live in water. Ponds and slow-moving streams are preferred.

Darwin's finches These are a group of birds living in the Galapagos Islands. They are not particularly attractive, but they are famous because they helped Charles Darwin to arrive at his theory of evolution by natural selection. Several species live in the various islands, all derived from a few finches that were blown to the Galapagos from South America. Before they arrived there were few birds on the islands, so

Many breeds of dogs now exist, but all have been derived from a wolf-like ancestor.

the newcomers were able to evolve into many different forms, each having a different way of life.

Dasyure The dasyure is the 'native cat' of Australia. It is not a true cat, but a pouched mammal. There are five species and they look something like civets, with short legs and long bushy tails. They feed mainly on invertebrates and small mammals, stalking them rather like a cat. Dasyures range from about 1 ft to 4 ft in length.

Deer mouse Deer mice are American rodents very similar to the European field-mouse, although the two are not closely related. There are 55 species, varying from 5 ins to 15 ins in length (including the tail). The feet are white, and the animals are often called white-footed mice. They are found all over North America, and one species reaches south to Colombia. Seeds, berries and insects are their main foods.

The great northern diver

Devil fish Related to skates and rays, these fishes live in the warmer seas of the world. The front fins are greatly enlarged as 'wings' and they also have forward-pointing extensions at the front of the head. These look like horns and give the fishes their common name. The largest of the species is the manta ray, which may be 22 ft across and which may weigh nearly two tons. The devil fishes swim mainly near the sea surface, in contrast to the other rays and skates which live on the sea-bed.

Dhole This is the wild dog of India. It is not a true dog, and can be distinguished by its rounded ears and short muzzle. The fur is greyish brown and the bushy tail is tipped with black. Dholes range from India and Malaysia to China and parts of the Soviet Union. They hunt in packs of 100 or more individuals, running down deer and other animals. They will even

attack tigers and leopards, but do not seem to attack men.

Dik-dik Dik-diks are small antelopes, reaching up to 16 ins at the shoulder. There are six species, distributed over most of Africa. The males have short horns, partly hidden in a tuft of hairs on the head. The tail is a mere stump, and the animal also has a longish snout.

Dingo This is the wild dog of Australia—one of the few placental (non-pouched) mammals to arrive there before the Europeans. It arrived with the Aborigines when they reached Australia from southern Asia. It is a reddish brown animal, about 20 ins high at the shoulder. The ears are erect and pointed, and the tail is bushy. Dingoes live in small packs and hunt sheep, wallabies and kangaroos.

Dipper There are four species of dipper, all very similar wren-like birds with short wings and tails. They live by the sides of swift-flowing, boulder-strewn streams and feed mainly on insect larvae. They walk down into the water and search for their food among the stones of the stream bed. They can also dive into the water. The dippers live mainly in the Northern Hemisphere, with one species in the northern part of South America.

Diver Divers or loons are streamlined water birds with very short tails and straight pointed beaks. Their legs are set well back on the body and so, although they are good swimmers, the birds can only shuffle on land. The four species of diver live in the northern parts of the world. They spend the winters in coastal waters, but move to inland waters to breed. They feed mainly on fishes.

Dog, Domestic Dogs have been domesticated for at least 8,000 years. They have probably been derived from mainly wolf stock with a possible jackal admixture. Domestic dogs of all shapes and sizes have been bred by man over the centuries ranging from the St Bernard weighing as much as 200 lb to toy dogs of only one lb.

Dogfish This name is given to several species of small sharks and cat-sharks. One of the commonest, certainly around the coasts of Europe, is the lesser spotted dogfish. This fish is about 3½ ft long, including the slender tail, and it has extremely rough skin—so rough that it has been used as sandpaper and also for scrubbing the decks of trawlers. Fishermen do not like the dogfish, but large numbers are caught and eaten. When fried, it is known as rock salmon.

Dormouse Dormice are rodents living in the Old World. There are several species, of which the best known is the hazel dormouse— the only native species in Britain. It is a most attractive animal with brown and white fur and a length of about 6 ins including the tail. The largest dormouse is the edible dormouse or glis glis, a grey squirrel-like animal about 1 ft long. Most of the species live among the woodland undergrowth, although the glis glis prefers taller trees and is sometimes a pest in orchards. The animals feed mainly on fruits and nuts. They hibernate during the winter when large numbers are believed to be killed by predatory species such as weasels and foxes.

Douroucouli This is the world's only nocturnal monkey, and it is also called the owl monkey because of its large eyes and rather owl-like face. It is about 2 ft long, including the tail, and has short greyish fur. It can move absolutely silently through the tree-tops at night, but it can also make unbelievably loud calls and it has a very large range of sounds. Insects, small birds and mammals are the main items in the douroucouli's diet. The animal lives in the rain forests of South America.

Drongo Drongos are birds of the Old World tropics. There are about 20 species, ranging from 7 ins to 15 ins in length. They have long pointed wings and their tails often have long trailing feathers. Most of the species are black with a bluish or greenish sheen. There are conspicuous bristles around the nostrils. Drongos live in the dense forests and not a great deal is known of their habits. They feed on insects and seem to prefer bees and wasps.

Dugong The dugong is a seal-like mammal about 8 ft long, with flipper-like front limbs and a broad paddle for a tail. Dugongs live around

Dugong

the tropical coasts of the Indian Ocean, from Africa to northern Australia. They feed on eel grass and other aquatic vegetation and they often come up the rivers. They can haul themselves partly out of the water. They usually live in small family groups. Dugongs are closely related to the manatees and both groups are often known as sea cows.

Duiker Duikers are small short-legged antelopes with arched backs. The tallest of the dozen or so species is only about 2 ft high at the shoulder. Both sexes have short, pointed horns as a rule, and there is a tuft of hair between them. Duikers are found throughout most of Africa, but they are very shy creatures and quickly take cover when disturbed. They are mainly vegetarian but also eat insects and other small animals.

Eel Eels are elongated fishes found throughout the warmer seas of the world. Most species are naked and slimy, although some have small scales imbedded in the skin. Eels are largely marine animals, with the exception of a few species which live in fresh water. Most make long migrations to the sea during the breeding season. European and eastern North American eels migrate to an area near the Sargasso sea to spawn, a journey which may take as long as three years to complete. Eels are carnivorous creatures ranging in size from a foot to 6 feet or more, weighing up to 100 lbs. The moray eel is probably the best known for its voracious disposition and as a danger to divers. Most of these accounts are unfounded in fact. However, morays are fierce animals with formidable teeth. Some species possess

venom glands which secrete a potentially lethal venom.

Egg-eating snake Many snakes eat eggs, but the six species of egg-eating snakes (found in Africa and India) live almost exclusively on them and have a remarkable device to deal with them. They are slender snakes, reaching up to about 3½ feet in length. They are active mainly at night, gliding through the tree-tops and seeking out birds' nests by smell. The eggs are swallowed whole and are then cut open by a row of teeth, or pegs, in the roof of the throat. The contents of the eggs are swallowed, but the shells are compacted and ejected through the mouth. Egg-eating snakes have no venom.

Egret The egrets are a group of birds related to the herons. They are mainly white birds. Most of them live around lakes and marshes and they feed in shallow water. Frogs, fishes, and various invertebrates are eaten. The cattle egret, however, feeds on land. It forages in small flocks and eats the insects it disturbs in the vegetation. It often associates with cattle and other large animals and it has increased its range remarkably in recent decades. Egrets are found all over the world except for the northern regions.

Eider Famed for the soft warm down that it provides, the eider is a duck a little smaller than a mallard. There are four species, all living around the northern coasts of the Pacific and Atlantic Oceans. The male common eider is a striking black-and-white bird in the breeding season, but it is brown (like the female) at other times. The birds usually move south for the winter and tend to stay further out to sea

A young emu

than during the summer. They dive for shell-fish and crabs on the sea-bed.

Eland The eland is the largest antelope and stands up to 6 ft at the shoulder. It is a heavy ox-like creature and both sexes carry spiral horns up to 4 ft in length. There is a short mane and there is a tuft of long dark hair on the throat. There are two species: the common eland, which lives in Central and Southern Africa, and the Derby eland of Central Africa. Both have fawn coats with faint white stripes on the back, but the Derby eland is richer in color and less grey. The animals live in herds of up to 100 and roam the open plains. They

Elephants usually live in small herds led by a female.

are now being farmed in several places and they have many advantages over cattle.

Elephant The two species of elephants are the largest living land mammals. The African elephant is the larger of the two, reaching 11½ ft in height and 6 tons in weight. The great size of the elephant necessitates a massive skeleton, but the weight of this is kept down because the marrow cavities are replaced by light spongy bone. The trunk is used for carrying food and water to the mouth, for spraying water over the body, for lifting things, and for smelling. It is immensely strong, and yet is also very sensitive, especially near the tip. The two species differ mainly in the size of their ears and tusks, and in the shape of the tip of the trunk. The African elephant is found over most of the southern half of Africa, while the Indian elephant lives in the forests of South-East Asia, from India to Sumatra. They eat a variety of plant material.

Elephant seal Also called the sea elephant, this creature is the largest of the seals. The male can reach a length of 22 ft and a weight of 3½ tons. Females are much smaller and weigh less than one ton. The name refers to the peculiar drooping snout of the male. There are two species, both very much alike. The northern elephant seal lives off the coast of Mexico and California, while the southern elephant seal lives mainly in the South Atlantic and the Antarctic. Like most other seals, they feed on fishes and squids.

Elephant shrew Elephant shrews are found only in Africa. They belong to the same order of insect-eating mammals as the ordinary shrews, although they are not closely related. They get their name from the greatly elongated snout. They range from 4 ins to 12 ins in length, excluding the tail, and they have long back legs. There are about 18 species.

Emu The second largest living bird, the emu stands about 6 ft high. It lives in Australia in a wide variety of habitats. There were several species until the Europeans arrived, but only one now remains. It is related to the cassowary and shares its drooping, downy plumage. It cannot fly, but can run at 40 mph over short distances. It is a vegetarian and is often a nuisance on farms during the dry season.

Featherstars

Fairy fly This name is given to various minute wasp-like creatures, not true flies at all. They belong to a group called chalcids and they spend their early lives inside the eggs of other insects. They are parasites. The name fairy fly refers to the tiny wings, little more than a couple of veins and a few hairs.

Fennec fox

Red fox

Fairy shrimp An inch or more long, the fairy shrimp is transparent and almost colorless. It is a rather primitive relative of the true shrimps and other crustaceans and it turns up now and then in temporary pools. It swims on its back by means of rhythmic movements of its many legs, which also function as gills and as food collectors. The animal feeds on tiny plants and animals that it filters from the water. The fairy shrimp's eggs can survive for many years without water, and this is why the creatures can appear quite suddenly in small puddles.

Fallow deer This attractive deer is now found nearly all over the world because man has taken it about with him as an ornament for his parks and gardens, but it originally came from the Mediterranean region. The coat is reddish brown, dappled with white spots in summer, and there is a prominent white patch on the rump, extending on both sides of the black tail. The fallow is a small deer, never more than 3 ft at the shoulder, and the buck carries broad *palmate* antlers. It feeds mainly on grass.

False scorpion This is a minute greyish animal only distantly related to the true scorpions. Most of the 1500 known species live in leaf litter and other vegetable material and they feed on small insects and mites which they catch with their relatively huge pink claws. A few species eke out a living between the pages of old books or in the bodies of stuffed museum specimens. They feed on the book-lice that they meet there. False scorpions often 'hitch-hike' by grabbing hold of the bristles of flies and other insects.

Fanworm This name is given to various marine bristleworms that live in tubes and filter their food from the water with a crown, or fan, of stiff bristles. Some fanworms live in little tubes made of lime, others make tubes of sand grains. One of the best known is the peacock worm, whose sandy tubes often stick out of the mud at low tide.

Featherstar A relative of the starfish, this creature has five basic arms, but they branch freely and it appears to have many feathery arms. The body is covered with little plates of limestone and the arms are supported by little bone-like cylinders joined together like a backbone. The body is anchored by little claws on the underside, but the animals can swim gracefully when they release themselves. The arms filter food particles from the water.

Featherstars are related to the sea lilies, in which the body is carried on a stalk. Sea lilies were very common about 350 million years ago, and certain limestone rocks consist almost entirely of their cylindrical 'bones'.

Fennec Also known as the fennec fox, this little animal is the smallest of the foxes, yet it has the largest ears. The head and body measure about 15 ins in length and the ears are 6 ins long. The animal lives in the deserts of North Africa and Arabia and does not need to drink. Its pale sandy color blends well with the desert sands, and its huge ears help it to pick up the slightest sound. The fennec feeds on a variety of invertebrates and small mammals.

Fer de lance A relative of the rattlesnakes, the fer de lance is found in coastal districts from Mexico to Argentina. It also lives on various islands in the Caribbean. It averages about 5 ft long and is brownish, with a diamond pattern on the back. It is a poisonous snake and many accidents occur because it frequents sugar plantations and houses in its search for rats and mice.

Ferret The ferret is a pale domesticated version of the polecat and it has been used for centuries to catch rabbits, although it is not used much now. The ferret is sent into a rabbit hole to chase the rabbits out into nets. The black-footed ferret is a rare North American animal—sometimes stated to be the rarest mammal of that continent. It is about 2 ft long and, like the ordinary ferret, related to the weasel. It feeds on prairie dogs.

Fiddler crab This is a little crab, only about one inch across, found in great numbers on tropical beaches. The males have one very enlarged claw and one normal claw. They are brightly colored and the large claw is waved about as a call sign to attract a mate. Some fiddler crabs can make sounds by rubbing their claws against a row of teeth on the shell. This stridulation produces a sound rather like that of a cricket.

Fighting fish Many fish fight, but the most famous is the Siamese fighting fish. It has been reared for its fighting ability, and fish-fighting is a traditional sport in its native Thailand (Siam). Two males will go on fighting for an hour or more before one comes out as the winner. The wild fish is yellowish brown, but many other color varieties have been produced now.

Flamingoes in flight

Firebelly This is a toad with an unusual way of scaring its enemies. It is less than 2 ins long, and dark grey on the back. The underside is bluish grey but spotted with red or orange. When alarmed, the animal rears up and exposes this 'firebelly'. It can also exude a strong smelling poison from its skin. The firebelly lives in Eastern Europe and Russia, extending into Germany and southern Sweden.

Firefly Fireflies are named for the brilliant light they give out when they are flying. This comes from the underside of the abdomen and is produced by a chemical reaction. The light appears much stronger than it really is because our eyes are especially sensitive to the firefly's wavelength. Fireflies are not really flies at all. They are beetles. The more brilliant ones live in the tropics, but several occur in the cooler regions.

Fisheagle Described as the most handsome of Africa's eagles, this relatively small species has a white head, back and chest, together with black wings and a chestnut belly. It lives all over the southern half of Africa and is also called the river eagle. It fishes along the shores of rivers and lakes. Other kinds of fisheagle live in Asia.

Fish-eating bat This is a more or less naked bat living in America from Mexico to Argentina. It has long narrow wings and it flies low over the water, impaling fishes on its sharp claws. It is also called the bulldog bat. Two more bats also feed on fish—one in northern Mexico and California, and one in South-East Asia.

Flamingo There are four species of these strange but beautiful birds. They have ex-tremely long legs and neck. The plumage is a delicate pink except for the flight feathers, which are black. The beak is strangely curved, and the birds feed with their heads upside down in the water—often very foul water, too. They live and breed in shallow lakes and lagoons. Commonest of the four species is the greater flamingo, a 4 ft bird found in many parts of South America, Africa and Central Asia. It also lives in southern Europe, and is an occasional visitor to S.E. United States.

Flicker Admired for their beautiful plumage and wide variety of calls, the flickers are small American woodpeckers. There are six species, found from Alaska to Chile. Most of them feed on ants, which they collect on tree-trunks or on the ground, often tearing open the nests to get at the insects.

Flying squirrel

Flounder A flat fish of the shallow seas of the Atlantic Ocean, the flounder may reach a weight of 6 lb. It is at home in both salt and fresh water and it frequents river mouths. It is greyish brown on the upper side and pearly white underneath.

Flying fish There are two types of herring-like flying fishes. One type has only one pair of wings—formed from the pectoral fins—and the other has two pairs of wings, formed from both pectoral and pelvic fins. They all live in the tropical seas and feed on plankton. They do not flap their 'wings' in flight: they swim quickly upwards and break surface, the fins are spread and, with a few flicks of the tail, they take off in a glide which may well last for over half a minute.

Flying fox The flying foxes are actually large fruit-eating bats, which have fox-like faces. There are about 60 species and, unlike the insect-eating bats, they depend upon their large eyes to find their way about. They are found in India and Indonesia, Australia, and many of the islands of the Indian and Pacific oceans.

Flying squirrel There are 37 species of flying squirrel, most of them living in southern Asia, although one lives in North America and one in northern Eurasia. All are similar in build and habits and all have a furry flap of skin stretching along the sides of the body from front legs to back legs and then on to the tail. They are nocturnal creatures and they glide from tree to tree on their outstretched flaps of skin. They range from about $5\frac{1}{2}$ ins to 4 ft long, including the tail. Some of Australia's flying phalangers (marsupials) are also called flying squirrels.

163

Fossa This is a cat-like carnivore from Madagascar, related to the civets and mongooses. It has short soft reddish brown fur and it measures about 5 ft in length, half of which is the stout tail. The fossa is a nocturnal animal, feeding on lemurs which it catches in the trees.

Frigatebird Frigatebirds, or man-o'-war birds, live around the coasts in the warmer parts of the world. They are not really seabirds, for their feet have only traces of webbing and their plumage is not very waterpoof. The plumage is mainly black, shot with blue or green, and the males have a red throat pouch. The wings are large, spanning 7 ft in some species, and flight is good. The frigatebirds get a large amount of their food by harrying other birds until they drop the food they are carrying. They spend a lot of time in the air and they can walk only with difficulty. They usually leap from trees or rocks to become airborne.

Frilled lizard This is a pale brown lizard, sometimes bearing yellowish or dark patches, named from the large frill around its throat. The frill normally lies like a cape around the shoulders. but it can be raised as a warning display when the animal is disturbed. The frilled lizard reaches a length of about 3 ft. It lives mainly in the sandy, semi-dry areas of northern Australia and it feeds on insects and small mammals.

Fritillary This is the name given to a group of butterflies, nearly all belonging to the family Nymphalidae, which have a chequered pattern rather like that of the fritillary flowers. Most of the fritillaries have orange brown wings marked with black, and the undersides often bear silvery spots. The majority of the species live in woodland or on rough grassland.

Fruit bat The 160 species of fruit bats live in and around the tropics of the Old World. About 60 of them are known as flying foxes. They vary in size from a tiny 2½ ins long to large bats with wingspans of up to 3 ft. Although called fruit bats, not all of them

Gannet

Fur seals breed on land or havens, with one bull to several cows.

actually eat fruit. They differ from the insect-eating bats in several ways, notably in the complete or almost complete absence of a tail.

Fulmar A member of the petrel family, the fulmar's name is derived from 'foul bird'—a name based on its musky scent and its habit of spitting a foul-smelling oil at intruders. The

bird has a stocky body and long narrow wings. It swoops low over the sea and catches young herrings, squids, and crustaceans. The fulmar has been increasing in numbers over the last 100 years and is now one of the commonest birds in the North Atlantic. The cause of the increase is the increased number of fishing boats. The fulmars follow them to pick up the discarded fish and offal.

Fur seal Fur seals belong to the eared seal family, a group which also includes the sea lions. They differ from true seals in having small ear flaps and in being able to turn their hind limbs forward when on land. This means that they can lollop along, whereas the true seals can do little more than wriggle. Fur seals also have a dense fur, and this is what makes them valuable. Most of the fur seals live in the Southern Hemisphere, but one important species lives around the shores of the North Pacific, especially around Japan and Siberia.

Gall wasp The familiar oak apples and marble galls on oak trees, and the robin's pincushion on wild rose bushes are made by insect larvae. These various swellings are known as galls. They can be produced by various kinds of insect larvae, but a large number of them are produced by the larvae of gall wasps.

The gall wasps are little ant-like creatures, only distantly related to the true wasps. Most of them produce galls on oak trees and they often have a complicated life history.

Gannet Gannets are white goose-sized birds that live around the temperate oceans of the world. They are oceanic birds, coming ashore only in the breeding season. They feed on fish and will actually swim after their prey under water when once they have dived in. Gannets tend to nest on steep cliffs and offshore rocks, and their nests are often packed tightly together—perhaps with only a couple of feet between them.

Gar The seven species of gar, also called gar-pike, live in rivers and lakes of North and Central America. They are rather lazy fishes, but they can move very rapidly to snap up an unwary fish in their long toothed jaws. Gars have rather slender bodies ranging from 2 ft to about 12 ft in length. The body is covered with tough diamond-shaped scales like those of the early fishes. The gar is in fact a rather primitive fish, closely related to various jurassic fishes. Like the coelacanth, it is a living fossil.

Garter snake Garter snakes are the commonest snakes in the United States and Canada. There are many species, ranging from Mexico to the north of Canada. They

A gecko

reach a length of about 3 ft as a rule and are not poisonous. The body is slender and marked with longitudinal stripes. Worms are the major items in the garter snake's diet, but frogs, newts and mice may also be taken. The snakes often enter water, and they may catch fish.

Gazelle Gazelles are dainty, slender antelopes in which the males have sweeping horns that diverge from each other to form a lyre-shaped outline. The females have only short spikes or no horns at all. There are 10 species of true gazelles, most of them living in Africa. Grant's gazelle and Thomson's gazelle are among the best known of the African species. The true gazelles have a white streak on each side of the face, usually underlined by a dark streak. Three other species are also known as gazelles.

Gecko The geckos form a family of lizards noted for the broad toes of many of its members and for the microscopic hooks on the toes. These hooks can hook into all but the smoothest of surfaces, and so the geckos can climb up almost anything. Even window panes can be climbed. Geckos live in all warm countries and there are many species. They vary from about 1 in to 14 ins in length. Most of them live in trees, but quite a number have made their homes in houses. They do useful service by catching flies and other insects.

Genet   The genet looks like a cross between a tabby cat and a mongoose. It is a carnivorous mammal related to the mongooses and civets. There are six species, of which three are rare. The body is about 20 ins long in the commoner species and the tail is about the same length. Genets are elegant spotted creatures, with white marks on the face. They feed on small rodents, birds and insects. The six species are all found in Africa, and one species—the feline genet—extends into Southern Europe and Arabia.

Gerbil   Gerbils, or sand rats, live in the desert or semi-desert parts of Africa and Asia. There are many species, and two genera are more accurately known as jirds. In some respects, the gerbils resemble the jerboas, although the two groups are not closely related. The fur is fawn or brown, often tinged with black, and the underside is pale. The tail is usually long and the animals also have long back legs. Most of the species can jump, although the jirds jump less well than the others. Gerbils eat seeds, roots, and other parts of plants. They will also eat insects.

Gerenuk   The gerenuk is a long-necked, gazelle-like antelope, although not a true gazelle. The male carries short thick horns and stands about 3½ ft high at the shoulder. Gerenuks feed by browsing on trees and they often stand up on their hind legs, lifting their heads to a height of about 7 ft. Gerenuks live only in Somalia and neighbouring parts of north-east Africa.

Gharial   The gharial is a long, slender-snouted crocodilian living in the rivers of northern India and Pakistan. It is also known as the gavial. It can reach a length of about 20 ft and keeps to the water more than other crocodiles. It feeds mainly on small fishes, which it catches with rapid movements of its slender snout.

Giant snail   The giant snail is a large land-living snail native to East Africa. The pointed shell may reach a length of 8 ins and the whole animal may weigh ½ lb. The snail is not particularly remarkable in itself, but it is remarkable for the way in which it has spread, with man's help, to many new areas in recent decades. It is a serious pest in crop-growing areas, although it can be eaten as a nutritious food.

Gibbon   Gibbons are among the most agile of mammals, and the smallest of the apes. There are six species and they normally range from 3 ft to 4 ft in height when standing upright. The arms are extremely long, perhaps 1½ times as long as the legs, and they are used for swinging through the trees. Gibbons live in the forests of South-East Asia and feed mainly on fruit. They will also eat leaves, insects, and small birds if they can catch them.

Gila monster   This is an ugly lizard living in the deserts of the south-western United States. It is up to about 2 ft long and it is one of only two poisonous lizards. The other poisonous species is the similar looking beaded lizard which lives in Mexico. The gila monster is pink and yellow, with black markings, and it has an unusually stout tail in which it stores reserves of fat. It is a rather sluggish animal, feeding on birds' eggs and young mammals. The bite is very painful and may be fatal. The

A gibbon (above), the acrobat of the forests.

Giraffes (right), the tallest living animals.

gila monster is now rare and it is a protected species.

Giraffe   The tallest animal in the world, the giraffe has remarkably long legs and a remarkably long neck. An old male may reach 18 ft in height. There are many races, differing in the pattern of the coat, but they all belong to the one species. Giraffes live in the savanna areas of Africa, mainly in the east and south. They browse the tree-top foliage with their long tongue and mobile, hairy lips. They have few enemies when adult: lions may attack them, but the giraffes have a powerful weapon in their long legs and hooves.

Glass snake   Glass snakes are legless lizards found in many parts of the world. They reach lengths of up to 4 ft, but two-thirds of this length is tail. All glass snakes have a deep furrow running along each side of the body. They live in open country, hiding under leaves and stones. They feed mainly on insects. The fragile tail is easily broken off if seized. The tail also breaks into pieces and so it looks as

The grasshopper (above) can be distinguished from the bush cricket (left) by the length of the feelers.

if the whole animal has broken up—hence the name of glass snake.

Gliding frog The gliding frogs belong to a family of tree frogs and they get their name from their habit of gliding from tree to tree, supported by the webbing of the large feet. The most common gliding frogs live in Malaya and Borneo.

Glow-worm The glow-worm is a beetle belonging to the same family as the fireflies. The female is wingless and it is she who is responsible for the name glow-worm. There are several species. All stages of the life cycle can give out light, but the female shines most brightly. She climbs up a grass stem and shines to attract the flying males. Young glow-worms, which look very much like the females, feed on snails.

Goshawk A magnificent bird of prey, the goshawk is closely related to the sparrowhawk. It has the same long tail and rounded wings when seen in silhouette, but it is much larger— the head and body measuring about 2 ft. The plumage is dark brown, with a pale underside marked with bars. The goshawk ranges over much of the higher latitudes of the Northern Hemisphere. It is mainly a woodland bird and it takes a wide variety of birds and mammals as food.

Grasshopper Grasshoppers are jumping insects which live among the grass and herbage close to the ground. They are active only when the sun is shining and their usual method of getting about is to jump with the aid of the long and powerful back legs. Most of the species have wings and they can fly. They are best known for their 'songs', which are produced by rubbing the hind legs against prominent veins on the wings. Usually only the male sings. True grasshoppers have short antennae. A related group of insects, known as bush crickets, have very long antennae and are sometimes (wrongly) called long-horned grasshoppers. Bush crickets sing by rubbing

their wings together and they feed mainly on other insects. Grasshoppers are vegetarians.

Grass snake The common grass snake, also known as the ringed snake, extends from the British Isles (not Ireland) across all but the most northerly parts of Europe to Central Asia. It is a non-poisonous snake reaching a length of $2\frac{1}{2}$–3 ft as a rule, although specimens twice this size have sometimes been recorded. The snake is an olive brown or greenish color and it can be recognised by the yellowish collar around the neck. It lives in damp places and is a good swimmer. Worms, slugs, frogs and small mammals are the main items on the grass snake's menu.

Grebe Grebes are long-necked water birds, many of which have plumes on the head. The feet are not webbed, but the toes have horny fringes that act as paddles. There are 18 species, distributed all over the world. They live on lakes, reservoirs and slow-moving rivers. They do not fly much, although some species migrate to the coasts for the winter.

A grass snake, showing the characteristic yellow collar.

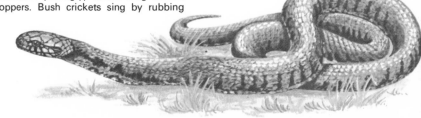

166

Green turtle Once abundant in the tropical seas, the green or edible turtle is much rarer now because its flesh and eggs are good to eat. The adult turtle has a dark brown shell marbled with yellow. It reaches a maximum length of 4 ft, and the adult animal weighs up to 400 lb. The name green turtle comes from the green tinge of its fat. Like all marine turtles, it comes on to the shore to lay its eggs in the sand. Young green turtles are carnivorous, but the adults feed only on the vegetation growing in the shallow water.

Grizzly Grizzly bears are a closely related race of North American brown bear, although a smaller form. They inhabit the western high timberlands of the U.S., Canada and Alaska. Grizzlies are known for their ferocious disposition, particularly when females are followed by cubs. Their diet includes fish, insects, fruits and berries, small animals and carrion. The cubs are born to the female while she rests half asleep during the coldest winter months, and remain at her side for about two years. Males are solitary animals and associate with the females only during the spring breeding season.

Ground squirrel Ground squirrels look very much like tree squirrels, but they do not have such bushy tails. The body and tail together measure between 8 ins and 30 ins, of which the tail may account for up to half. The animals are brownish grey or yellowish grey, often with stripes or rows of spots. There are 32 species, most of them living in North America. Some live in Africa and some in Eastern Europe and Asia, where they are known as susliks or spermophiles. They live in burrows or under logs and they eat seeds, fruits, bulbs, insects, and many other items. Those animals living in the northern regions hibernate, but the others remain active throughout the year.

Green turtle

Grouper Groupers, also known as sea bass or sea perch, are heavy-bodied fishes living mainly in the tropical seas. There are about 400 species, although not all are large. The largest is the Queensland grouper, which reaches some 12 ft in length and has been said to swallow divers. Groupers have many strong needle-sharp teeth and feed on a variety of other fishes. Many species lurk on the bottom, or among coral reefs, and wait for food to come along. Some groupers are brightly colored and they can also change their colors to match various backgrounds.

Grouse The grouse family contains 18 species of game birds, ranging in size from a domestic hen to a turkey. Most of them live in the northern regions and they feed mainly on leaves, shoots and berries. One of the best-known species is the ruffed grouse, a woodland game bird with a repeated thumping call. Other species include the blackgrouse (blackcock), the capercaillie, and the ptarmigan. The males are much larger than the females in all the species.

A ground squirrel

Guillemot Guillemots are birds of the northern seas. They belong to the auk family and are related to puffins. On the water they look like ducks, but they stand upright on land and look more like penguins. They fly with a rapid whirring sound. There are various species, all black-and-white. Out of the breeding season, the birds stay at sea, diving and swimming underwater to catch fishes and crustaceans. They breed on steep cliffs and pack their nests tightly on to the ledges.

Guinea pig Guinea pigs belong to a large family of South American rodents. The wild form is usually called a cavy, and it is eaten by the natives of South America. The Incas used to rear them specially for food, and the Spaniards eventually brought them to Europe. They are now kept as pets and as laboratory animals, and a large number of varieties have been reared. Wild cavies live in burrows and come out at night to eat a variety of plant material.

Gundi This is a long-furred rodent said to be the commonest mammal in North Africa. It is about 10 ins long and it looks rather like a sandy-colored guinea pig. It lives on dry mountain slopes and other rock areas, basking in the sunshine and seeking shelter only during the hottest part of the day. It also shelters from the rain, which damages its silky fur. Gundis eat grass and other plants, getting all their water from this source.

Guppy The guppy is a small freshwater fish from the Caribbean region and the northern parts of South America. It is probably the commonest aquarium fish after the ordinary goldfish, and it has also been released in many places to keep down the mosquito population. The male guppy is a colorful little fish, about one inch long, but the female is rather drab and reaches a length of two inches. She brings forth active young, but frequently eats them as soon as they are born. Fish breeders have produced many attractive strains of guppy.

Gurnard The gurnard is distinguished from all other fishes by its wing-like pectoral fins, in which the front two or three rays are separated off to form 'feelers'. The fish has a heavy, box-like head protected by bony plates and spines, but it can lift itself up by digging the 'feelers' into the sea-bed. It can also drag itself along in this way. There are several gurnard species, generally living in shallow coastal waters.

Gymnure The gymnures, of which there are five species, are related to the hedgehogs and they are sometimes called hairy hedgehogs, although they look more like shrews. They have thick underfur and a dense covering of coarse guard hairs. All five species come from South-East Asia and the largest, often called the moon rat, is about 2 ft long including its almost naked tail. Gymnures are nocturnal and,

like all insectivores, they feed mainly on worms and insects.

Haddock A close relative of the cod, the haddock, or finnan, is another important food fish. It is greyish brown above and it has a white belly. It is easily recognised by its three dorsal fins and by the dark patch just behind the gills. It reaches a length of about 44 ins— much smaller than the cod—and weighs up to 36 lb. It lives in the North Atlantic and the Arctic Ocean.

Hagfish The hagfish is a degenerate descendant of the jawless fishes that lived many millions of years ago. It bears little resemblance to other fishes except the lamprey. Its scaleless body is up to 2 ft long and has no paired fins. It has a large mouth, surrounded by tentacles, but no jaws. Its powerful tongue carries many horny teeth, and the animal feeds by attaching itself to some other fish—usually a dying or weak one—and rasping its flesh away. There are 15 species of hagfishes, living in all the temperate seas of the world. They are often called slime fishes because of the huge amount of slime they exude from their skin.

Hairy frog This is a greenish brown frog, about 4 ins long and flushed with pink on the underside. It does not have any hairs and the name refers to the tufts of skin filaments that develop on the sides and thighs of the male during the breeding season. These filaments are thought to increase the breathing surface of the frog. The hairy frog lives in the fast-flowing streams of West Africa.

Hake A close relative of the cod, the hake lives in the Mediterranean and the North-Eastern Atlantic. It is a streamlined fish with one short dorsal fin and one very long ·one. The anal fin on the lower side is also very long. The hake is brownish grey above and silvery white on the sides and belly. Its scales are large. Hake feed mainly on squids and other fishes.

Hartebeest

Hedgehog

Halibut This is a fish that is in some ways half way between an ordinary fish and a flat-fish. It lies on its left side, but its body is not as flat as that of a plaice and its jaws are not distorted like those of the true flatfishes. It feeds on bottom-living invertebrates and also on other fishes. There are two species of halibut—one in the North Atlantic and one in the North Pacific. They may reach lengths of 12 ft and they are fished for food and for the oil in their livers.

Hammerhead shark There are five species of these fishes, in which the sides of the head are drawn out to form a hammer-head shape. They live mainly in the warmer seas and feed on bottom-living fishes and invertebrates. The largest species may reach a length of 20 ft and span 3 ft across the head from eye to eye.

Hippopotamus

Hamster There are about 14 species of hamsters, short-tailed rodents related to the voles. They range from Western Europe to China and usually live in dry regions such as sand dunes and the grassy steppes. They eat various plant materials but cereals are their staple diet and they are sometimes a nuisance to farmers. The golden hamster probably came into being from a chance hybridization between two wild species in Syria. All the golden

Honey ant

hamsters in the world have descended from a single family found there in 1930.

Harrier Harriers are a group of hawks generally living in open country. They are usually brown, although some species are largely black. They have owl-like heads, long wings, and long legs. They are found in all the continents, the hen harrier (marsh hawk in America) being the most widely distributed species. Harriers take a variety of prey, which they spot as they fly low over the ground.

Hartebeest This is a large and rather un-gainly looking antelope standing up to 5½ ft at the shoulder. The head is long and narrow and the back slopes sharply down from the shoulder to the hindquarters. Both sexes bear horns. There are three species which, between them, cover most of the southern half of Africa. They are among the commonest of the antelopes.

Harvester The harvesters or harvestmen are related to the spiders. They are often called harvest spiders or daddy-long-legs. Harvesters do not have the narrow waist of the spider, nor do they have the silk glands. They have four pairs of long legs, and a pair of eyes mounted on a turret on the top of the body. Most of the world's 2000 species of harvesters are brown-ish in color and the majority live on the ground or in low growing vegetation. They feed mainly on other small animals.

Harvesting ant The harvesting ants belong to various species and they get their name because they all collect the seeds of grasses and store them in their nests. They make huge nests and they go out to forage in large bands. Small workers gather the seeds, and larger workers crush them with their jaws. Harvesting ants live in the drier regions of the world, mainly in the tropics and sub-tropics.

Harvest mouse The harvest mouse, one of the smallest British mammals, is about 5 ins long. Its prehensile tail accounts for about half of this length. The animal has thick soft fur and weighs only 6 or 7 grams. It is light enough to be able to swing about on grass stems and make its nest there. In winter the animal lives in a burrow. It feeds mainly on seeds and is found right the way across Eurasia to Formosa. The American harvest mouse is a different creature and is really a vole.

Hatchet fish The 15 species of hatchet fishes live in the deep sea. The largest is only 3½ ins long. The crinkly, shiny body is flattened from side to side and has a sharp lower edge. This feature, together with the slender handle-like tail, is responsible for the name hatchet fish. The animals go down to 1500 ft during the day, but they come nearly to the surface at night when they feed on various floating creatures.

Hawkmoth Hawkmoths, or sphinx moths, are large, thick-bodied moths found through-out the world, although most of the 900 species are tropical insects. The wings are long and narrow and the moths can fly very fast. Most of them fly at night. The caterpillars are stout creatures and they usually have a curved horn at the hind end.

Hedgehog There are 15 hedgehog species, all quite similar, living in Eurasia and Africa. They are insectivores, related to the moles and shrews, and they feed on a variety of inverte-brates and small vertebrates. They also eat plant food, especially fruits. The most obvious thing about a hedgehog is its prickly coat, com-posed of numerous hardened hairs. These spines are found only on the upper surface, the lower surface being clothed with perfectly ordinary soft hair. The hedgehogs of cooler regions hibernate for the winter.

Hellbender This is a giant salamander living in North America and reaching a length of up to 30 ins. It has a flattened, brownish body and a broad head with small eyes. The tail, which accounts for one-third of the animal's length, is oar-like at the end. The hellbender lives in swift streams in the eastern part of the United States. It feeds on any animal it can catch. Similar species live in China and Japan, the Japanese species reaching a length of up to 5 ft.

Hermit crab Hermit crabs live in empty sea shells, such as those of winkles and whelks. In place of the short squat body of the typical crabs, the hermits have a soft banana-shaped abdomen. Only the front end of the animal has a hard skeleton, and the right claw (larger than the left) is used to close the opening of the shell when the animal retires. The robber crab of the South Pacific is a land-living hermit crab which does not live in a shell, although it still has a twisted abdomen.

Howler monkey

Herring The herring has for long been the most important food fish in the world. Some 3000 million herrings are caught in the Atlantic and neighboring seas every year. It is a greyish green color on top and silvery beneath. It lives near the surface of the sea and swims in huge shoals. It feeds on tiny planktonic animals which it strains from the water with its gills. There are several races of herring.

Hippopotamus Distantly related to the pigs, the hippopotamus rivals the Indian rhinoceros as the third largest living land animal behind the two species of elephants. It reaches a length of 14 ft and may be nearly 5 ft at the shoulder. It may weigh 4 tons. Hippos spend most of their time wallowing in rivers and lakes, but they come on to land at night to feed. They eat large quantities of grass, but they also consume plenty of aquatic vegetation. They are found over much of the central part of Africa. The pigmy hippopotamus of West Africa is a separate species.

Hoatzin One of the strangest of all living birds, the hoatzin lives in the dense forests of the Amazon basin. It is related to the chickens and other game birds. It is basically brown and cream, with long spiky feathers forming a crest on the head. The young hoatzin has claws on its wings and it uses these to climb about in the trees. It does not fly well for it has only small flight muscles. The hoatzin feeds almost entirely on leaves and fruits.

Honey ant Honey ants belong to several species and they live in the drier regions of the world. They are famous for the development of 'living honey pots'. The ants feed largely on the honey dew exuded by aphids, and much of the honey collected in time of plenty is stored in the bodies of special workers called *repletes*. These become greatly swollen and they hang from the roof of the nest chamber, giving out honey when stroked by other ants. The honey-filled repletes are eagerly sought by local people and are considered a great delicacy.

Honeyguide The honeyguides are small birds related to the barbets. They are brown or grey above and pale beneath, sometimes with yellowish markings. The eleven species live in Africa and southern Asia and they feed mainly on bees and wasps, especially the honeybees. They seem to be immune to the stings and they even enter the bees' nests to eat the wax combs. The birds get their name

The humming bird hawkmoth sucks nectar while hovering at a flower.

because some species actually attract honey badgers and lead them to the bees' nests. The badgers rip open the nests to get at the honey, and the birds feed on the scraps.

Hornbill Hornbills are large birds—up to 5 ft long—found in Africa and South-East Asia. They have huge bills, often topped by large bony swellings called casques. Most of the species live in the forests and feed on fruit. They nest in hollow trees and, after laying her eggs, the female shuts herself in with a wall of mud and other materials. Only a small slit is left, and the male feeds his mate through this.

Hornet This name is commonly applied to almost any large wasp, but the true hornet is a particular species living in wooded areas of Europe. It differs from the ordinary wasps in being brown and yellow instead of black and yellow, but it leads just the same kind of social life. It builds its nests in hollow trees. Hornets have a powerful sting, but are less ready to use it than the wasps.

Horsefly This name is given to certain large flies of the family Tabanidae. The females feed on the blood of animals, especially cattle and horses, and the males suck nectar from flowers. The body is dull in color, but the eyes have beautiful rainbow colors in life. Cleg-flies are related to the horseflies and cause great distress to picnickers because they fly so silently that one rarely notices them until they have sunk their sharp mouths into our skin.

Hoverfly Hoverflies are true flies, belonging to the family Syrphidae. They are wonderful aeronauts and they can move in all directions as well as hover. Many of them mimic bees and wasps, but the hoverflies can always be distinguished by their single pair of wings. Each wing has a 'false margin' formed by a vein running near the edge. Adult hoverflies

take nectar, while many of the larvae help us by eating greenfly and other pests.

Howler monkey The largest of the South American monkeys, the howlers are named from their loud calls which are produced in a large 'echo chamber' in the throat. The monkeys have a thick and heavy neck to accommodate this apparatus. There are five species, up to 4 ft long with the tail, and ranging from red to black in colour. They live in the forests and feed mainly on leaves and fruit.

Humming bird There are over 300 species of these little birds, distributed throughout the New World. The largest is a mere $8\frac{1}{2}$ ins long, and the smallest has a body no bigger than a bumble bee. Most of them have brilliant plumage. They get their name from the noise made by the wings as the birds hover in front of flowers. Most of them feed on nectar, sucking it up through the tubular tongue and narrow, pointed beak. Some hummingbirds make incredible migrations for such small animals—from Alaska to South America in some instances.

Hyena There are three species of hyena: the spotted or laughing hyena, the brown hyena, and the striped hyena. The latter extends from North Africa to India, but the other two are found in the southern half of Africa. They are brownish or greyish yellow creatures, up to 3 ft high at the shoulder. The back slopes sharply down to the hindquarters. The head is large and the jaws are very powerful. Hyenas feed mainly on carrion, although they often kill weak animals. Their strong jaws and teeth can even chew up the bones.

Hyrax The dozen species of hyraxes or conies are so unlike any other animals that they are given a whole order to themselves. They are greyish brown, tail-less mammals with short muzzles and small rounded ears. The largest species is only 18 ins long. The skeletons show resemblances to rodents, rhinos, hippos and elephants! It is thought that the elephants are probably the nearest relatives of the hyraxes. Hyraxes live in Africa and South-West Asia, some in the forests and some in rocky country. They feed on grass and other vegetation.

Ibex This name is given to several species of goats that live on high mountains. They range from Spain to Mongolia. Ibexes are sturdy animals, usually with long backward-curving horns. The horns, which are shorter in the females, generally carry prominent knobs or 'knots' on the front edge.

Ibis Ibises are birds related to the herons and flamingos. There are about 25 species,

Spotted hyena

found in all the continents. They all have long legs, fairly long necks, and long curved beaks. Many are brightly colored, such as the scarlet ibis of tropical America. Ibises live in marshes and around shallow lakes, feeding on fishes, frogs and other small animals. They nest in trees, however, just like the herons.

Iguana Iguanas are lizards. There are many species, including the basilisk, the ground iguana, the green iguana and the marine iguana. Many of the species are eaten by men, for their flesh is very tasty. Various species live on the Galapagos Islands, but some are now rare as a result of human interference.

Impala The impala is a graceful antelope about 3 ft high at the shoulder. It is chestnut brown, with a sharply defined white belly. The male has horns up to 30 ins long, but the female is hornless. The species can be recognised by a tuft of black hair on the 'heel' of the hind leg. It inhabits a large part of eastern and southern Africa, usually keeping near to water.

Indri The largest of the lemurs, the indri reaches 30 ins in length. It has very long hind limbs, but its tail is only one inch long. The fur is basically black-and-white. The indri lives only in the northern parts of Madagascar, where it dwells in the trees and feeds on leaves.

Jacana Also called the lily trotter, the jacana is a water bird related to the curlew, although it looks more like a long-legged coot. It walks over the floating leaves, supported by its long spreading toes, and feeds on plants and small animals. There are seven species, spread throughout the warmer parts of the world.

Jackal Jackals belong to the dog family. Two species live in Africa, and a third lives in North Africa, Asia and south-eastern Europe. They range from black to dirty yellow in color and stand about 16 ins at the shoulder. They feed on the left-overs from lion kills, but also hunt gazelles and other small animals.

Jackdaw A member of the crow family, the jackdaw is well known for its habit of stealing bright objects. It is about 13 ins long and has black plumage shot with blue on the back and head and a grey nape. It has a short strong bill and feeds on insects, small vertebrates, and fruits and seeds. It ranges across Europe and Central Asia.

Jaguar The largest of the American cats, the jaguar is about the same size as a leopard but more heavily built, weighing up to 300 lb. The coat is yellow with black spots arranged in rosettes of 4–5 around a central spot. It is a carnivorous animal ranging from extreme

southwestern United States to Patagonia. It lives mainly in the dense forests.

Jaguarundi Although a member of the cat family, the jaguarundi is more like a large weasel in shape and habits. Its body is long, up to 50 ins, but with a shoulder height of only 12 ins. It has short, rusty-red or grey fur. It is found in grasslands or dense brush on the edge of forests from the southern border of the United States to northern Argentina. Little is known of its habits but it is said to feed on rodents, ground birds and fruit.

Jay Colorful members of the crow family, jays make up a varied subfamily of over 40 species, 32 of which are in America. The common jay ranges over Europe and Asia and has reddish brown plumage, darker on the back, with black and blue barred wings. It lives in woods and forests but feeds in open scrubland on a variety of foods including seeds, fruits and insects.

Kagu An unusual bird found only in the forested highlands of New Caledonia, the

The kiwi, one of many flightless birds in New Zealand.

kagu has no very close relatives the nearest being the cranes and rails. The kagu is somewhat larger than a domestic fowl and has pale grey plumage. Its head bears a loose crest that it can raise in anger or display. It eats mainly snails, worms and insects, coming out to feed at dusk. Like the domestic fowl, the kagu seldom flies, running fast with wings outstretched when alarmed. Kagus are becoming increasingly rare and may soon be found only in zoos.

Kakapo The kakapo is an extremely rare, owl-like parrot living in the rain forests in parts of New Zealand. It is a flightless bird about 20 ins long with greenish upper parts streaked with black, brown and grey. It is nocturnal and feeds on fruit and grass.

Kangaroo rat This is a rodent, named for its long hindlegs and tail, short forelegs, and leaping gait. It resembles a jerboa and, like the jerboa, it lives in deserts. There are several species in North America, ranging up to about 20 ins in length, including the long tufted tail. They hardly ever drink, and get all their water from the seeds and plants they eat.

Katydid This is the name given to various American bush crickets. They have remarkably loud 'voices'. (See Grasshopper.)

The great kudu male (above) and the female (right).

Kookaburras and a koala, two of Australia's famous animals.

Kestrel Kestrels are small falcons, closely related to the American common sparrow hawk, noted for their hovering flight. There are several species, distributed almost all over the world. All are brown and grey, with black spots. The common kestrel lives in Eurasia and Africa. It nests and roosts in open woodland, but hunts over open country. It hovers in the air and drops down on various small animals which it takes for food.

King crab This marine creature is not a crab, but a 'living fossil' more closely related to the spiders. It is also called the horseshoe crab because of its horseshoe-shaped body. The front end is covered with a horny brown or olive green shield, and the hind end bears a long moveable spike. The whole animal may be up to 2 ft long. There are five species, one on the east coast of North America and four in East and South-East Asia. They live in shallow water and feed on other small animals.

Kingfisher There are over 40 species of kingfishers, living in the temperate and tropical regions. They are stocky birds with long beaks and short tails, and often with brilliant plumage. Not all of them actually catch fish. Many of the species sit on perches and dart after lizards and insects on the ground. The kookaburra or 'laughing jackass' of Australia is a kingfisher.

Kinkajou A relative of the panda, the kinkajou or honey bear ranges from South America to southern Mexico. It has large eyes, small ears, and is about 12 ins long with an 18 ins prehensile tail. It lives in the tree-tops and uses the tail for extra support. The kinkajou has soft woolly fur coloured dark gold to brown. It feeds mainly on fruit, but will also eat insects and small mammals.

Kite The kite is a bird of prey with long pointed wings and a forked tail. There are three groups of kites which occur in Europe, North America, Asia and Africa. Several species are found in the United States. Kites may resemble swallows in displaying a deeply forked tail during flight. Kites live mainly in wooded valleys and they feed on fish as well as on many other animals and on carrion.

Kiwi The three species of kiwis are small flightless birds of New Zealand. They are about the size of a chicken, but they have long curved beaks with the nostrils at the tip. Kiwis live in the forests and come out at night to search for worms and insects. The eyes are small, and the ears and nose are the main sense organs.

Koala The koala, Australia's most famous animal, is a pouched mammal related to the opossums. It looks like a small grey bear, with a prominent beak-like snout and tufted ears. It is about 2 ft high and weighs up to 33 lb. It has only a stump for a tail. Koalas are nocturnal creatures and feed largely on the leaves of eucalyptus trees. They rarely come down to the ground. They live mainly in eastern Australia.

Krill This is a Norwegian word, used to describe the food of the whalebone whales. It refers especially to the little shrimp-like creatures called *Euphausia* that float in millions in the colder seas of the world. They are about 2 ins long, but they feed the world's largest animal—the blue whale. The krill might be one source of food for the world's growing human population.

Kudu The kudu are among the largest of the antelopes. There are two species—the greater and the lesser kudu—and both have long twisted horns in the male. The greater kudu stands up to 5 ft at the shoulder, while the lesser kudu is up to 3½ ft. Both have white stripes on the flank. The greater kudu ranges from the Sudan to South Africa, but the lesser kudu is found only in East Africa. Both live in the open bush.

Lacewing Lacewings are delicate insects with two pairs of thin, netted wings. The best known ones are the green lacewings, or goldeneyes, but there are also many brown species. Both adult and larva feed on aphids and they are therefore very useful insects. The larvae suck the juices from the aphids and then often

171

Ring-tailed lemur

the origin of back-boned animals. It feeds by straining food particles from the water. There are several species, found all over the world. Most live near the coasts.

Langur Langurs are brightly colored monkeys with crests of hair on the head and patches of naked skin. There are 20 species in South-East Asia and Indonesia. Most are slender animals, about 2 ft long and sporting a tail of about the same length. With one exception, they all keep to the trees, and they feed mainly on leaves. Some species are rare now because the forests are being cut down. The proboscis monkey, with its long drooping nose, is a langur.

Leathery turtle The leathery or leatherback turtles are the largest of the sea turtles, growing to a maximum of 9 ft. The foreflippers are also very large and a leathery turtle of 7 ft may have flippers spanning 9 ft. Unlike other turtles, the shell is made up of hundreds of bony plates covered with a leathery skin. The animals are usually dark brown or black and are found mostly in the warmer seas of the world. They feed on jellyfish and other soft-bodied animals.

Leech Leeches are relatives of the earthworms, but they have a strong sucker at each end of the body. Most of them live in water and many live as parasites on fishes and snails. They suck blood and often attach themselves to people who enter the water. Some leeches eat whole insects and other small creatures.

Lemming Lemmings, of which there are 12 species, are small rodents up to 6 ins long. They live in the cold regions of the Northern Hemisphere, where the vegetation consists mainly of lichens and stunted shrubs. They continue to live under the snow during the winter. Like the related voles, the lemmings periodically build up to huge numbers. They then emigrate in all directions.

Lemur Lemurs are primates—relatives of men and monkeys—confined to Madagascar. There are several species, ranging from 6 ins to 40 ins in length. In addition, they have long bushy tails. Leaves, fruit and insects are the main foods of the lemurs. One of the best-known species is the ring-tailed lemur, which runs about with its black-and-white ringed tail held vertically.

cover themselves with the empty skins as a form of camouflage. Many lacewings lay eggs with stalks on them.

Ladybug Ladybugs are small, brightly colored beetles, most of which are oval or circular in outline. They nearly all have red-and-black or yellow-and-black patterns, warning birds that they are unpleasant to taste. When handled they often exude a bitter fluid. The adult and larva both consume greenfly and other pests.

Lammergeier This is a graceful vulture found from the Mediterranean region to South Africa and China. It has a wingspan of up to 9 ft and, unlike other vultures, it has feathers on its neck. It is greyish black on the back and rusty brown underneath. The tail is long and diamond-shaped. Lammergeiers are cowardly birds and usually have to make do with the scraps after other vultures have finished with corpses. They will eat small bones, and will carry large ones up in the air and drop them until they break.

Lamprey This is an eel-like creature, but it has no jaws and its nearest relative is the hagfish. It is not a true fish, therefore. There are about 30 species, in fresh and salt water, in the temperate parts of the world. Young lampreys, and some adults, feed by sucking up debris from the sea or river bed. Others are parasites and feed by rasping flesh from other fishes.

Lancelet The lancelet is a small marine creature about 2 ins long. It looks like a fish, but it has no paired fins and it is not a back-boned animal. It has some similarities with the vertebrates, however, and is probably close to

The leopard, a solitary hunter of Africa and Asia, is a good climber.

172

Leopard One of the big cats, the leopard is a handsome animal up to 8 ft long, including the tail. The fur is basically tawny yellow, with black spots. Black individuals, known as panthers, are quite common. Leopards live in Africa and southern Asia, and feed on a wide variety of antelopes and other animals. They often catch fish.

Limpet This name is given to various water-living snails in which the shell is more or less tent-shaped. Some live in the sea, some in fresh water. When covered by water the animals glide about on the broad foot, and they scrape algae from the rocks and stones. When the tide goes out the seashore limpets return to their own spots on the rocks and pull their shells down tightly around them. The rock and the shell are gradually worn so that they fit perfectly and it is then very hard to remove the limpet.

Linsang This is a very slender relative of the mongoose. There are two species, both living in South-East Asia. They have long bodies and short legs. The fur is short and velvety, brownish grey with darker bands or spots. Linsangs are active at night, preying on a variety of small animals. They are excellent climbers.

Although often called the king of the jungle, the lion actually lives in open country.

Lion Once found all over Africa and southern Asia, lions are now restricted to Africa south of the Sahara and a small area north of Bombay in India. Lions feed chiefly on zebra and antelopes.

Llama This is a South American relative of the camel. It is not really a wild animal, but a domesticated descendant of the guanaco. The alpaca is another domesticated animal, probably derived from the same ancestor. The llama has a long coat and has long been used as a beast of burden in the Andes, especially in Peru. The alpaca has been bred more for its fine wool.

Lobster Lobsters are large crustaceans, weighing as much as 40 lb. They live in the sea and are blue when alive (turning red when boiled). They have large claws. Lobsters live on the sea-bed and feed on a variety of living and dead animals. They are active mainly by night, and fishermen catch them by lowering baited pots into the sea.

Locust Locusts are large grasshoppers whose populations periodically build up to plague proportions. At such times the insects fly out in great swarms and do a great deal of damage to crops. Africa suffers particularly from locusts, but various species are found throughout the warmer and drier parts of the

The African lungfish

173

world, extending to northern China and Canada.

Lorikeet Lorikeets are colorful little parrots found from Malaya to Australia and the South Pacific. There are 31 species, some of them with several sub-species. They are mainly green, with other bright colors on the head and breast. They range from 6 ins to 15 ins in length. Lorikeets live in flocks and feed mainly on nectar. They are also called honey-parrots.

Loris The three species of loris are slow-moving relatives of the lemurs. They live in South-East Asia. They are nocturnal animals, with large forward-looking eyes. They have broad grasping hands and feet, which can cling tightly to branches for long periods. Lorises have rather woolly fur and live mainly in the rain forests. They eat fruit, leaves and insects.

Lovebird Lovebirds are small colorful African parakeets. There are six species, generally green with patches of red or black on the face. They eat fruit, nectar and insects. Lovebirds pair for life and the pairs spend much of their time 'kissing' each other—hence their common name.

The llama, a beast of burden in the Andes.

Lugworm This is a large worm living in U-shaped burrows in the sand of the sea-bed. Its coiled castings are a familiar sight at low tide. The animal is up to 12 in long and looks rather like an earthworm. The central part bears tufts of reddish feathery gills. Lugworms swallow sand and extract food particles from it. A current of water passes through the burrow, bringing oxygen to the gills.

Lungfish There are only six living kinds of lungfishes, one in South America, one in Australia, and four in Africa. All live in fresh water. They have air-breathing lungs as well as gills and they can live in very stagnant water. Apart from the Australian species, they have to gulp a certain amount of air even when in well-aerated water. The Australian species has broad scales and strong fins which allow it to walk about on the river bed. The other species have eel-like bodies and small scales. Their fins are very narrow. The African and South American species can survive when the rivers dry up. They burrow into the mud and aestivate, surrounded by slime which keeps them moist.

Lynx Lynxes are short-tailed, long-legged members of the cat family. They have tufted ears. The European lynx, up to 3½ ft long, lives in the forests of northern Europe and Asia. It has a sandy coat and black spots. The Spanish lynx is a little smaller and more heavily spotted. The Canadian lynx, found all over Canada, is larger and has few spots.

Lyrebird The two lyrebird species are the largest of the perching birds or song birds. The male has a body about the size of a bantam and, when he is three years old, he grows the distinctive tail which gives the birds their name. The tail is 2 ft long and its two outer feathers are very broad. They take on the shape of a lyre. Lyrebirds live in the mountain forests of Australia. They feed on seeds and small invertebrates which they collect on the ground.

Macaw The 18 species of macaws include the largest and the most colorful of the parrots. They live in tropical America. They have hooked beaks, and tails up to 2ft long. Macaws live in large flocks, except at the breeding season. They feed on fruit and seeds, cracking hard nuts with their strong beaks and cleverly extracting the seeds with their tongues.

174

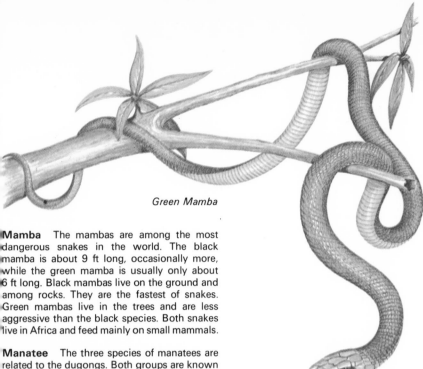

Green Mamba

Mamba The mambas are among the most dangerous snakes in the world. The black mamba is about 9 ft long, occasionally more, while the green mamba is usually only about 6 ft long. Black mambas live on the ground and among rocks. They are the fastest of snakes. Green mambas live in the trees and are less aggressive than the black species. Both snakes live in Africa and feed mainly on small mammals.

Manatee The three species of manatees are related to the dugongs. Both groups are known as sea cows. Manatees are seal-like mammals, up to 15 ft long and 1500 lb in weight. One species lives around the West African coast, one around the Caribbean and Florida. The third lives in the Amazon and Orinoco rivers in South America. They all feed on aquatic vegetation.

Manatee

Mandrill This is a forest-living baboon with a sturdy body up to 3 ft long. The male's face and rear end are decorated with naked patches of pink and blue. The rest of the body is dark brown, with a golden collar and beard. Mandrills live in Central and West Africa and are mainly vegetarian. They spend most of their time on the ground, but sleep in the trees.

Golden Marmoset

Mantis Mantises, often called praying mantises, are voracious insects related to cockroaches. They feed on other insects which they catch with their large spiny front legs. The mantis gets its name because it holds its front legs folded in a praying attitude. Most of the 1800 species live in the tropics. It was introduced into the United States to control the Japanese beetle.

Marabou This is a rather large and ugly stork of Central and East Africa. It stands 4 ft high and has a wingspan of over 8 ft. The back, wings and tail are dark grey, while the underparts are white. The head and neck are pink and more or less naked, with a fleshy pouch dangling from the throat. The bill is large and heavy. Marabous eat carrion and other rubbish.

Markhor This is a wild goat of the Himalayan region. Males stand about 3 ft at the shoulder and have fine spiralled horns. The hair is sandy colored in summer and grey in winter. Both sexes have a long beard and heavy mane. Most markhors live between 4000 and 8000 ft.

Marmoset These South American animals are the smallest of the monkeys. There are several species, most of them about 8 ins long with 12 ins tails. They have claws instead of the more usual nails, and these help them to get a grip as they scurry through the tree tops. Like most monkeys, they feed mainly on leaves, fruit and insects.

Marmot Marmots are mountain-dwelling rodents living in North America and Eurasia. There are about a dozen species. One of the best known is the alpine marmot of Europe. It is about 2 ft long and may weigh 18 lbs. It has coarse brown and grey fur and lives near the tree-line in the Alps. Marmots feed on grasses and other plants. They hibernate in burrows during the winter.

Marten Martens are relatives of the badger and stoat. They have long bodies and short legs. There are several species in North America and Eurasia. The largest is the fisher marten, or pekan, about 3 ft long. This species hunts mainly on the ground, but the others live more in the trees and hunt squirrels and birds. Most of the species are brown or black and some, notably the sable, have valuable fur.

Mayflies Mayflies are rather primitive insects whose nymphs live in water. The adults have one or two pairs of delicate wings and two or three long filaments on the end of the body. The adults do not feed and live for only a short time—a few hours in some species. The nymphs, however, live for about a year before they mature. There are about 1000 known species.

Millipede

Merganser Mergansers are colorful ducks with saw-edged bills. They are closely related to the smew and the goosander. There are several species, breeding in freshwater and spending the rest of the year around the coasts. The commonest species, the red-breasted merganser, breeds throughout the northern latitudes and moves south for the winter.

Mongoose

Midwife toad This small toad, under 2½ ins long, gets its name because the male assists the female when she is laying the eggs and then carries them around attached to his back legs until they hatch. The upper surface of the animal is greyish or light brown and it is covered with warts. The feet are only partly webbed. Midwife toads live in Western Europe and, like other toads, feed mainly on slugs, earthworms and insects.

Millipede The millipedes are arthropods in which nearly every segment of the body carries two pairs of legs. Most of them are long and slender, but some species, known as pill millipedes, are short and fat. They look more like

Moray eel, a voracious feeder.

175

The mud-puppy

woodlice and they can roll themselves into a tight ball. Millipedes live in the soil and leaf litter, eating the decaying vegetation as a rule, although some species attack growing crops.

Mink   The mink is a relative of the stoat and weasel. There are two species: the American mink and the European mink. The two species are very similar, the males being 1½–2 ft long including the tail, and the females about half the size. Wild mink are brown, but many color varieties have been produced in captivity, for the mink has a valuable fur and is reared on large farms. Mink feed on small mammals and ground birds. They will also dabble in streams for fish, for they normally make their homes in river banks.

Mitten crab   The three species of mitten crabs are natives of the rivers and estuaries of China and Japan. They are about 3 ins long and they get their name because the pincers are clothed with short hair and look as if they are wearing gloves. Mitten crabs reached Europe early in this century and now live in most of the rivers of Western Europe. They interfere with the fishing industries by taking bait from hooks and traps, and they have also damaged river banks in places by burrowing into them.

Moccasin   Properly called the cottonmouth water moccasin, this rather large North American pit-viper may reach lengths of up to 6 ft. Cottonmouths are found in and around swamps, ponds and streams from southern Virginia to Florida and west to Texas. Although unaggressive by nature, this snake is quick to bite and possesses a potent venom which may result in death if not promptly and properly treated.

Mocking bird   There are several mocking bird species in the Americas. They are about 1 ft long, with slender beaks and long tails. They are related to the catbirds. Mocking birds are popular creatures in the United States because of their ability to mimic other birds' songs. They can also make passable imitations of human voices and other man-made sounds.

Mynah bird

They eat mainly fruit and insects. They defend their nesting territories with song and threatening displays, and also with violent pecking attacks against intruders.

Mole   The several species of moles all live in the Northern Hemisphere They are small mammals related to hedgehogs and shrews. They spend almost all their lives under the ground and have very strong front limbs with which they excavate their tunnels. Their eyes are almost useless, and the animals rely on the senses of smell and hearing. They feed mainly on earthworms, but also take insects. When worms are abundant the mole will collect large numbers, bite their heads off, and store them in underground chambers. The worms are not killed, but they are paralysed and they cannot move away. They form an important larder because the mole cannot go for more than a few hours without food.

Moloch   The moloch is an Australian lizard which is covered with sharp spines. It is also called the thorny devil, although it is quite

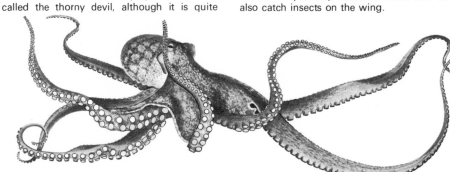

Octopus

harmless. The animal is about 6 ins long and it lives in the dry parts of the continent. It eats almost nothing but ants.

Mongoose   Mongooses are carnivorous mammals living in Africa and Asia. They are long-bodied creatures with short legs and sharp muzzles. The largest of the 48 species is about 4 ft long, including nearly 2 ft for the tail. Most of the species have a uniform brown or grey coat. They are very alert and quick animals, and they are well known for their snake-killing abilities. Mongooses severely interfered with the wildlife of the West Indies when they were taken there to combat the rats that were destroying the sugar cane.

Monitor lizard   The monitors include the largest of all living lizards. Several species exceed 5 ft in length and the Komodo dragon reaches a length of 10 ft. Monitors live in the warmer parts of the Old World, from Africa and Arabia to Australia. They have long forked tongues and have lost the ability to shed their tails when attacked. Monitors spend a lot of time basking in the sun, but they can be nasty opponents if disturbed. They lash out with the long tail and bite with their powerful jaws. They feed largely on small vertebrates, but some species take insects. Some live in desert regions, while others never go far from the rivers.

Moose   The moose is the largest of the deer, standing nearly 8 ft at the shoulder (although the rump is considerably lower) and weighing up to 1800 lb. The male carries flattened antlers more than 6 ft across. The moose lives in the northern forests of North America and also ranges from Scandinavia to Siberia. In Europe it is known as the elk. It is found mainly in the wetter regions, where there are plenty of ponds and lakes and plenty of willows for it to feed on. The animals also feed on water plants and

often wade into the lakes right up to their necks.

Moray eel   There are about 120 species of these fishes, living in the warmer seas of the world, especially around coral reefs. Some species reach lengths of 6 ft or more. The head and front part of the body are usually more bulky than the rest. There are no paired fins and no scales. Many are brightly colored. They hide by day and come out at night to feed on any kind of animal that they can catch. The mouth has numerous sharp teeth, which can inflict a nasty wound, but the morays are not normally aggressive fishes.

Motmot   Motmots are colorful birds related to kingfishers. There are eight species, all in Central and South America. Their beaks have saw-like edges, and most species have two long racquet-shaped tail feathers. Motmots live in wooded country and perch on branches while scanning the ground for large insects or lizards. They swoop down on these and they also catch insects on the wing.

Moufflon   This is the only wild sheep in Europe. It is about 27 ins high at the shoulder and reddish brown in color, with a whitish 'saddle'. The males usually have large curving horns, but the females are usually hornless. Truly wild moufflon are probably confined to Corsica and Sardinia, but flocks exist in the mountains of several countries.

Mudpuppy   This is a North American salamander that never really grows up—it retains its gills throughout life and lives permanently in weedy ponds and streams. It is usually about 12 ins long and it feeds on intertebrates and small fishes.

Muntjac   Muntjac are small deer, 16–25 ins high at the shoulder and somewhat higher at the rear. Male muntjac have short antlers and long tusk-like canine teeth in the upper jaw. Muntjac come from South-East Asia, but several species have been introduced to other countries. They live in wooded regions and feed on low-growing vegetation.

Musk ox   This creature, more closely related to sheep and goats than to oxen, lives on the bleak tundra of Canada and Greenland. The bull is a bulky creature, 5 ft high at the shoulder and weighing 700 lb. The cow is nearly as large and both sexes have thick shaggy coats. They stay out on the tundra throughout the year, nibbling at the sparse vegetation to keep alive.

Musk rat   This is a large vole, not a rat at all. It is up to 12 ins long, with a tail of about the same length. The fur ranges from silvery brown to black and, under the coarse guard hairs, there is a dense layer of soft hair. This is an important item in the fur trade and is known as musquash. Musk rats are natives of North

America and they live in and around streams, feeding on various water plants. When introduced to Europe they did great damage to river banks.

Mynah Mynahs, or mynah birds, are large starlings of southern Asia. The Indian hill mynah, 15 ins long, is a popular cage bird because it can mimic human voices. It is also known as a grackle. It is a glossy black bird with a yellow bill and yellow wattles behind the eyes. Most of the mynah species live in the forests and eat fruit and insects.

Narwhal This is a smallish whale, reaching about 16 ft in length excluding the long tusk. The tusk is normally found only in the males and it is formed from one of the upper teeth. It may be 9 ft long. The narwhal is greyish in color, with black spots. It lives in the Arctic Ocean and often forms great herds. It feeds mainly on cuttlefish and squids.

Natterjack The natterjack toad is easily recognised by the yellow line running down the center of the head and back. Apart from that, it is very like the common toad. It is about 3 ins long and greyish brown. Natterjacks are found only in Europe. They are very local in the British Isles, but common in France and Spain. They live mainly in sandy places. They have very loud voices.

Opossum

Nightjar The common nightjar, or goatsucker, is a nocturnal bird about 10 ins long. It is grey, brown and black and very difficult to see as it sits on the ground during the daytime. It produces a churring noise throughout the night as it flies about catching insects on the wing. It is a migratory bird, spending the summer in Europe and Asia and going to Africa for the winter. There are about 70 species of nightjar altogether. All are much alike and they are found nearly all over the world. The whip-poor-wills are American nightjars.

Nilgai This is the largest Indian antelope, the males being nearly 5 ft high and weighing up to 600 lb. The male's horns are only about 8 ins long and the female is hornless. The male is bluish grey and the female is tawny. They may be recognised by the white 'garter' below each fetlock. Nilgai live in hilly country and open plains, feeding on trees and grasses.

Noddy Noddies, of which there are five species, are terns living in the warmer parts of the world. They are between 12 and 16 ins long. The white noddy or fairy tern is white, but the others are brown or grey. Most of them

Orang-utan

nest on remote islands, but they spend most of their time at sea. They catch small fish at the surface and they do not dive.

Numbat This is a termite-eating marsupial, often called the banded anteater. It grows to a length of 18 ins, including a bushy 7 in tail, and has brownish fur with white bands around the body. It has prominent ears and 52 teeth—more than any other land mammal. The numbat lives in South-West Australia.

Nuthatch Nuthatches, of which there are 18 species, are small birds with bluish grey or bluish green plumage above. They have strong feet and claws and run up and down tree trunks with ease. They do not use their tails to support them. The birds eat insects for much of the year but in autumn and winter they eat nuts, often wedging them into bark crevices and then hacking them open with their sharp beaks. Most of the nuthatches live in Eurasia.

Oarfish This marine fish, also called the ribbon fish, has a body up to 20 ft long yet it is only 2 ins across. It is about 12 ins deep at the front, but tapers to a point at the tail. It is silvery grey, with darker markings and deep red fins. The dorsal fin runs the whole length of the body and the front few rays form a spectacular

Black-naped oriole

crest over the head. The pelvic fins are long and slender but broaden at their tips, rather like oars—hence the fish's name. Oarfishes live in nearly all parts of the oceans and feed on tiny planktonic animals.

Ocelot This is one of the most beautiful members of the cat family. Its color ranges

177

The otter, a fish-eating carnivore

from grey, through yellow, to deep brown, and it is decorated with spots and streaks. It is about 4 ft long and has rather long legs. Ocelots live in the warmer parts of America, from the southern United States to Paraguay. They are forest-dwellers mainly and feed on birds and small mammals.

Octopus The octopus is a marine cephalopod with eight arms. These creatures range in size from a few inches across to more than 30 feet. They feed mainly on crabs and shellfish which they catch with their tentacles, poison with venom, open with their beak and scoop out with their tongue. Octopuses are found in all oceans of the world. They live on the sea-bed in nooks and crannies.

Okapi This African mammal is the only close relative of the giraffe but the neck is not so long. The top of the head is only about 6 ft from the ground. Its coat is a rich reddish brown and it is marked with white on the legs and haunches. Okapis live in the rain forests of the Congo and were not discovered by Europeans until less than 100 years ago. They eat leaves and fruit.

Olm The olm is a blind white amphibian that lives in underground rivers and lakes. It is related to the newts and salamanders but it does not grow up completely. It retains its red gills throughout its life. It has an eel-like body up to 12 ins long. Olms live only in the limestone caves of Yugoslavia and Italy.

The giant panda

Opossum Opossums are American marsupials. The best known species is the Virginia opossum, common in many parts of the United States and South America. This animal is up to 20 ins long and looks rather like a large rat. Like the other opossums, it lives in the woods and forages for invertebrates and small vertebrates on the ground. The young are carried in the mother's pouch for about 10 weeks and for a further month they are carried about on the mother's back. Not all opossums actually have a pouch.

A macaw, a long-tailed member of the parrot family.

Orang utan This is one of the apes. It lives in Borneo and Sumatra and is now in real danger of becoming extinct. The male stands about 4½ ft when upright. The animals have rather sparse reddish brown hair. They live in the forests and stay mainly in the trees, swinging from branch to branch with their arms. Orangs feed mainly on fruit.

Oriole Orioles are forest birds living in the Old World. Only the golden oriole is found in Europe. This is a very striking black and yellow bird a little larger than a song thrush. Orioles eat fruit and insects, being among the few birds to accept hairy caterpillars. The Old World orioles, of which there are 28 species, belong to a quite different family from the 'American orioles'.

Oryx The three species of oryx are among the most beautiful of the antelopes, with white or fawn coats and long slender horns. They all live in the desert areas of Africa and Arabia, the Arabian oryx being very nearly extinct in the wild. The scimitar oryx has distinctly curved horns and lives in the Sahara. The gemsbok has a fawn coat, with black patches on the head and black stripes on the back and flanks. It lives in South-West Africa, but there is also a northern race, known as the beisa oryx, in Somalia.

Osprey The osprey, or fish hawk, is a fish-eating bird of prey. Its narrow wings have a span of about 5 ft. It fishes in the sea and in lakes, diving from a height of up to 100 ft and scooping up the fish with its talons. Its toes have spiny scales, which help it to grip the prey. The head and under-parts are whitish but the rest of the plumage is brown. The osprey is found in all continents except Antarctica.

Ostrich The ostrich is the largest living bird. The male may stand up to 8 ft high, although half of this is taken up by the neck. The head, neck and legs are almost naked. The males have black plumage with white plumes on the wings and tail. The females are brown. Ostriches cannot fly, but they can run very fast. They live in the dry bush and desert. They once ranged over all of Africa and Arabia, but they are now found mainly in East and South Africa. The Asian race has not been recorded for many years.

Otter The otters are water-dwelling carnivores related to the weasels. There are several species, all much alike, ranging up to 6½ ft long, including the tail. They have slim bodies and short legs. The head is broad and flattened, and the ears are almost hidden in the sleek brown fur. Otters are wonderful swimmers and they catch most of their food in the water. They eat a variety of animals including eels and other fishes, crayfishes, frogs and small mammals. Otters frequently move from one lake or river to another and will often cover some distance on land.

Ovenbird Ovenbirds are South American birds named from the domed clay nests that some species make. The birds are between 5 and 11 ins long and most of them are dull brown. Most of the species live in the pampas and the mountain regions of Argentina and Chile and most of them eat insects. The North American ovenbird is not a true ovenbird and belongs to the wood warbler family. It makes an oven-shaped nest, using grass not mud.

Pack rat Pack rats are large voles living in North America. They are about 18 ins long, half of which consists of the tail. There are 22 species, found in almost every kind of habitat from woodland swamps to deserts. They are nocturnal animals and they collect all kinds of brightly colored and metallic objects to hoard in their nests. Some of the species make houses of twigs, and the desert-dwelling pack rats use pieces of prickly cactus. These homes provide safety from all enemies. Pack rats are mainly vegetarian, eating fruit, leaves and roots.

Paddlefish This is a large and rather primitive freshwater fish related to the sturgeon. It has a long shark-like body with few scales and a long flattened snout looking rather like a canoe paddle. The snout may account for half of the fish's length. There are two species, one living in the Mississippi region of America, and the other in the Yangtse basin of China. The American species is about 6 ft long, but the Chinese species is reported to reach a length of 23 ft. Both feed on plankton.

Panda The giant panda is probably the most famous of all animals. It is a black and white bear-like creature, up to 6 ft long and weighing up to 300 lb. It lives in the cold damp bamboo forests of Tibet and South West China, and is probably rather rare. Pandas eat mainly bamboo shoots, but they also eat other plants and various small animals, including fishes.

Pangolin Pangolins are strange mammals in which the hair of the back has been converted into large brown overlapping scales. The seven species—four in Africa and three in Asia—are placed in an order by themselves. They have slender bodies and short legs, with long tails and snouts. The eyes are very small. Pangolins live in the forests and feed on ants and termites. They use their strong claws to rip open the nests and they pick up the insects with their long tongues. They are often called scaly anteaters.

Paradoxical frog The adult paradoxical frog, which lives in South America, looks just like any other frog. It is basically greenish brown, with darker markings on the back and yellow and black markings on the hind legs. Its outstanding feature is the great size of the tadpole. When fully grown, the tadpole is more than 10 ins long, but then changes into a tiny froglet less than 2 ins long. The adult frog is only 3 ins long.

Parrakeet Parrakeets are small brightly colored parrots with long tails, but the name has been given to a wide variety of species and does not refer to any clearly defined group. The budgerigar is the best known of the parrakeets.

Parrot The parrot family contains more than 300 different species, but only 107 of them are strictly called parrots. The others are macaws, cockatoos, lorikeets, and so on. Parrots are generally stocky birds, with square or rounded tails. They all have large heads and strongly hooked beaks. One of the best known is the African grey parrot, which is a favorite pet because of its great ability to 'talk'—that is to mimic human voices. It is a remarkably intelligent bird, and it is also very long-lived. In the wild, parrots are forest-dwelling birds. They feed mainly on fruits and seeds.

Parrot fish Parrot fishes are brilliantly colored fishes in which the teeth are joined together to form beak-like structures around the mouth. There are numerous species living around the coral reefs of the tropics. They feed mainly on seaweeds, but they also nibble away at the corals with their sharp beaks. Parrot fishes range from 12 ins to 12 ft in length.

Peccary Peccaries are the South American equivalents of the Old World pigs although they are smaller than true pigs. The tail is vestigial and the body is covered with thick bristly hairs. There are two species. The collared peccary lives in deserts, woodlands and rain forests from Argentina to southwestern U.S. It is greyish and has a paler collar of hair around the shoulder. It feeds mainly on fruit and roots, together with some invertebrates. The white-lipped peccary lives in the rain forests, never far from running water. It is reddish brown or black, with a whitish area around the mouth. It feeds as much on animal material, including carrion, as on vegetable food.

Pelicans

Pelican The pelicans are among the largest living birds, some of the eight species reaching a length of 6 ft and having wingspans of 10 ft. They are rather ungainly birds on the ground, but they are expert swimmers and fliers. The birds have long necks, and the large beak carries an extending pouch which is used for catching fish. The bird swims on the water and dips its head under the surface, using the pouch as a net. The brown pelican is a sea bird, but the other species, which are mainly white, live around inland lakes. They are found in the warmer parts of the world.

Père David's deer This deer, which stands nearly 4 ft high at the shoulders, came originally from the plains of northern China, but it no longer exists in the wild. There is a herd at Woburn Park in England, and there are quite a number of the animals in zoos. The male carries large antlers which fork near the base into two more or less equal arms. The hooves are large and spreading, and it is believed that the animals originally lived in swampy areas. They certainly like water.

Peregrine The peregrine is a large falcon, up to 19 ins long. The upper surface is slaty blue and the underparts are white, both regions having dark markings. Peregrines are most numerous in rocky areas, especially around the coast. They are swift birds and feed mainly on other birds, especially pigeons. They have been used to frighten birds away from airfields to prevent damage to aircraft. Peregrines live in all continents except Antarctica and have several races, or sub-species.

Phalanger Phalangers are tree-dwelling Australian marsupials. They fill the niches occupied by monkeys and squirrels in other parts of the world. There are several species and they all feed mainly on fruit and leaves, although some take insects as well.

Piddock This is a marine bivalve mollusc that bores into stone and wood. It has a rather delicate long white shell which is decorated with numerous spines and ridges, especially at the broad front end. The hinge is so arranged that the two halves of the shell can rock backwards and forwards, and it is this motion that gradually bores a hole into the stone. The animal can make quite a long burrow and it is well protected there. It feeds like most other bivalves by drawing a current of water in through a siphon and filtering out food particles. The largest of the piddocks is about 6 ins long.

Pika Pikas are small mammals related to rabbits and hares. There are two species in North America and 12 in Asia, the largest being about 12 ins long. They look like rabbits except that they have short rounded ears and all four legs are about the same size. There is no tail. Pikas are generally greyish brown and they live in a variety of habitats. One species even lives more than half way up Mount Everest. The animals cut grass and dry it in the summer. They then store it and use it during the winter.

Pike The pike is the fiercest freshwater fish in the Northern Hemisphere and it is aptly nicknamed the freshwater shark. It is a long-bodied fish, sometimes reaching 5 ft in length, with a flattened head and large jaws. The latter bristle with teeth. The fish is greenish, and it blends

Pipe fishes, relatives of the sea horse

well with the weeds among which it hides. When another fish swims close enough the pike will dart out and snap at it. Pike have been known to swallow fishes almost as large as themselves.

Pilot fish Pilot fishes are small black and white striped fishes that swim about in the company of whales, sharks, and other large fishes. They do not guide their larger companions in any way and they do not even feed on the scraps left by the sharks. It is a mystery why pilot fishes should associate with the larger animals.

Pipefish The pipefish is a long, thin fish with a long snout. It has an unusual way of swimming. Sometimes, instead of swimming headfirst through the water it swims upright. The male pipefish keeps the eggs and the young in a pouch on his body. Slow swimmers, the pipefishes depend upon bony armor for protection.

Piranha A ferocious fish found in South American rivers. It travels in schools and has been known to attack and devour large animals within a matter of minutes. Piranhas are between two and eight inches long and have extremely sharp teeth.

Polar bears

Pit-viper Pit-vipers are poisonous snakes found throughout North and South America and Asia. They have a characteristic deep pit on each side of the face and long movable front fangs which fold up out of the way when the mouth is closed. The pit is a sensitive heat receptor, capable of detecting the elevated body temperatures of mammals and birds. Some Asian and South American species are small tree climbing snakes while most are ground dwelling. The largest species are the eastern diamondback rattlesnake of southeastern U.S. and the tropical bushmaster. Both may be more than 8 ft long and weigh 20 lbs. The bite of a pit-viper is painful, and in some species, fatal. Pit-vipers inhabit jungles, swamps, deserts, prairie grasslands and temperate forests. They are represented in the U.S. by the cottonmouth water moccasins, copperheads and rattlesnakes.

Pocket gopher This is a small hamster-like rodent living in North America. There are about 30 species, up to about 18 ins in length including the tail. They get their name from the fur lined external pouches on their cheeks. These are used for carrying food. Pocket gophers are burrowing creatures with small eyes and ears and powerful front legs. They feed on bulbs and roots, often pulling whole plants down into their burrows.

Polar bear One of the largest and most carnivorous of the bears, the male polar bear may reach a length of 9 ft. Females are a little smaller. The animal has a long head with small ears. It has powerful legs, with broad feet and hairy soles which help it to grip the ice and snow. Polar bears live along the southern edge of the Arctic pack ice and occasionally venture as far south as Japan. They eat seals and fish and will also take vegetable food when they come on land.

Polecat A member of the weasel family, the polecat is about 20 ins long, including the tail. The upper surface has a dense golden underfur with coarse dark brown guard hairs. The underside is blackish. The animal prefers wooded country, but lives more or less anywhere that it can find small mammals and birds to eat. Polecats are found throughout most of Europe, but do not extend far into Scandinavia. In the British Isles they are confined to Wales at present.

Pond skater The pond skater is a small bug with a flat narrow body and two pairs of long legs. These have hairy tips which enable the insect to rest on the water surface. The legs also 'row' the insect about. The front legs are short and are used for catching small insects that fall on to the water surface.

Poorwill Poorwill or whip-poor-will is an American nightjar about 8 ins long and mottled brown in color. Like the other nightjars, the poorwill is nocturnal and catches insects on the wing. It is the only bird known to hibernate.

Porbeagle This is a shark, normally about 6 ft long, living in the Atlantic Ocean. It is a plump fish, with a sharply pointed snout and a large dorsal fin. It feeds mainly on other fishes and is a rather aggressive creature.

Porcupine fish This animal looks much the same as any other kind of fish when it is relaxed. It is normally about 12 ins long, with numerous spines lying flat on the body. When alarmed, the fish swallows water and puffs itself up. The spines stand out from the body then and the fish is very difficult for any enemy to swallow. Porcupine fishes live in tropical seas and feed on corals and shellfish.

Porpoise Porpoises are small whales, never more than 6 ft long. They have blunt snouts and broad rounded flippers. These features distinguish them from the other major group of small

Potto

whales, the dolphins. There are several species generally found in coastal waters. They swim in the surface waters, catching herrings and other small fishes.

Portuguese man o' war This is a marine animal—actually a colony of minute animals—related to the jellyfishes and sea anemones. There is a gas-filled float at the surface, and below it are numerous long tentacles with stinging cells. These have a venom almost as powerful as that of a cobra, and the man o' war is something to be avoided. It lives in the warmer seas of the world and is often brought to British waters by the Gulf Stream. The tentacles catch and digest fishes.

Potto This is a slow-moving relative of the lemurs and lorises, up to 16 ins long with a 3 in tail. It lives in the forests of West and Central Africa. It has thick brown fur and large round eyes, enabling it to see as it moves around at night. It grips the branches tightly and never lets go with more than one foot at a time. Pottos eat leaves, fruit, invertebrates, and small birds and eggs.

Prairie dog This is a hamster-like, short-tailed ground squirrel which has a barking call. It is about 12 ins long, yellowish brown or grey. The five species live on the plains of North America. They are burrowing creatures and they live in 'towns' of many individuals, with numerous burrows for each family group. Prairie dogs feed on grass and other plants.

Pronghorn This is an antelope-like creature living on the grasslands of Western North America. It stands about 3½ ft at the shoulder.

Rabbit

The upper parts are reddish brown, with a black mane and white underparts and rump. Both sexes bear horns covered with a sheath of hair. The horns have a short prong at the front—hence the name—and they are shed each year.

Ptarmigan   This game bird is distinguished from its grouse relatives by its white wings and underparts. It ranges right across the northern parts of the world. The birds have a brownish mottled color in spring and summer. In autumn they become greyish, and in winter they are white except for the black tail and the black eyepatches of the male. They eat berries, leaves and insects.

Puffin   This is a small black and white sea bird about 12 ins long, with a triangular bill of red, blue, and yellow. These colors develop only during the breeding season. The birds eat small fishes and they swim under the water scooping up the fishes as they go. Puffins live around the North Atlantic and breed on cliffs. Outside of the breeding season they may go far out to sea. Puffins are related to guillemots and razorbills.

Puma   Also called a cougar or mountain lion, the puma is a large cat looking rather like a lioness. The body is generally some 4 ft long. The animal is found throughout the Americas, especially in the mountains and plains. It also

Black rat

lives in forests. It feeds mainly on deer, but may also take domestic animals.

Python   Pythons are the Old World equivalents of the American boas. They are large constricting snakes, of which the Asian reticulated python is the largest. It reaches a length of about 30 ft. Pythons live in various habitats, but the larger ones prefer to be near water. Many climb trees. They eat all kinds of small mammals and birds.

Quail   Quail are game birds rather like partridges in many ways. There are nearly 100 species, living all over the world. They do not fly much, and keep within a few feet of the ground when disturbed. They keep to open ground, feeding on seeds and insects.

Quelea   The quelea, or red-billed quelea to give it its full name, is a sparrow-like bird with a stout red beak. It lives in vast flocks on the grasslands and savanna of Africa and does immense damage to grain crops. Whole fields of grain may be eaten.

Quokka   This is a small wallaby, looking like a large rat. It is only 3 ft long, including the tail, and it is greyish brown with a reddish tinge around the front part. The quokka lives in the swamps and brush of southwest Australia, feeding on grasses and other plants at night.

Rollers

Rabbit   The rabbit can be distinguished from the hare by its shorter ears and limbs. Rabbits live in extensive burrows or warrens with several emergency exits. Their diet includes a wide variety of plant food such as grass, corn and kale. During the summer the female may have a number of litters at intervals of 5–6 weeks. There are over 50 races of domesticated rabbits.

Raccoon   Raccoons are carnivorous mammals related to pandas and coatis. There are seven kinds all living in the Americas. They are between 24 and 40 ins long including the tail. The fur is generally greyish with dark marks on the face and tail. Raccoons originally lived in the forests, especially around streams, but they have adapted themselves to the changes brought about by man and they often come into towns in search of food. They will eat more or less anything, and even raid garbage cans in the towns.

Ragworm   This is a marine relative of the earthworm, living in shallow coastal waters all over the world. There are many species. Each segment bears broad lobes and tufts of bristles and, unlike the earthworm, the ragworm has a fairly distinct head with eyes and jaws. Ragworms range from 1 in to 3 ft in length. Most of them are carnivorous.

Rail   Rails are long legged, short-tailed birds, most of which live in and around water. Coots and moorhens are rails. They are generally rather shy birds and they run for cover when

Male rattlesnakes in a combat dance.

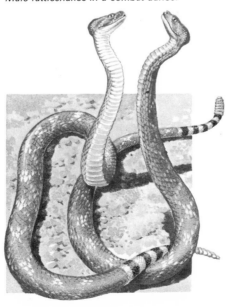

disturbed. They do not fly very strongly. Many species have loud and rather eerie calls. They eat a variety of plants and small animals.

Rain frog   This very rounded brownish frog is about 1½ ins long. It lives in the dry regions of southern Africa and sleeps (aestivates) for much of the year. It comes out after the rains and feeds for a couple of months or so on worms and termites. The eggs are laid under the ground and the tadpole stage is passed entirely in the egg. When the egg hatches, a tiny frog emerges.

Rat   Black rats originated in South-East Asia but are now found all over the world. They feed almost exclusively on plant food. Brown rats are larger and omnivorous. They are a serious agricultural pest in many parts of the world and spread infectious diseases.

Rattlesnake   The rattlesnake is a rather heavy-bodied snake, named because of the horny rings on the tail. These are the remains of cast-off skins and they rattle when the snake is disturbed. There are 25 species in North America, some of them dangerously poisonous. One species may exceed 8 ft in length. They feed on a variety of small mammals. They all give birth to live young.

Raven   This is the largest member of the crow family, about 25 ins long and completely black. It ranges over much of Eurasia and North America. Similar species live in Africa and Australia. Most ravens live in mountainous and coastal regions and they feed largely on carrion and small mammals. They nest in trees or on rocky ledges.

Razorbill   This is a black and white sea bird related to the puffins and guillemots. It is a plump bird about 16 ins long, with a white stripe crossing the black beak. Razorbills are confined to the North Atlantic and spend much of the year out at sea. They come inshore during the winter to breed on cliff ledges in the spring. They feed mainly on small fishes and swim after them underwater.

Reed frog   There are many species of reed frogs, all living in Africa. They are only about 1 in long, but many species are beautifully colored and marked. The markings often vary from individual to individual and depend to some extent on the background. The animals generally bask in the sunshine, although they are always ready to leap back into their ponds. They feed on flying insects, especially mosquitos.

Reindeer   The reindeer of northern Europe and Asia is a semi-domesticated race of the

181

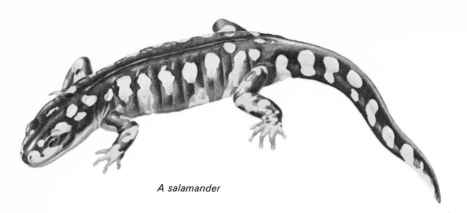
A salamander

American caribou. They are true deer, but both sexes bear antlers. Reindeer stand about 3½ ft at the shoulder. They have been domesticated for hundreds of years. Some Siberian tribes merely herd them and kill some for meat. Others, including the Lapps, use the reindeer for meat, milk, transport and clothing. Its dung is also used for fuel.

Remora This is a fish in which the dorsal fin has been converted into a large oval sucker. The remora attaches itself to the undersides of other fishes and gets a free ride in this way. Sharks commonly carry remoras in the warmer seas of the world. The remora does no harm to the shark and often moves away to feed on passing small fishes. Remoras vary from about 7 ins to 3 ft in length.

Rhea This flightless ostrich-like bird is the largest bird in America. It stands about 5 ft high and, like the ostrich, it has an almost featherless head and neck. The long, powerful legs have three toes, compared with only two in the ostrich. There are two species of rhea: the common rhea and the slightly smaller Darwin's rhea. Both live on the grasslands of America and eat plants and small animals.

Rhinoceros The rhinoceros is a bulky, thick-skinned mammal related to the horse. Both animals bear their weight on the central toe of each foot, although the rhino does have two extra toes on each foot. Rhinos bear one or two horns on the snout, composed of densely compacted fibers like hair. There are five species: two in Africa, one in India, and two more in South-East Asia. The last three are very rare. They range between 4 ft and 6½ ft in height and the Indian rhino weighs up to 2½ tons. All are vegetarians, grazing or browsing on the vegetation.

Rice rat Rice rats are actually voles. There are about 100 species, up to 18 ins long including the long tail. They have brownish fur. Rice rats live in the Americas in all kinds of habitats. They are pests because they eat rice and other crops, but they are less of a pest than the true rats.

Right whale Right whales are whalebone whales. There are five species and they got their name because they were the right whales to catch when men first started whaling. None of them is common today because they have been hunted for a long time. The Greenland right whale is one of the rarest of the whales.

Roadrunner This is an American member of the cuckoo family, famed for its speedy running and its ability to catch snakes. Its wings are short and it rarely flies, but it has a long tail which helps to balance the bird as it runs. The roadrunner is about 2 ft long, from the beak to the tip of the tail, and is mottled brown in color. It lives in the deserts of North and

Central America and catches a variety of animals as well as snakes.

Roller Rollers are attractive birds related to kingfishers and bee-eaters. They are sturdy, jay-like birds from 9 to 13 ins long. They have long curved beaks and long tails. Their plumage is largely blue, green, and red. Most of the rollers live in Africa, but one species breeds as far north as Scandinavia. They get their name from their rolling courtship flight. Rollers feed mainly on insects, but they also eat lizards and small snakes.

Rorqual Rorquals are whalebone whales related to the blue whale. There are four species —the fin whale, sei whale, Byrde's whale, and the minke whale or lesser rorqual. Byrde's whale is found in the warmer seas, but the others live in the polar seas and are the mainstay of today's whaling industry.

Rove beetle Rove beetles have very short wings and wing covers which reach only a short way down the back. They look rather like earwigs, but without the forceps. Most are very small, but some, such as the devil's coach horse, may exceed an inch in length. Many are black or black and red. Most of them live in damp places, feeding on decaying matter and smaller insects.

A scallop showing the eyes on the edge of the mantle.

Ruff The ruff is a sandpiper in which the male has an ornate ruff or collar of feathers around the neck. The bird is about 12 ins long (the female, or reeve, is a little shorter) and is very variable in color. Most of the birds are mottled brown. They breed in northern Europe and Asia, inhabiting damp meadows and marshes. Out of the breeding season they migrate as far as South Africa and Australia. They feed mainly on insects.

Saiga The saiga is an antelope with a noticeably swollen snout. It has features

of both sheep and gazelles. It is about 2½ ft tall and it has a woolly coat which is buff in summer and white in winter. The male has lyre-shaped horns about 12 ins long. The saiga lives on the cold steppes of Central Asia, and feeds on wormwood and other low-growing shrubs.

Salamander The salamanders belong to the group of amphibians with tails—the Caudata. They are found in most parts of the Northern Hemisphere in brooks and ponds and in moist spots on land. Most salamanders are under six inches in length but the giant salamander of Japan reaches 6 feet from snout to tip of tail.

Sand dollar This creature, also known as a cake urchin or sea biscuit, is a rather flattened form of sea urchin with a bilateral symmetry brought about by the unequal development of the 'arms'. The flattened shell is purple or black and is about 4 ins across. It is strengthened by internal pillars. There are several species, found mainly in the warmer seas, especially in the Pacific. The animals live buried in the sand below low tide level and feed by trapping food particles in mucus produced by the spines.

Sapsucker This is a North American woodpecker, so called because of its habit of drilling holes in trees and sucking the oozing sap. There are two species: the yellow-bellied sapsucker and Williamson's sapsucker. They are both like other woodpeckers in form, being about 8½ ins long and brightly colored. The birds also catch the insects that are attracted to the sap.

Sawfish Sawfishes are large fishes, occasionally up to 20 ft or more, with a long drawn out snout flattened into a blade with a row of strong teeth sticking out of each side. This blade may account for as much as one third of the fish's length. There are 6 species, found in all warm coastal waters and sometimes venturing up rivers into fresh water. They feed on molluscs and crustaceans, using the saw to search in the mud. They also eat other fishes.

Sawfly Sawflies are related to ants, bees, and wasps, but do not have the narrow 'wasp waist'. Their name is derived from the fact that they have a saw-like edge to the ovipositor with which the female cuts holes in plants to lay her eggs. Some sawflies are called wood wasps because they lay their eggs in wood. Many sawfly larvae are very destructive to plants.

Scale insect Scale insects are small bugs distantly related to greenfly. They are among the most serious of crop pests. The males normally have one pair of wings and they fly freely, but the females are wingless and they merely sit on plants and suck sap. They are flattened creatures, never more than 1 in long and usually much smaller. They are covered with waxy or resinous secretions and many are also protected by a horny scale which makes them look like tiny limpets.

Scallop Scallops are bivalve molluscs in which one valve of the shell is normally flat and the other is convex. The valves are ridged, and they bear 'ears' near the hinge. Scallops are among the few bivalves that can swim. They do it by opening and closing the shell quickly and shooting water out. They thus move about by jet propulsion. The mantle edge, just inside the shell, bears numerous

Sea horses

though and they generally feed on small insects or decaying material. The head is prolonged into a long 'beak'.

Scrub-bird The two scrub-birds look like wrens, but they have their own family. They are between 6 and 8 ins long and they live in Australia. They live in the dense scrub and are rarely seen. They feed on the ground and take worms, snails and insects. Nests are built on the ground and consist of grass and twigs lined with 'papier mache' made from chewed-up wood.

Sea hare Sea hares are molluscs, halfway between a sea slug and a sea snail. They have a soft body and a thin transparent shell. The foot has a crinkly flap on each side and these flaps are folded up over the shell. They wave gently in the water and force a current of water over the gills. The sea hares of European waters are usually about 3 ins long and olive-green, brown or reddish in color. Similar species are found in temperate or tropical parts of the world but are much larger. They feed on sea lettuce and other seaweeds.

Shearwaters

burrows in search of food which consists of decaying animal matter.

Sea slug Sea slugs are marine snails of various kinds that have lost their shells at some time during their history. They are rather

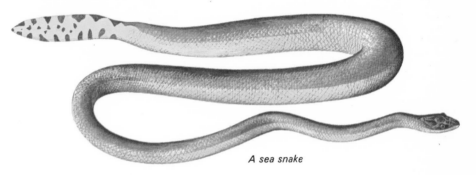

A sea snake

tentacles and many small but beautiful blue eyes. Scallops live mainly around the coasts and they feed like any other bivalve.

Scarab Scarab beetles are stoutly built blackish beetles with brilliant metallic sheens in many species. The head and the thorax often bear ornamental spines, especially in the male. Most scarabs feed on the dung of mammals and they are called dung beetles. They usually bury lumps of dung and feed on it underground. They also lay their eggs on it. Some species, such as the sacred scarab of Egypt, roll balls of dung about until they find a suitable place to bury them. Scarabs belong to the same large group as the stag beetle and the huge Hercules and Goliath beetles. There are hundreds of species.

A sea slug

Scorpion See page 107.

Scorpionfish Scorpionfishes include some of the most beautiful, the most ugly, and the most poisonous of fishes. They nearly all live in the temperate seas. Their fins are normally broken up into numerous ribbons and spines, the latter often being very poisonous. Striped color patterns add to the strange appearance of the fishes. Members of the group include the highly poisonous stonefish and the Norway haddock. Although the fishes have poisonous spines, their flesh is quite pleasant to eat.

Scorpionfly Scorpionflies belong to a small but very ancient group of insects. They get their name because the males of many of the species have turned-up abdomens rather like those of scorpions. They are quite harmless

Sea horse The sea horse is about 8 ins long and has a large head with a tubular snout, a rotund body, and a very long tapering tail. It is almost entirely covered with bony bumps which give it the appearance of a wood carving. Despite its name and appearance, the sea horse is a fish. There are about 20 species, mostly brownish in color. They are found almost all over the oceans, except for the colder regions. They eat small fishes and crustaceans.

Sea lily Although very plant-like in shape, the sea lily is an animal related to the starfishes and sea urchins. In most species the body has five arms and is carried at the top of a slender stalk up to 2 ft long. At the bottom of the stalk there are branching tentacles which anchor it to the sea bed. Sea lilies are found in deep water in most of the world's oceans. They feed on particles of decaying matter that float down from above.

Sea lion Sea lions are relatives of the fur seals. Both groups have small external ears and in this respect they differ from the true seals. The most familiar sea lion is the Californian sea lion, which is seen in zoos and circuses. This is the smallest of the five species, with a maximum length of about 7 ft. It lives around the coasts of California and Mexico, and also on the Galapagos Islands and the Japanese coast. Sea lions are beautiful swimmers. They feed almost entirely on fishes and squids.

Sea mouse This is a stout marine bristle worm, related to the ragworms. It looks rather like a mouse, hence the name, and moves over the sand with the creeping movements of a mouse. The animals live in shallow seas and those found around the coasts of Europe are about 7 ins or more long. The sea mouse

flattened creatures, but the upper surface bears numerous flaps and filaments which act as gills. Many of the species are brightly colored. They are several inches long and they feed on seaweeds, sponges, sea anemones, and other small sea creatures. Most sea slugs crawl on the sea bed, but some swim freely, often by 'rowing' with the flaps on the sides of the body.

Sea snake Sea snakes are reptiles that have returned to the sea from the land. Sea snakes still have to rely on air for breathing and will drown if held under water too long. Most sea snakes are between 4 and 5 ft long although 9 and 10 ft specimens have been recorded. There are about 60 species and they are found in tropical waters, mainly in the Indo-Pacific region. They live mainly near sandy shores and river mouths. All sea snakes are venomous with cobra-like fangs at the front of the mouth. They feed on eels and small fish.

Sea squirt Sea squirts are marine animals which, in the adult state, live permanently fixed to the sea-bed or to rocks. They look little more than bags of jelly and they feed on tiny particles which are drawn in with water currents. Many live in clusters or colonies. Nothing could look less like a vertebrate animal, but the sea squirts are in fact related to vertebrates. The young sea squirts are tadpole-like creatures with a primitive backbone.

Sea urchin A small round animal with a spiny shell which lives off rocky coasts. The shell or test of a sea urchin is made up of hundreds of interlocking plates. These chalky plates carry little knobs to which the spines are attached. The sea urchin moves about on small water-filled tube-feet which stick out through pores in the test. The sea urchin's test is not an external skeleton like a crab shell. It is covered with a layer of skin and the spines protrude through this. Some urchins have poisonous spines. Sea urchins feed mainly on decaying matter and algae which they scrape from the rocks with an apparatus called Aristotle's lantern. This consists of five bony teeth worked by a series of levers and muscles which acts like the jaws of a grab. There are many species including the purple sea urchin, the heart urchin and helmet urchin.

Secretary bird The secretary bird is a rather unusual-looking bird of prey. Its name comes from the crest of black-tipped feathers which hang down behind the head in the way 18th century clerks carried their quill pens stuck into their wigs. Adult male secretary birds stand about 4 ft high at the shoulder and have a wingspan of 7 ft. The plumage is grey, with black on the wings and legs. They are found in Africa, south of the Sahara. They feed on insects, small mammals and snakes.

Serval The serval is one of the most beautiful members of the cat family. It is a slender animal, 15–20 ins at the shoulder and up to 3 ft long with a 10–12 in tail. The serval is long-legged, with large pointed ears. It has smooth, short hair and is a yellowish brown color with a pattern of bold black spots and stripes. The serval is found in most of Africa, south of the Sahara, and lives almost entirely on rodents.

Shearwater The 15 species of shearwater are petrels, related to the albatross and fulmar. They are medium sized sea birds mostly between 10 and 15 ins long. The plumage is dull, usually black or brown above and whitish underneath. Outside the breeding season shearwaters live out at sea, and are found in many parts of the world excepting the polar seas. They feed on fishes and squid.

Shelduck The shelduck is a brightly colored, rather goose-like duck common in

A skink, showing the rather smooth cylindrical shape of these largely burrowing creatures.

estuaries and on coasts. It is about 23 ins long and the plumage is white with bold patterns of chestnut and black. The shelduck ranges from the British Isles, across Europe and Asia to China. The shelduck's main diet consists of small shore living animals such as winkles, whelks, crabs and sandhoppers. It also eats grasses and other plants.

Shrew Shrews are the smallest of the mammals. There are 170 species, found everywhere except Australasia and the polar regions. The common shrews of North America and Europe are about 4 ins long, including 1½ ins for the tail. They are mouse-like animals, but they have a much more pointed snout, full of sharp red-tipped teeth. Shrews live in undergrowth and leaf litter. They feed on insects, slugs and worms, with some vegetable matter. They have to feed every two or three hours.

Sidewinder This is a rattlesnake living in the deserts of western North America. It gets its name from its peculiar method of moving along the sand. It throws its body into curves, but only two points are in contact with the sand at any one time. The only tracks left in the sand are unconnected parallel grooves. Sidewinders are up to 2 ft long, with stout bodies. They are brownish in color and they feed mainly on small rodents.

Siren The three species of siren are North American amphibians. They are related to newts and salamanders, but they are eel-like with no back legs and only weak front ones. There are feathery external gills behind the head. The sirens are thus amphibians that have not properly grown up, because they still have some of the features of young amphibians. The

The Arctic skua

three species range from 8 ins to 30 ins in length. They live in shallow streams and marshes and eat plants and small animals.

Skate Skates are cartilaginous fishes, related to sharks and rays. There are several species and they are all flattened from top to bottom. The body is more or less diamond-shaped, with a slender spiny tail. The eyes are on the top of the head, but the nostrils and mouth are right underneath. Skates reach a length of 8 ft. They live in most temperate and tropical seas, especially in the Northern Hemisphere, and feed mainly on crustaceans.

Skimmer The three species of skimmers are relatives of the gulls and terns, but they have very long beaks with the lower half much longer than the upper. They are up to 20 ins long,

with stout bodies and long wings. They are black (or brown) and white. Skimmers fly low over water and scoop up fish with their beaks. The black skimmer lives around the coasts of the Americas. The other two species live in Africa and South-East Asia and feed mainly inland.

A skunk

Skink Skinks are lizards with very few frills or decorations. They are perfectly ordinary lizards and most of them are burrowing creatures. The largest of the 600 or so species is about 2 ft long, but most species are much smaller. They live in a variety of habitats and feed mainly on insects.

Skua Skuas are sea birds related to gulls. There are four species, and they normally feed by attacking other sea birds and forcing them to give up the fish they are carrying. They will catch fish for themselves, however, and they also catch and eat smaller birds. The beak is hooked and the toes have sharp claws. The great skua breeds on the North Atlantic coasts and also around Antarctica. The other skuas nest around the Arctic. All are brownish and between 20 ins and 24 ins long.

Skunk This animal, related to the weasel and badger, is famed for its ability to squirt foul-smelling fluid at its enemies. Its bold black-and-white pattern acts as a warning to these enemies. There are several species, all with long shaggy fur. They live in the Americas and feed mainly on insects. They also take birds and mice, and they eat nuts in autumn.

Sloth Sloths are strange South American mammals related to ant-eaters and armadillos. They spend most of their lives hanging upside down in the trees by means of their huge

A 3-toed sloth

curved claws. The hair runs from the belly to the back, so that water runs off easily. The hairs are clothed with tiny algae and the animals thus look green. There are seven species, between 2 ft and 3 ft long. They move slowly and feed on leaves and fruit.

Slow-worm This creature is not slow and it is not a worm. It is a legless lizard. It is also called a blind-worm, but it is not blind either. The body is brownish, often with a bronzy sheen and with a black line down the back of the female. Adults are generally about 1 ft long, although they often lose their tails and appear much shorter. Slow-worms eat slugs and insects. They are found over much of Europe and Western Asia.

Snapping turtle There are two kinds of snapping turtles in the U.S. with several races extending through Mexico and into South America. The alligator snapping turtle of the southeastern states is the largest, weighing up to 200 lbs. The common snapping turtle is found throughout our swamps, ponds, and streams. Snapping turtles defend themselves viciously, inflicting severe wounds with their lightning-like strike. Stories of their severing a broom handle with one bite are unfounded. Snappers are largely scavengers, but do take small animals on occasion.

Snipe Snipe are short-legged wading birds with long straight beaks. The plumage is mottled brown. There are several species, ranging from 7 ins to 12 ins long. They live in marshes and other wet places and use their beaks to probe for worms and other creatures in the mud.

Sole Soles are tongue-shaped flat-fishes usually up to about 12 ins long. They lie on their left sides and are almost completely surrounded by a fringe formed by the fins. Most species live in shallow seas in the warmer parts of the world.

Solenodon The two species of solenodons are shrew-like mammals found only in the West Indies. They are about 12 ins long with a tail nearly as long. The head is relatively large and bears a long snout. The animals are brown and have remarkably large front feet and claws. They are rather clumsy creatures and feed on a variety of invertebrates and plant food. Both species are now very rare.

Spadefoot Spadefoots are toads in which there is a horny projection on the side of each hind foot. This is used for digging burrows

They have rather moist skin and may be up to 4 ins long. Spadefoots are widely distributed in the Northern Hemisphere, the best known species living in the North American deserts. They start to breed as soon as the rains come and the tadpoles grow up very quickly.

Sparrowhawk This handsome bird of prey lives in Europe, Africa and Asia. It has short, rounded wings and long legs. The plumage is brown above and white with brown bars below. The bird lives mainly in woodland and feeds on small birds. It has a characteristic flight consisting of bursts of rapid wingbeats alternating with long glides.

Spadefoot toad

Spectacled bear Named from the whitish markings around the eyes, the spectacled bear is the only South American bear. It is 5–6 ft long and stands only 30 ins at the shoulder. It has a dark shaggy coat much less dense than that of bears from the temperate regions. The animal lives in the forests of the Andean foothills from Panama to Bolivia. It feeds mainly on vegetable material.

Sperm whale This is the largest of the toothed whales, the males being about 60 ft long and the females about half the size. The head is huge but the lower jaw is relatively small. It contains up to 56 conical teeth, each up to 8 ins long. Sperm whales are found in all the oceans, but are most common in the warmer seas. They live mainly on squids and many of them carry scars where giant squids have gripped them in a struggle. These whales will also eat fishes and seals.

Spider monkey Spider monkeys are South American monkeys with slender bodies and sparse, rather wiry fur. They are up to about 2 ft long and have a tail up to 3 ft long. The tail is remarkably prehensile and is used as an extra hand. There are two species—the common spider monkey and the woolly spider monkey. The latter lives in southern Brazil. It is more robust than the common species and has woolly hair. Both species appear to eat nothing but fruit.

Spiny ant-eater Also called the echidna, this creature is one of the egg-laying mammals. It looks rather like a large hedgehog. There are several species, ranging from 18 ins to about 3 ft in length. They are found in Australia and New Guinea and they live in all sorts of habitats. They can burrow very rapidly with their strong front feet and claws, which are also used for ripping open termite nests. The animals

A spider monkey, showing its prehensile tail

feed almost entirely on termites, which they pick up with their long tongues.

Spiny lizard Spiny lizards or fence swifts are found throughout the warmer regions of North America. They range in size from 4 to 12 ins long. Males of most species have bright blue patches on the underside of the throat and body. All are harmless creatures, although erroneously considered poisonous by local people who may call them "scorpions." They consume large quantities of insects.

Spoonbill Spoonbills are long-legged wading birds related to ibises. The beak is flattened and broadens out at the tip to form a 'spoon'.

A spoonbill and a scarlet ibis

The six species are up to 30 ins long and are generally white. They are found in all continents, but not in the northern regions. They live in marshes and other damp places and feed on small aquatic creatures.

Stork The name given to a family of large black or white birds that inhabit marshes and plains in most parts of the world. Adult storks have no voice but they can clap their beaks and make a loud noise. They hold their long beaks downwards when resting but extended horizontally when in flight. Storks feed chiefly on snakes, lizards, frogs, fishes and worms.

Sun bear Also called the honey bear, this animal is the smallest of the bears. It is about 4 ft long and about 2 ft high, but it is a stocky creature and may weigh up to 200 lb. It has short black hair, with a grey muzzle and a distinct white or yellowish crescent across the chest. The sun bear lives in the forests of South-East Asia and spends most of its time in the tree-tops. It eats both fruit and small animals and is very fond of honey.

Surgeonfish Surgeonfishes are colorful fishes that get their name from the razor-sharp bony keels at the base of the tail. These are used as weapons by the fishes. The fishes are found around coral reefs of the Indian and Pacific oceans and they browse on the algae and other creatures. They grow to lengths of about 2 ft and have deep bodies flattened from side to side.

Surinam toad This very flat toad is about 4 ins long and it is covered with small warts. It lives in Brazil and spends almost all of its time in the water, eating any creature (dead or alive) it finds on the stream-bed. It has neither teeth nor tongue. The eggs stick onto the spongy skin of her back. A pouch and lid develops around each egg and the whole larval development takes place in the pouch. After about 4 months tiny toads crawl out of the pouches and swim away.

Swallow The name given to a group of birds numbering some 100 species found in all parts of the world. The commonest species are the barn swallows. The tree swallow is a North American species about 5 inches long.

Swan The six species of swans are long-necked water birds closely related to geese. Most familiar is the mute swan, which is all white with an orange bill. It is a native of parts of Eurasia but has been taken to various other parts of the world. Whooper swans and Bewick's swans also occur in Europe and sub-species occur in North America. They have black or black and yellow bills. Black swans live in Australia, and black-necked swans live in South America. Swans feed on water plants and small aquatic animals. They rarely visit deep water.

Swift Swifts are swift-flying birds with long, narrow wings. Perhaps no other bird is so well adapted to life in the air, for the swift feeds, mates, bathes and even sleeps in the air. Their legs are short and weak and they rarely land on the ground. All swifts have dark plumage. Most of them live in the tropics and feed on flying insects. Some species migrate to cooler regions for the summer.

Swordfish The swordfish is a striking blue and silvery fish in which the upper jaw is carried forward to form a sword-like beak. The

The tree swallow

dorsal (back) fin is very tall. The fish is normally between 6 ft and 11 ft long, including the sword which is up to about 4 ft long. Sword-fishes live mainly in the warmer seas and feed on smaller fishes. The prey is snapped up or else stunned by rapid blows with the sword.

The white-rumped swift

Swordtail This is a popular and brightly colored aquarium fish related to the carp. The sword is an extension of the lower part of the male tail fin. Females are up to 5 ins long and males are about 3 ins long excluding the tail. Swordtails live in fresh water in Central America and feed on a variety of small animals and plants. They bear their young alive.

Tailor-bird Tailor-birds are warblers living in Southern Asia and Australia. They are small birds with inconspicuous plumage, and they

The marabou stork (left) and the saddle-bill stork (below).

feed entirely on insects. Their nests are made by stitching leaves together to form a pouch which is then filled with soft fibers or grasses.

Takahe This is a flightless bird of New Zealand and it is very rare, only about 300 birds being known. It is related to the coot and stands about 18 ins high. The wings and body are green and blue and the beak is pink and scarlet. The bird lives in swampy regions and feeds entirely on the leaves and seeds of a plant called snow grass.

Takin This is a heavily built mammal, half goat and half antelope. It stands about $3\frac{1}{2}$ ft high and weighs 500–600 lb. The male is usually yellowish, with a dark stripe along the back. Females are greyish and the young are black. Takins are browsing animals living in the mountains of western China. They feed mainly on bamboo.

Tanager There are over 200 species of tanagers, medium sized birds related to the buntings. Most of them are brilliantly colored and they nearly all live in the forests of tropical America, although five species live in North America. They feed on fruit and insects.

Tapir This is a stoutly built mammal related to both the rhinoceros and the horse. It is about 6 ft long and stands up to 40 ins at the shoulder. The snout is prolonged into a short trunk and the body is covered with short, bristly hair. Tapirs live in the wet tropical forests, one species

Mute swans (above) and black-necked swans (right).

being found in South-East Asia and three in South and Central America. They are browsing animals, feeding on young shoots and leaves as well as a certain amount of aquatic vegetation. The animals spend a lot of time in the water.

Tarantula This name is popularly applied to almost any large spider, especially to the hairy bird-eating spiders of South America. The true tarantula, however, is a wolf spider found in southern Italy. It is a greyish spider about 1 in long and it has a painful, but not particularly serious bite.

Tarpon This is a marine and freshwater fish with many common names, including big-scale, silverfish and silver king. It is up to 8 ft long and feeds mainly on crabs and fishes such as mullet, bream and needlefish. It lives along the tropical and sub-tropical coasts of the Atlantic. The tarpon reacts violently to being hooked and it is a big-game fish for the sea angler.

Tarsier The three species of tarsiers are found only on certain islands in the East Indies. They are small primates, the largest being only about 6 ins long, with thick woolly fur and enormous eyes. They have large rounded

sucker-like discs on their fingers and toes and they spend much time clinging to vertical tree-trunks and branches. Tarsiers are carnivorous animals and they feed on insects, frogs, lizards and small birds.

Tasmanian devil This is a carnivorous marsupial now thought to be confined to Tasmania. It is stockily built, rather bear-like and up to 3 ft long including the 1 ft tail. Its coat is black with white patches on the throat, shoulder and rump. It has large pink ears and its powerful teeth give it a rather fierce expression. It feeds on any flesh, dead or alive, but mostly on small mammals and birds.

187

A tapir

Tenrec Tenrecs are small insectivorous mammals, up to 15 ins long, related to the shrews and hedgehogs. They live only in Madagascar. There are 20 species, some looking like hedgehogs and the others looking like shrews or moles. They behave very much like the animals they resemble and they eat a wide variety of invertebrates. They also eat some vegetable food. Tenrecs are not common creatures.

Tern Terns are rather slender sea birds related to the noddies and gulls. There are over 30 species. Their wings are narrow and tapered and their tails are forked. The bill is narrow and the legs are short. Most of the species are black and white and range from 10 ins to 15 ins in length. Most terns live in the tropics, but there are terns in all parts of the world.

Terrapin This name is loosely used for any edible tortoise or turtle living in fresh water. More strictly, however, it means the diamond-backed terrapin found in the coastal waters and estuaries of the United States and Mexico. This animal gets its name from the diamond-shaped pattern on its shell. It is up to 8 ins long and weighs up to 2 lb. It feeds on plants and small invertebrates and it hibernates during the winter.

Thresher Thresher sharks, also called fox sharks, are found mainly in the warmer seas of the world. They reach lengths of up to 20 ft, but half of this is tail. The sharks feed on small fishes such as herrings and pilchards, which they 'round up' and then stun by fierce lashing of the long tail.

Thrips This is a tiny black insect with a narrow body and narrow fringed wings. Thrips are extremely common and can be found in almost any flower in summer. They fly about and often get in our eyes and hair. They hibernate for the winter and often come into houses where they hide behind picture frames, loose wallpaper, and in other crevices. Thrips feed by sucking plant juices and some are real pests.

Thylacine This is a dog-like flesh-eating marsupial, also called the Tasmanian Wolf. It is about 3 ft long, with an 18 in tail. Although the front half is dog-like, the hind quarters and tail are much more like those of a kangaroo. The fur is light brown, with darker stripes across the hind part of the back. Thylacines were once common in Australia, but now survive only in the wilder parts of Tasmania. They feed mainly on small wallabies and birds.

Tick Ticks are small blood-sucking creatures related to spiders and, more closely, to mites. There is a small head and a larger, leathery body with little sign of segmentation. The head

The matamata a South American terrapin.

carries mouthparts which pierce the host—usually a mammal—and withdraw blood. The tick stays put for long periods and gradually swells up as it sucks blood. More serious than the actual blood sucking is the fact that the ticks spread diseases.

Tiger The largest of the cats, the tiger is a magnificent striped beast ranging from Bali in Indonesia to Persia and Siberia. The tigers of Siberia and northern China are very large and may reach 12 ft in length, although 9 ft including tail is a good average for tigers from

Tiger

other areas. There are several races, differing in color, pattern and size. Tigers live in a variety of habitats, from mountains to dense forests. They do not climb well, but they are good swimmers. They eat a variety of animals, from insects and fish to pigs and antelopes.

Tilapia This name is given to about 100 kinds of freshwater fishes in Africa and the Middle East. They belong to the cichlid family and have large heads and deep laterally compressed bodies. Most of them are 8–12 ins long and they live in lakes and sluggish rivers. Tilapias eat mainly plant food and they themselves are eaten by men in many places. Most species carry their eggs and young in their mouths.

Toucan Toucans are large birds of tropical America with bright colors and huge bills which may be as long as the rest of the bird. The birds live in the forests, usually going about in small flocks. They feed almost entirely on fruit, supplemented by some insects. The bills are probably used for display and they are not connected with the diet.

Trapdoor spider Trapdoor spiders are related to the large, hairy bird-eating spiders, but instead of hunting for food they lie in wait for their prey. They construct silk-lined tunnels on the ground and fit them with lids of silk and soil particles. The spiders then sit peering out of the doors waiting to pounce on passing insects. Trapdoor spiders live only in the warmer regions of the world. They are not dangerous, although some species can give a painful bite.

Tree frogs The name given to a family of tree-dwelling frogs widely distributed with many different species. They have suckers on their toes and some have webbed hind toes.

Tree shrew There are about 20 species of tree shrews, small squirrel-like mammals living in South-East Asia. They are 8–16 ins long, about half of which is tail, and generally brownish. They live mainly in woodland and feed on a variety of plant and animal food. Tree shrews are generally included with the primates and it is thought that some creature of this type gave rise to the ancestors of monkeys, apes and man.

Tree squirrel Many squirrels live in trees, but the true tree squirrels are members of the genus *Sciurus*—the grey and red squirrels and their relatives. They live in most parts of the Northern Hemisphere. They differ from the ground squirrels in having bushy tails about as long as the head and body, although the tail is sometimes more feathery than bushy. They are remarkably agile in the trees, although they often wander over the ground as well. They feed mainly on fruit and buds, but also eat birds' eggs and fungi. They open hazel nuts very neatly by splitting them clean down the middle, although young squirrels have to learn how to do this and often find it difficult at first.

Tree frogs showing the little suction pads on their toes.

Tropic bird Tropic birds are graceful sea birds with two very long tail feathers. They look rather like terns, but are actually related to cormorants and pelicans. They are white with black markings and are up to 2 ft long without the tail plumes. These add up to 16 ins to the length. The three species all live in the tropical regions and come to land only in the breeding season. They fly over the oceans and dive for fishes and squids.

Trout The classic game fish of North America and Europe, they are closely related to salmon. Most species prefer cold-water streams or deep lakes. Rainbow, Dolly Vardin, brook, golden, cutthroat and lake trout are well-known American species. The European brown trout, "salmon trout," spends part of its adult life in salt water. Numbers of trout are raised in commercial fish hatcheries. Once lake trout were abundant in the Great Lakes, before the advent of water pollution and the parasitic lamprey fish.

Trunkfish Also known as boxfishes, the trunkfishes are almost completely encased in bony boxes. Only the tail and the fins are unenclosed. The head is more or less conical and has a small mouth armed with strong crushing teeth. The eyes are large and placed near the top of the head. Trunkfishes live in the warmer seas of the world and generally live on or near the bottom. Many of them feed by nibbling at coral reefs.

Tsetse fly Tsetse flies are African insects which carry the dreaded disease called sleeping sickness. This is caused by a tiny protozoan called a trypanosome. The flies also carry a disease of cattle and horses. It is called nagana and it prevented men from keeping cattle in many parts of Africa. Tsetse flies look rather like house-flies, but they suck blood. Unlike mosquitos and horse-flies, both male and female suck blood. The female does not lay eggs. Every now and then she gives birth to a fully developed maggot. This immediately turns into a pupa.

Tubeworms are bristle worms that live in tubes of various kinds. They are found buried in the sand or attached to rocks and weeds on the sea-shore. The chalky white tubes sometimes found on the surface of stones are made by tubeworms.

Tuna Tuna are large oceanic fishes of which there are several species. The bluefin tuna or tunny of the Atlantic and Mediterranean is said to reach lengths of 14 ft and weights of 1800 lb, although most specimens are not much more than half of this size. Tuna are very streamlined, deep-bodied fishes, dark blue above and silvery beneath. The hind part of the body bears a number of small finlets on the top and bottom surfaces. Tuna swim near the surface and feed on smaller fishes such as herrings. They are important as human food.

Turaco Turacos are lively fruit-eating birds related to the cuckoos. Most of the 18 species are about 18 ins long, with short rounded wings and longish tails. They have strongly curved bills and all but one species have large crests on the head. Turacos live only in the African forests. Some species are called plantain-eaters.

Turnstone Turnstones are small wading birds that get their name from their habit of turning stones over in their search for food. They are about 9 ins long, with short legs and a short, stout bill. The common turnstone, which breeds all around the Arctic regions, has a 'tortoiseshell' coloring in summer—black and white and chestnut. The black turnstone, which breeds only in Alaska, is nearly all black. Both species migrate far south for the winter.

Vampire Vampires are bats. They live only in tropical and sub-tropical America and they feed on the blood of birds and mammals. Contrary to popular belief, they are very small creatures—less than 4 ins long and weighing only an ounce or two. They have razor-sharp teeth, which aided by a somewhat anti-coagulating and anesthetic saliva, enables the bat to effect a painless wound which bleeds freely. The blood is simply lapped up as it wells to the surface. There are three species and they are quite dangerous creatures because they can carry the dreaded disease called rabies.

Tube worms

Velvet ant The velvet ant is not an ant at all: it is a kind of wasp. It gets its name because the female is wingless and looks like a large, hairy ant. There are hundreds of species, most of them living in the warmer regions. Velvet ants lay their eggs in the nests of bees and wasps, and the young velvet ants eat the larvae and pupae of their hosts.

Viperfish The deep-sea-viperfishes are slender pencil-like fishes up to 10 ins long. The lower jaw is relatively huge and both jaws carry long fang-like teeth. There are three species and they live in all the oceans except the Arctic and Antarctic. They live down to depths of 9,000 ft and feed on smaller fishes and crustaceans. The viperfishes are equipped with light organs which probably attract their prey.

Viscacha Viscachas are South American rodents related to the chinchilla. The plains viscacha is up to 2 ft long and may weigh 15 lb. It is dark grey, with black and white markings. The four species of mountain viscachas, sometimes called mountain chinchillas, are only about half as long as the plains species and weigh only about 3 lb. They are paler in color. All viscachas live in colonies, with the plains viscacha digging extensive burrows. Mountain viscachas live among rocks and boulders. They all feed on grasses and other plants and they can be pests in farming areas.

Vulture Vultures are scavenging birds of prey. True vultures live only in the Old World, but the name is also applied to the condors and their relatives in America, although these belong to a different family. All the vultures have more or less naked heads and most have naked necks as well. This is a useful asset because the birds plunge their heads and necks into carcases when feeding. Vultures rarely kill and they have rather weak talons. They soar around until they find a dead or dying animal and then they swoop down for a feast.

Wallaby Wallabies are kangaroo-like marsupials living in Australia and New Guinea. Most of the two dozen or so species are about the size of hares, although the brush wallabies may reach lengths of 3 ft excluding the tail. They are all vegetarians and feed mainly on grass. Since the Europeans introduced foxes and other carnivores to Australia many wallabies have become rare and some have become extinct.

Walrus The walrus is a heavily built seal in which the upper canine teeth are enlarged to form the long tusks. It feeds on shellfish which it digs out of the sea-bed. Walruses live in the North Pacific and the North Atlantic, those in the Pacific usually being somewhat larger. Adult bulls may weigh more than a ton.

Wapiti This is a large North American deer, rather like a large red deer. It reaches a height of 5 ft at the shoulder. The antlers may be more than 5 ft long. The wapiti was once an abundant animal, found right the way across the continent, but numbers have been much reduced by hunting and the remaining animals are nearly all in reserves.

Warbler There are about 300 species of warblers, named after the melodious song of many species. They are generally 4–5 ins long and have fine-pointed beaks. The plumage is rather dull, with greens, browns and greys predominant. Warblers are found throughout the Old World, from Europe to Australia. Many of them migrate. The American wood warblers belong to a different group.

Warthog One of the ugliest of animals, the warthog is a relative of the domestic pig. The head is long and armed with warts and a pair of upward-curving tusks. The male is about 5 ft long and about 28 ins high at the shoulder. The animals have an 18 in tail which is carried erect when they run. Warthogs are found in open country throughout much of Africa south of the Sahara. They feed mainly on grass, but also eat berries and occasionally take a meal of carrion.

Waterbuck This is a large antelope always found near rivers, although it does not live in marshy ground. It is about 4 ft high and has a coarse brown coat. The males have long, slender horns which produce a horse-shoe shape when viewed from the front. There are

189

two sub-species, one with a white patch on the rump and one with a white ring round the rump. Waterbuck range over much of Africa south of the Sahara.

Water scorpion This is a flattened aquatic insect, in no way related to ordinary scorpions. The front legs are thickened and strongly hinged for grasping prey. The tail end of the insect is drawn out into a long breathing tube which can be pushed up to the surface of the water. Water scorpions live on the bottom of shallow muddy ponds and resemble dead leaves when seen from above. They eat insects and other small creatures. The water stick insect is a long slender relative of the water scorpion.

Water snake Water snakes (not sea snakes) are worldwide. Many species are native to the U.S., particularly the southern lowlands. The northern banded water snake is common along woodland ponds and streams. Water snakes are quick to bite if touched, although not venomous.

Water spider Many spiders can live temporarily under water, supported by the film of air trapped around the body, but only one species lives permanently under water. It is a perfectly ordinary-looking spider, although the male is normally larger than the female—the reverse of the normal situation in spiders. It is reddish-brown with a grey abdomen. It is able to live under water by constructing a silken diving bell into which it releases bubbles of air collected at the surface. The water spider ranges through temperate Eurasia.

A male wapiti

Wattlebird Wattlebirds have rounded wings, long tails, and long hindclaws, with wattles, usually orange, at the corner of the mouth. They feed mainly on fruit leaves and insects. There are two species, both living in New Zealand and both very rare.

Waxwings Named after the red tips of their secondary flight feathers, which look like blobs of sealing wax, the three species of waxwings are found in the northern half of North America and northern Eurasia. Waxwings are about the size of starlings, about 6–7 ins long, with prominent pointed crests. Their main diet is berries and insects, and even during winter, waxwings have been seen to feed each other as they do during courtship.

Weasel The weasel, found throughout Eurasia and North America, is similar in form to its near relative the stoat, although somewhat smaller, averaging about $8\frac{1}{2}$ ins plus a $2\frac{3}{4}$ in tail. It has a slender body, short limbs and long neck, giving it a snake-like appearance. The fur is reddish-brown with white throat and underparts. Weasels are carnivor-

Opposite: A young white-tailed deer

ous animals sometimes killing animals larger than themselves, although they feed mainly on voles and mice.

Weaver bird The weavers are small, mainly seed-eating birds which live in Africa and Asia. There are about 70 species and many of the males have bright plumage during the mating season, reverting to the drab streaky plumage of the females outside it. Weaver birds live in various types of tree country and make flask-shaped nests by weaving leaves and grasses together.

Weevil Weevils are beetles in which the head is drawn out into a snout. Most weevils are about one-eighth of an inch long, although there are tropical species up to 3 ins. Weevils are found all over the world and usually feed on some particular species or genus of plant, a notable example being the cotton 'boll weevil' of American cotton plantations. There are already 40,000 species of weevils known to man and several hundred new ones are being discovered every year.

Whale shark The largest of all sharks, the whale shark grows to a length of 50 ft, and even 70 ft monsters have been seen. The skin is about 6 ins thick. The animal is dark grey, with whitish spots all over the upper surface. The head is broad and flattened, and the huge mouth contains hundreds of tiny teeth. Although the mouth could easily engulf a man, the whale shark is harmless. It lives in tropical seas and feeds on plankton.

Whelk This name is given to many kinds of sea snails, but the true whelk is a widely distributed species with a thick whitish or yellowish shell up to 6 ins long. It lives all over the sea-bed below low tide level and has long been used as food. It has a long breathing siphon, which draws water over the gills, and it feeds by rasping away at various living and dead creatures. Whelk egg cases are often washed up on the shore. They look like lumps of sponge.

Whirligig beetle This is a small, shiny beetle that lives on the surface of still water. It skates round and round searching for small insects that fall on to the water. The eyes are in two parts, one part being used for seeing on the surface, and one part for looking down into the water. There are about 800 species of whirligig beetles.

White-tailed deer These deer were almost exterminated in North America through years of slaughter by settlers. Now they are so abundant that their numbers are thought to be greater than prior to the arrival of Europeans in North America.

Wallabies

Whiting An important food fish, this member of the cod family reaches a length of about 28 ins. It is easily distinguished by its silvery sides and the black spot at the base of the front fin. The whiting is found in the North Atlantic, the North Sea, and the Mediterranean. It eats crustaceans and small fishes.

Whydah Whydahs or widow birds are African seed-eating birds that lay their eggs in the nests of weaver finches. There are about 11 species. The females are drab and sparrow-like, but the males have striking breeding plumage, often with very long tail feathers.

Wild cat The European wild cat resembles a heavily built tabby cat, about 2 ft long excluding the tail. The latter is thick and bushy and distinctly ringed. Wild cats live in the upland regions of Europe and Asia Minor. The American wildcat, or "bob-cat," is so called because of its short stumpy tail. Grouse, rabbits and hares are the main food of these fierce carnivores.

Wobbegong Also called carpet sharks, the wobbegongs are quite unlike the normal shark shape. They have stout flattened bodies and broad heads, with a rough mottled skin. They rest on the sea-bed and look like seaweed-covered rocks—perfectly camouflaged to pounce on unsuspecting fishes. There are five species around Japan, China and Australia. Most are small fishes, but some reach a length of 10 ft.

Wolf Possibly an ancestor of the domestic dog, the wolf once covered almost all of the Northern Hemisphere. It probably had a greater range than any other land mammal. Today it has disappeared from all but the wildest parts of Western Europe, but it still ranges over some

Waterbuck

of Asia and North America. Its body is about 4 ft long and it is about 38 ins high at the shoulder. Wolves live in forests and in open country and they hunt by day, usually in small packs. Their endurance allows them to wear down all kinds of animals, and wolf packs have been known to bring down a moose.

Wolf spider These spiders are so-named because, instead of making webs, they run down their prey like wolves. They are smallish spiders, less than 1 in long in the body, and they are all rather drab. Females carry their eggs around in spherical sacs of silk which are attached to the hind end of the abdomen. Wolf spiders are very widespread. The famous tarantula of Southern Europe is a wolf spider.

Wolverine The wolverine or glutton is the largest of the weasel family, although it looks more like a badger or a small bear. A fully grown male may reach 4 ft in length, including about 1 ft for the tail. It may weigh up to 60 lb. It has a shaggy coat of dense dark brown fur, powerful limbs, and strong claws. It lives in the coniferous forests of the Arctic and Sub-Arctic regions and feeds on any animal it can catch. It will also drive other carnivores away from their kills and take over the food.

Wombat This is a marsupial related to the koala bear. It is a stocky, short-legged creature up to 4 ft long and has strong claws. The two species live in southern Australia and behave rather like badgers, hiding in their burrows by day and coming out to feed at night. Wombats feed mainly on grass and roots.

Wood ant This large reddish brown ant lives on heathland and in dry woodlands in Europe. It is noted for the great piles of pine needles and other debris that it builds up over its underground nest. Wood ants are carnivorous creatures, feeding on a wide variety of other insects. They also 'milk' aphids on the trees around their nests.

Woodchuck The woodchuck is a large North American rodent related to the squirrels. It is up to about 2 ft long and weighs between 5 and 10 lb according to the season. It is a golden or reddish brown animal and lives in deep burrows in woods and farmlands. It is a solitary creature and feeds mainly on grasses and other low-growing plants. It will also eat fruit and it does some damage to crops.

Woodcock This is a wading bird related to the snipe. It is a brown and black bird, about 13 ins long, with a straight 3 in bill and a short neck. The wings are rounded. There are two

Spectacled weaver bird

species: the Eurasian woodcock, which extends from the British Isles across to Japan, and the American woodcock, which lives in the eastern half of North America. Although they are wading birds, woodcock live mainly in woodland undergrowth. They feed mainly on earthworms and other ground-living invertebrates.

Woodlouse Woodlice, also called sowbugs, bibble-bugs, sink-lice, and many other quaint names, are crustaceans not too distantly related to the crabs. They are the only group of crustaceans that have successfully invaded the land. Never more than about three-quarters of an inch long, they have a small head and abdomen, but the thorax is large and makes up most of the body. It has seven segments, each with a pair of legs and a hard shield on the upper side. Some species can roll into a ball when disturbed. Although they live permanently on land, few species can withstand dry air and they live in damp places. They feed on a variety of decaying materials.

Woodpecker There are about 200 species of woodpeckers, spread over most parts of the world except the Australian region. No birds are better adapted for life on tree-trunks and branches. They range up to 2 ft in length and are usually brightly colored. The bill is straight and pointed. The legs are short and the tail is made up of stiff feathers which act as supports when the bird clings to a trunk. The birds have harsh calls and they also give themselves away by their rapid hammering on the wood. Woodpeckers spend most of their time searching for insects on tree-trunks. The bill is used to pierce pieces of bark and the long tongue is then used to pick up the insects. Nearly all woodpeckers nest in holes which they drill with their bills.

Woolly monkey A relative of the spider monkey, the woolly monkey lives in South America. It is about 18 ins long, with a tail of about 2 ft, and it weighs up to 20 lb. The animal is covered with close, woolly fur and it has a black, rather human-looking face. The

Wombat

rest of the body is greyish brown. There are two species, but one is very rare. The monkeys live high in the trees and eat leaves and fruit.

Wryneck This bird is a relative of the woodpecker, but it looks more like a thrush, especially when feeding on the ground. It is about 6 ins long, with mottled brown plumage. Although in the woodpecker family, it does not have a strong bill and it does not have stiff tail feathers. Wrynecks feed on insects and are particularly fond of ants. They nest in holes in trees, but they have to find ready-made holes as they cannot drill their own. There are two species, one breeding in Europe and Asia and one breeding in the southern half of Africa. The Eurasian species migrates south for the winter.

X-Ray fish This little fish from the equatorial rivers of South America is almost completely transparent. Much of its skeleton can be seen from the outside. It is an attractive fish and popular for aquaria. Although only 2 ins long, it is fiercely carnivorous and it is actually closely related to the famous piranha.

Yak

Yak This is a large member of the cattle family that lives high in the mountains of Tibet—one of the most inhospitable regions of the world. A wild yak bull may stand well over 6 ft high and weigh $\frac{3}{4}$ ton. The animals have long, thin horns which extend sideways and may reach 3 ft in length. They also have long, shaggy black coats which reach almost down to the ground. Yaks graze on the sparse grasses and other plants of the mountains. They have been domesticated as beasts of burden and also to provide milk. Domestic yaks, smaller and paler than the wild ones, are found in many parts of the Himalayan region.

Zebra Closely related to horses and asses, zebras are easily distinguished by their stripes. There are three species, all living in Africa. The commonest species is Burchell's zebra, in which the stripes reach under the belly. There are several races of this zebra. In the mountain zebra and Grevy's zebra the stripes do not extend under the belly. A fourth zebra, known as the quagga, was common until about 150 years ago, but it is now extinct. Like horses, the zebras are grazing animals. They are the favorite prey of lions.

Zorille The zorille looks and acts like a skunk, although it is only distantly related to the American skunks. It is more closely related to the weasels and polecats. It is a slender black and white animal about 2 ft long, the bushy tail accounting for nearly half the length. Like the skunk, it ejects a foul-smelling fluid when disturbed. The zorille lives in most parts of Africa and feeds mainly on small rodents and insects.

PART 4

omes and Habitats

ants and animals are not scattered haphazardly over the surface of the Earth. Each ecies has its own preferred *habitat*. Habit is just another way of saying surroundgs. *Environment* is another word meaning ich the same thing. The habitat is made up the climate, the soil, the air or water, and e other plants and animals. The climate d the soil are the most important factors cause they control the vegetation. The getation in turn affects the distribution the animals, because animals will go ere they can find food and shelter. The idy of habitats and of the plants and anials that live in them is called *ecology*.

Some of the Earth's major habitats are scribed in the following pages together th examples of how the plants and animals e adapted for life under various conditions.

The Earth provides a variety of homes for living things ranging from the cold, bleak wastes of the tundra (above), through forest retreats (left) and open grasslands (right) to sun-scorched deserts (below). In all types of environment plants and animals are found adapted to the prevailing conditions.

Life on the Grasslands

The major grasslands of the world occur on each side of the desert belts, in regions where there is a certain amount of rain but not enough for the growth of forests. There are two main types of grassland: temperate and tropical.

Temperate Grasslands

The temperate grasslands are outside the tropics and most of them are found in the inner regions of the continents, where rainfall is low throughout the year. The summers are generally very hot and the winters are generally very cold. Among the major temperate grassland areas are the prairies of North America, the pampas of South America, the veld of South Africa, and the steppes of Eurasia. Smaller areas are found in eastern Australia.

The nature of the grassland depends upon the amount of rainfall. Where the rainfall is moderately high, the grasses are tall and lush. Areas with less rain support shorter grasses and fewer of them. A large proportion of the temperate grassland has been taken over for agriculture and stock rearing.

Tropical Grasslands

Tropical grasslands, also called savanna, are found in and around the tropical areas, between the tropical forest and the deserts. The largest areas are in Africa and South America, and smaller areas occur in India and northern Australia. Rainfall is markedly

Above: The sable antelope a rather rare animal of the African grasslands.
Below: the distribution of the major grasslands.

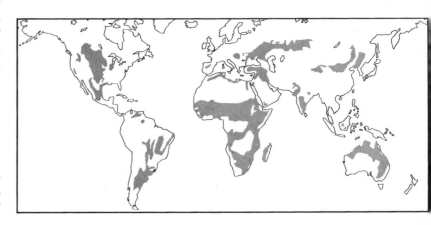

seasonal and falls only for a short period during the summer. The rain and the high temperature produce rapid growth and the land quickly becomes covered with lush grass. Some of the grasses reach heights of ten feet. Then, as the dry season progresses, the plants die down and the land takes on a rather barren appearance for several months. Acacia and baobab trees are dotted here and there over the savanna. They get more common towards the forest regions and rarer towards the deserts.

The Animals of the Grasslands

Grasses are very hardy plants, able to withstand or recover from grazing, trampling, fire, and drought. They produce abundant leaf and they can therefore support a large number of animals. The dominant animals of the grasslands are the large grazers. These include horses, antelopes, zebras, bison, and kangaroos. Not all of them live in the same place, of course. The savanna lands, where there are trees and bushes, also

support elephants, giraffes, and rhinos. Grasslands do not provide much cover for the animals, and we find that most of the large grazers are fast runners. This is their way of escaping from their enemies. Most of them also live in herds, allowing some individuals to feed while others keep a watch for enemies.

The grasslands also support many smaller mammals, such as voles and prairie dogs. These feed largely on seeds. They escape from their enemies by burrowing instead of running away. There are also many insects, notably the grasshoppers and locusts. These feed directly on the grass and other plants and they themselves are eaten by birds, lizards, and small mammals.

Best known of the carnivorous mammals of the grasslands are the various large cats —the lions and leopards and tigers. The lions and leopards of Africa feed mainly on

The elephant is found over much of Africa, in both grassland and forest. It is entirely vegetarian.

Above: The water hole is very important for the grazing animals. Zebras and wildebeest have come to drink but some individuals are always on the look-out for danger.

Right: A pride of lions. Lions are rather lazy animals and spend a lot of time lying in the shade They kill only when they are hungry.

Right: The coyote or prairie wolf of the American prairies. It feeds mainly on hares and small rodents.

Left: Termite nests are a dominant feature of some areas of tropical grassland. Made of soil cemented by saliva, they may rise 18 feet above the ground.

the antelopes and zebras. Another famous hunter is the cheetah. The fastest land animal in the world, the cheetah can reach 70 mph as it bounds after its prey. It cannot keep this up for long, however, and it gives up if it does not catch its prey within about 400 yards. Other well known hunters include the coyote or prairie wolf of North America and the Cape hunting dog of South Africa. The grasslands also have their full complement of scavengers waiting to pounce on a dead or dying animal or to clear up after another carnivore. These scavengers include the jackals and hyenas of Africa and the various kinds of vultures.

Water is always a problem to animals living in the grasslands, especially those in the tropical grasslands where rain falls for only a short time each year. Water holes, formed where water-bearing rocks come to the surface, are very important in the grasslands. Many of the animals come to drink daily and they often come at definite times, each species or group of species coming in turn. When the water holes eventually dry up, as they often do towards the end of the dry season, the herds move on to other regions. One hundred years ago there used to be mass migrations of antelopes in herds many miles long. There are far fewer animals now and the need to migrate is not so pressing, although one can still see migrating animals. The smaller animals often burrow down into the ground and sleep for the driest part of the year. In the temperate grasslands, where the winters are very cold, many of the smaller animals hibernate.

195

Life in the Rain Forests

The low-lying equatorial regions of the Earth have high temperatures and heavy rainfall throughout the year. Such conditions are ideal for plant growth, and we find here the richest of all habitats—the tropical rain forest. Rain forest develops where the temperature is always above 65°F and where the rainfall is at least 80 inches per year. Most rain forest regions actually receive between 100 and 500 inches of rain each year, and it generally rains every day. These figures compare with about 40 inches of rain per year for most parts of the British Isles.

There are three main regions of tropical rain forest, shown on the map.

Hundreds of Species

Because there is no cold season, the rain forest plants can grow throughout the year and they are all evergreens. The abundant moisture means that the plants grow rapidly. The plants do not need special features to withstand drought or cold, and so a great many kinds of plants are able to grow in the rain forests. The woods and forests of Europe and North America contain relatively few kinds of trees and they are usually dominated by one or two species. Oak woods, for example, may contain little but oak trees. Rain forests are much more variable, however, and there may be as many as 300 different kinds of trees in one square mile of forest.

Although the rain forests in different regions contain different species of trees, all rain forests look much alike. The main roof,

The jungle fowl is a native of South-East Asia. It is probably the ancestor of domestic chickens.

Below: Tapirs are shy plant-eating forest dwellers of South America and South-East Asia. They live near to water and are excellent swimmers.

or *canopy*, of the forest is generally about 100 feet high. There are scattered taller trees, reaching 200 feet perhaps, and there is also a lower layer at about 50 feet. Even the trees themselves look much alike, although they belong to many different families. They generally have straight trunks, with smooth grey bark and buttress roots to support their great weight. Many of the trees bear glossy, oval leaves rather like those of the laurel. They usually have sharply pointed tips, called 'drip-tips', which allow the water to run off easily. The flowers also tend to look alike, most of them being small and rather pale in color. Visitors to the rain forest might well think

Temperate Rain Forests

Several parts of the world experience very high rainfall without the high temperatures that are found in the tropics. As long as the temperatures are not too low, these areas will support temperate rain forest. The best examples of this type of vegetation are found on the west coast of North America, the west coast of New Zealand, and along the coast of southern Chile. Like the tropical forests, the temperate rain forests are largely evergreen, but the trees are nearly all cone-bearing species. They include the largest of all trees—in fact, the largest of all living things. These are the giant redwoods and sequoias, which reach well over 300 feet in height. Epiphytes are abundant in these forests, but they are mainly ferns and mosses. They reach further down the trunks than the tropical epiphytes, and the whole forest looks much greener than the tropical rain forest. Evaporation is not rapid because of the lower temperatures, and so the forest literally drips with water. The low temperatures also slow down the rate of decay, and the forest floor is littered with dead trunks and branches, all covered with ferns and mosses. The temperate forest is much more difficult to walk through than the tropical rain forest.

that there are only a few kinds of trees there, and even the botanist is sometimes fooled by the similarities of the trunks and leaves.

A Gloomy Place

Along the banks of rivers, and where man has interfered by cutting down trees, the rain forest is thick and often impenetrable. This is the 'jungle' that we often hear about. Further into the forest, however, much of the undergrowth disappears because there is not enough light for it to grow. The dense overhead canopy cuts off almost all the light and only a few large-leaved plants can survive on the ground. This is the true rain forest. One can walk through it without much trouble, but it would be a rather gloomy walk, with little to see apart from the tall straight tree trunks. There is not even a layer of dead leaves to kick about, for the warm damp atmosphere ensures that leaves decay almost as soon as they fall.

Treetop Orchids

Most of the rain forest's abundant life is to be found in the branches of the trees, 50 feet or more above the ground. This is where the light is and, as well as the leaves and flowers of the trees themselves, there are thousands of small herbaceous plants growing on the branches. These plants have their own leaves and they make their own food. They merely use the trees as perches to enable them to get some sunlight. Plants which grow in this way, without taking any food from the trees on which they live, are called *epiphytes*. The most famous of them are the orchids, whose beautiful flowers festoon all the tropical rain forests. Other common epiphytes include ferns and bromeliads (relatives of the pineapple).

The roots of many epiphytes are thick and spongy and they absorb water directly from the moist air. Other epiphytes, such as the bromeliads, trap water in their funnel-shaped leaves. These water traps provide water for the plants, and they are also homes for mosquito larvae and for the tadpoles of various tree frogs. The bromeliads

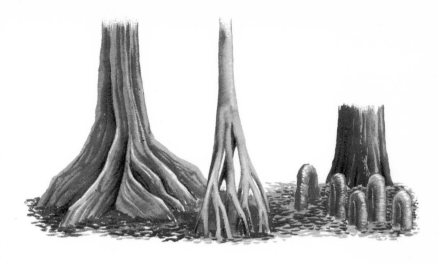

Some trees in the swampy rain forests have roots which push above the wet ground and take in oxygen (right). These are called pneumatophores. Other trees are lifted into the air on stilt roots (center). The large trees have buttress roots to support their great weight.

The African grey parrot.

The distribution of the rain forests.

live in the South American rain forests and they are often so numerous that their communities have been described as 'treetop marshes'.

The epiphytes get their mineral requirements from the droppings of the many birds and other animals that live in the canopy.

Another very common type of plant in the rain forest is the liana. There are many different kinds, but they are all climbing plants and they reach the light by twining up the tree trunks. Many of them have drooping branches which sometimes reach ground level.

Animals of the Rain Forest

Although the visitor to the rain forest is usually struck by the quietness of the place, it does support an abundance of animal life, especially among the upper layers of the vegetation. Each layer has its characteristic animals. Large birds of prey, such as the South American harpy eagle, live in the uppermost branches and swoop down to catch smaller birds and monkeys. The latter live in the main canopy, but they wander through most of the layers in their search for fruit and other food. The brilliant colors of the birds, together with the epiphytic flowers, make a vivid contrast with the lower levels of the forest. Snakes, such as the boas, also live in the trees, and so do many amphibians. Small antelopes and deer live on the ground, together with various cats. The latter are also very good at climbing the trees. Rain forests are the homes of the largest insects, including huge beetles and cockroaches, stick insects a foot long, and the beautiful morpho and birdwing butterflies. It may never be known just how many kinds of plants and animals there are in the rain forests, but we can be certain that the stability of the rain forest depends on its complexity. If man interferes with it too much, it will break down and never build up again.

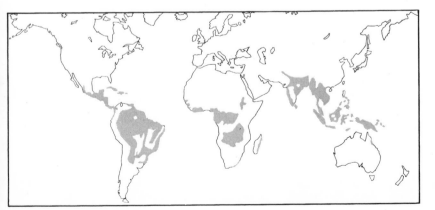

Desert Life

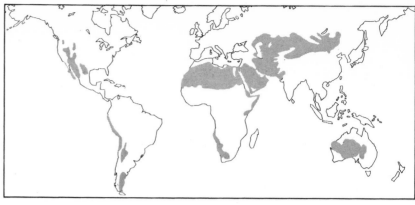

A desert is an area which receives very little rain—less than 10 inches per year on average. The sparse rainfall may be spread out over the year, but more often it comes down in irregular heavy showers. Some desert areas may go for several years without any rain. Although there are some deserts in the colder parts of the world, in Siberia for example, most of the deserts are found in and around the tropics. Because there is so little cloud to absorb the sun's heat, these desert areas get extremely hot during the daytime. They cool down rapidly at night and there may even be frosts from time to time.

The plants and animals living in the deserts therefore have two main problems —shortage of water and great heat. We tend to think of the deserts as rather lifeless places, but a surprising number of plants and animals manage to make a living there.

Desert Plant Life

Desert plants solve the water problem in several different ways. Many of the smaller plants are known as *drought evaders,* because they avoid the dry periods altogether.

Above: Nearly all desert regions have some form of jumping rodent. They all have long tails which they use for balance when leaping through the air. This is a jerboa.

The dry season, which may be as much as 11 months of the year, is passed in the seed stage. When the rain eventually comes these seeds start to grow very rapidly. They produce leaves and flowers very quickly and, by the time the dry season sets in again, they have already scattered a new crop of seeds. One small plant living in the Sahara can scatter its seeds only 10 days after growing from seed itself.

The other desert plants are *drought resisters.* A few plants, such as the creosote bush of the American deserts, can actually survive being shrivelled by the sun, but most of the plants need a supply of water. The mesquite tree and some others have very long roots that reach down to the water table way below the desert floor. These plants are never short of water and they do not need special features to help them to live in the desert. Cacti and many other desert plants have shallow, spreading roots which absorb large amounts of water as soon as the rain falls. The water is then stored in juicy leaves or stems and it is used sparingly during the dry season. Plants that store water in this way are called *succulents* and they are often covered with waxy deposits which cut down the evaporation of water. Some also have hairy coverings which reflect away some of the heat as well as cutting down water loss.

A plant loses most of its water through its leaves, and many desert plants cut down

The cacti are the best known of the desert plants. The great saguaros tower up to 50 feet above the desert floor. The spiny cacti provide a safe refuge for many small animals such as the cactus wren (below right).

the loss of water by having very small leaves or no leaves at all. The leaves are often modified into woody spines, which reduce water loss and also protect the plants, from browsing animals. Some desert plants, such as the ocotillo, produce leaves only in the wet season and drop them again as soon as it gets dry.

Desert Animals

The desert may seem quite empty of animals if it is visited during the heat of the day, but it 'comes alive' towards evening as its numerous inhabitants venture forth into the cooler air.

As well as bringing forth the short-lived annual plants, the periodic rains bring out the insects. Bugs, beetles, butterflies and other insects emerge from their eggs or pupae. The adult insects help to pollinate the flowers, while their young make rapid growth at the expense of the leaves. Within a short time the cycle is complete and another generation of eggs and pupae is left in the ground to await the next rain. Many of these young stages will never develop any

further for they will be found and eaten by the many lizards of the desert. The lizards, in turn, may be eaten by snakes or scorpions, or by some of the carnivorous mammals or birds.

Frogs and toads are usually associated with damp places, but even some of these creatures have managed to adapt themselves to the desert. The spadefoot toad, so called from a spade-like projection on its hind feet, digs itself down into the desert sand and spends up to 11 months in sleep. Woken up by the rains, it makes its way to a temporary pond where the eggs are laid. The eggs hatch almost at once and the tadpoles grow so quickly that they turn into little toads within three or four weeks. They then bury themselves away when the dry season starts.

Few birds live in the desert because, unlike the other animals, they can not burrow into the ground to get away from the hot sun. There is a burrowing owl in the American deserts, but it uses the old holes of other animals and does not make its own burrow. Several other birds such as

Right: The camel is often known as the ship of the desert. It can go for long periods without water because its hump or humps contain a lot of fat which releases water when broken down. The animal also sweats very little. There are two camel species —the one-humped and the two-humped. These are the two-humped camels of Central Asia.

Above: The prickly pear cactus in flower.

Left: The rattlesnake of the American deserts.

Right: Date palms are familiar and very important plants in the desert oases of Africa and the Middle East.

The spiky yucca, native to the North American deserts, is a member of the lily family.

the cactus wren, make use of the shade provided by cacti.

Among the desert mammals, the most abundant are the little rodents—the mice and voles and their relatives. They feed on seeds, insects, and occasionally leaves. Many of these rodents never drink. They get all the water they need from their food. Insects provide quite a bit of water, but seeds contain very little and the rodents have to be very careful about conserving water. Hiding by day and coming out at night helps to conserve water because the air is cooler and damper at night. The animals do not then lose water by evaporation. They sweat very little, if at all, and they pass very little urine. The kangaroo rat has a clever way of increasing its water intake. It eats seeds, but it does not eat them until they have been stored deep in its burrow for a time. During this period of storage the seeds absorb moisture from the damp burrow, and the animal thus gets extra water. Several of the desert rodents go to sleep in their burrows during the hottest and driest part of the year.

199

Life in Northern Forests

Cool forests occur mainly in the Northern Hemisphere—in Europe, northern Asia, and the northern half of North America. The equivalent latitudes in the Southern Hemisphere are generally too dry to support forests. Small patches occur in South America and Australia, but most of these areas are covered with grassland.

The northern forests are of two main types: deciduous and coniferous. The deciduous forests grow in the warmer regions—the temperate zones. Coniferous forests, which are mainly evergreen, grow in the colder parts. The two types of forest differ a great deal in the animal and plant life that they support.

The Deciduous Forests

Deciduous forests are forests composed mainly of trees that drop their leaves in the autumn. These include oak, beech, hickory, ash, and maple. They require fairly even rainfall throughout the year, and winters that are not too severe. Such conditions are found in Europe, in the eastern half of the United States, and in eastern Asia. The central regions of Asia are too dry for deciduous trees and the grasslands merge northwards into the coniferous forests.

There is plenty of light in these woodlands in the spring, before the trees have spread their leaves. The light allows a rich variety of plants to grow beneath the trees. The types of plants vary from place to place according to the soil, but they include primroses, bluebells, lesser celandines, wood anemones, and many other attractive flowers. There are also many ferns.

Distribution of coniferous and broad-leaved forests in the Northern Hemisphere.

Squirrels are ideally adapted to a forest life. Sharp, curved claws enable them to climb trees rapidly and powerful back legs help them to leap from tree to tree using their outstretched tail for balance.

Woodpeckers (left) feed on insect larvae. They drill holes in the bark of trees with their spear-like beaks and scoop up the larvae with their sticky tongues. Nuthatches (right) wedge hazel nuts in bark crevices and hammer them open with their beaks.

The moist conditions allow dead leaves and twigs to decay quite rapidly. The layer of decaying material, known as the leaf litter, is full of small animals. There are worms, mites, spiders, insects, slugs, snails, and centipedes. Shrews are very common and so are hedgehogs. Both find plenty of food in the shape of worms and insects in the leaf litter. Mice and voles are also plentiful. They feed on seeds and other plant material, together with some small animals.

Up in the trees there is a different community. Hordes of caterpillars and other insects feed on the leaves. Bark beetles and other wood-borers eat into the wood. These insects are snapped up by the huge flocks of birds that live in the woods. The insect-eating birds include titmice, warblers, flycatchers, and woodpeckers. Other birds eat the buds and seeds of the trees. These birds include wood pigeons, finches, and nuthatches. The latter cleverly wedge hazel nuts into bark crevices and hammer them open with their beaks. Then there are the predatory birds, including sparrow hawks and owls, that feed on the smaller birds and mammals. The trees also support plenty of squirrels, which feed on fruit, bark, and occasional birds' eggs.

200

Among the larger mammals, there are deer, bears, badgers, foxes, and beavers. The latter live mainly in the North American forests and are remarkable for their tree-felling activities.

The Coniferous Forests

The coniferous forests grow in the colder parts of the world, extending far into the Arctic Circle in parts of Europe and Asia. They consist mainly of pines, firs, spruces, and larches. Most of the trees are evergreens, although the larch is deciduous. The cold soil does not allow decay to proceed very rapidly, and so there is a deep layer of dead leaves on the forest floor. The rain water becomes acidic as it drains through the peaty deposits. It removes useful minerals from the surface layers and the soil becomes very poor.

The poor soil, together with the low level of light in the coniferous forests, means that few plants can grow under the trees. There are some mosses and lichens, and there are plenty of fungi, but there are hardly any flowering plants. The forests are therefore relatively poor in animal life, although they are not as poor as some people suggest.

Quite a number of insects feed on the leaves, and many beetle grubs tunnel into the bark and the underlying wood. These insects attract plenty of birds, such as goldcrests, titmice, and woodpeckers. A few birds feed on the seeds of the trees. The best known of these birds are the crossbills, whose strange beaks are perfectly adapted for tearing open the cones. Owls and other birds of prey are also plentiful in the forests.

Small mammals are less common than in the deciduous forests because there is not so much for them to eat in the leaf litter. There are some shrews and voles, however. Larger mammals include many squirrels, some ground-living and some living in the trees. The European red squirrel prefers coniferous forests and nowadays it leaves the deciduous forests to the introduced grey squirrel. Also in the trees there are martens and wild cats. Both prey on birds and squirrels as well as on other small mammals. Deer are common and they sometimes do much damage by browsing on the young trees. In the more northerly parts they are kept in check by wolves, but the forester's gun has to be used to reduce numbers in other places.

The badger is a shy woodland dweller in North America, Europe and Asia. Badgers build a large den with many entrances called a sett. They search at night for worms, frogs and mice which form their main diet. They also eat plants and roots.

Mountain Life

Climbing in the Andes in the year 1802, the German explorer Alexander von Humboldt reached a height of 18,096 feet above sea-level. The mountain was Chimborazo, a volcano in Ecuador, very close to the Equator. No man had ever climbed to such a height before.

But Humboldt's achievement was more than a mountaineering triumph. Throughout the climb he had taken detailed scientific readings with the most up-to-date instruments available to him. He found that as the height increased the temperature and the air pressure fell. What is more, he noticed how the changes in physical conditions were reflected in the plant life. Luxuriant tropical rain forest at sea-level gave way to forests of broad-leaved deciduous trees. The deciduous forests in turn gave way to the coniferous forests of pines and firs. Higher still it became too cold for even the conifers to survive. Here there were only stunted shrubs and small plants. Finally, a region of permanent snow and ice was reached. There were no plants growing here at all.

These different belts of plant life on a mountainside are just like those we find if we make a journey from the Equator to the polar regions. The similarity is due to the similar fall in temperature towards the poles and towards the summits of mountains.

Right: Rocky Mountain goat

Below: The mountain avens is a typical alpine plant, forming mats or cushions of tough leaves on the rocky ground.

Alpine Plant Life

Like the Arctic tundra, the mountain peaks experience long and bitterly cold winters, punctuated by brief summers. In addition, the slopes are exposed all the year round to strong gusts of wind which remove much of the warmth from the ground. These severe conditions allow only very specialized plants to grow on the upper slopes of the mountains. Lichens are common, of course, for they can grow almost anywhere. Grasses are common, too, and there are also many broad-leaved flowering plants which are collectively called alpines.

The alpine plants belong to many different flower families but, living under the same conditions, they tend to evolve along similar lines. They are small plants, generally forming mats or cushions very close to the ground. In this way they avoid the full blast of the wind and they also gain some heat from the earth. Despite their very short stems, they have long roots. These are

Man in the Mountains

Mountaineers climbing the highest peaks—more than 21,000 feet above sea-level—take oxygen with them to help them to breathe. The air is very thin (rarefied) at high levels and there is not enough oxygen for a man to work properly. Even at a height of 10,000 feet a lowlander would find it very difficult to move quickly. But some races of men live and work quite normally at heights of up to 18,000 feet. These races have lived in the mountains for generations and their bodies have become adapted to the conditions. They have barrel-shaped chests and large lungs, taking in more air and oxygen at a time. Their blood is also different from that of a lowlander in that it has more red corpuscles to carry the oxygen around the body. Mountain peoples, typified by the Sherpas of the Himalayas, tend to be fairly short and stocky. This cuts down the loss of heat from the body surface and it also speeds up the circulation, allowing heat and oxygen to reach all parts of the body with ease.

Ice Age Remnants

Many of the plants found in the European Alps are closely related to plants in the Caucasian Mountains and in the Arctic tundra. These regions are hundreds of miles apart and we have to look back to the great ice age for an explanation of the peculiar distribution of these plants. During the ice age there were sheets of ice over all of Northern Europe and glaciers also spread out from the mountainous areas. Large areas to the south of the ice sheets must have been rather like today's tundra, and alpine plants must have been widely distributed. As the ice melted and the land warmed up, the alpine plants shrank back to the far north or to the mountain tops where we now find them. The alpine plants are therefore survivors from the ice age.

necessary for secure anchorage on the sloping ground and also to draw up water, which drains rapidly from the stony ground.

The densely packed leaves of the alpines act rather like the fur of a mammal and they trap moist warm air around the plant. Coatings of wax and hair also help to cut down water loss and to protect the plants from the cold. The alpines cannot make food in the winter, when they are covered with snow, so they must make the best of the sunlight they receive in the summer. They tend to have large amounts of chlorophyll in their leaves, making them rather dark in color. The flowers also tend to have deep colors. Both features help the plants to absorb warmth from the sun.

Above: The Rocky Mountain goat (left) is a North American relative of the chamois. The Alpine ibex is a wild goat found in a few places in the Alps. The argali (right) which lives in the mountains of Central Asia is the largest of the wild sheep.

Some of the world's largest birds live in the mountains. The golden eagle has a wing span of 7 feet or more.

Many of the alpine flowers are very attractive and large numbers are cultivated as 'rock plants'.

Nearly all alpines are perennials. The conditions on the mountains are generally too severe for plants to complete their life cycles in one year. Growth is very slow because of the low temperatures and short growing season. Ten years may pass before a plant has stored enough food to flower.

Mountain Animals

A few insects, such as the Apollo butterflies, are found on the higher mountain slopes. Spiders are found there too, but only the warm blooded animals—the birds and mammals—are really successful in this cold habitat. They maintain their bodies at a constant high temperature, regardless of the outside temperature.

The most conspicuous of the upland creatures are the large grazing animals, such as the yak, the chamois, and the various kinds of sheep and goats. During the summer they wander high on the mountains, but most of them move down into the forests for the winter. Their large size and their shaggy coats are a protection against the cold. These animals, especially the goats and their relatives, move about on the rocky slopes with remarkable ease. They are helped by pincer-like toes which grip the surface, and also by 'suction cups' formed by soft pads under the feet.

As well as the larger grazing animals, there are also immense numbers of rodents, including ground squirrels, marmots, pikas and voles. They are all preyed upon by wild cats, including the snow leopard and the puma or mountain lion.

Some of the world's largest birds live in the mountains. The Andean condor has a wing span of about 10 feet, and it soars around the mountains on the many up-currents of air. Like the vultures, it is a carrion feeder and it eats animals that have fallen down the slopes and died.

Polar Life

The polar regions occupy the extremities of the Earth's surface, bounded by the Arctic Circle in the north and the Antarctic Circle in the south. Because the Earth is tilted on its axis, the polar regions experience at least one day in summer when the sun never sets. They also experience at least one day in winter when the sun never rises. At the poles themselves, the sun never rises during the six months of winter and it never sets during the six months of summer.

Polar winters, with little sunlight, are intensely cold. Temperatures of −120°F have been recorded in the Antarctic, and temperatures of −60°F have been recorded in the Arctic. Even in the summer the temperature may remain below freezing point because, although the sun may be shining, its rays strike the Earth at a low angle and much of the heat is reflected. This is especially so if there is snow and ice on the ground.

The big difference between the Arctic and Antarctic regions is that the Arctic region is nearly all ocean, whereas the Antarctic region contains a continent considerably larger than Europe. Land does not hold heat as well as water does, and so the Antarctic is much colder and less hospitable than the Arctic. The Arctic Ocean absorbs a lot of heat in the summer and retains some of it during the winter.

Life in the Arctic

Although the Arctic Circle is a fixed boundary at 66½° north, biologists make use of another boundary—the line beyond which no trees will grow. This is partly within and partly outside the Arctic Circle. Between this *tree line* and the shores of the Arctic Ocean lies the *tundra*—a vast expanse of flat land which is frozen and bare for most of the year. During the summer the surface layers thaw out, but the deeper soil remains frozen. Drainage is impossible, and the melt water is trapped in little pools. This is the tundra's major water supply, for the snow and rain that falls each year amounts to only about eight inches. The tundra is in fact a cold desert.

A nesting skua shrieks a warning in Antarctica. Skuas are rapacious birds. In the Antarctic their diet consists mainly of the eggs and young of penguins.

The snowy owl hunts over the Arctic tundra. Small animals such as lemmings form its chief diet.

The most abundant of the arctic plants are the lichens. They encrust bare rock surfaces and, nearer the tree line, they cover large areas of ground. Reindeer moss, the main food of the reindeer, is one of these lichens. Dwarf species of birch and willow creep about on the tundra surface. They can survive for hundreds of years, but they grow very slowly and are rarely more than a foot or two high. Between them in the summer there are numerous flowering herbs which, for a few short weeks, turn the grey tundra into a colorful landscape. These herbaceous plants generally grow in little clumps or cushions very close to the ground. Here they escape the icy winds and obtain a little heat from the ground. The long hours of daylight during summer help to counteract the shortness of the growing season, but even then there is not much time for plants to store up food and produce flowers. Many have given up flowering and they reproduce themselves vegetatively, by sending out runners.

Animals are rarely seen on the tundra in winter, but they soon reappear when the ice melts. Insects are soon on the wing, and lemmings, shrews and voles appear from the burrows where they have spent the winter. During the summer they will collect grass and seeds and store them away in readiness for the next winter. Ground squirrels awake from hibernation in the spring, while many other animals such as reindeer, wolves, and weasels move north from the forests. Ducks and many other birds fly in from the south to nest beside the temporary lakes of the tundra.

Life in the Antarctic

Most of the Antarctic continent is permanently covered with ice. Only about 3,000 square miles of the coastal regions thaw out during the summer. Summer temperatures rarely rise much above freezing point, however, and freshwater pools are less common than they are in the Arctic. There are some 150 kinds of lichen in the Antarctic, together with a few mosses, but there are only two species of flowering plants. These flowering plants are confined to the Antarctic Peninsula, a narrow strip of land that reaches north and just manages to cross the Antarctic Circle. Such sparse vegetation can support very little animal life, and the largest truly terrestrial creature is a wingless fly about a quarter of an inch long. Various other insects, mites and crustaceans live in the pools or among the lichens.

The more characteristic animals of the Antarctic region are sea creatures. They include various seals and several species of penguins. These animals come on to the land to breed but, with the exception of one bird that scavenges around the penguin rookeries, they get all their food in the sea. The penguins eat fish, and the seals eat both fish and penguins. During winter the emperor penguin has the frozen shoreline to itself. The females each lay a single egg, which the male then picks up and supports on his webbed feet. His thick feathers surround the egg and keep it warm until it hatches eight or nine weeks later. During this time the male stands exposed to the bitter winds and snows and takes no food. This must surely be one of the strangest of all breeding habits.

The ptarmigan, like many other Arctic animals, is white in winter and brown in summer, its colors matching the changing landscape.

A young seal surfaces at a blow-hole in the polar ice. Like all aquatic mammals, seals have to come up to the surface to breathe at regular intervals.

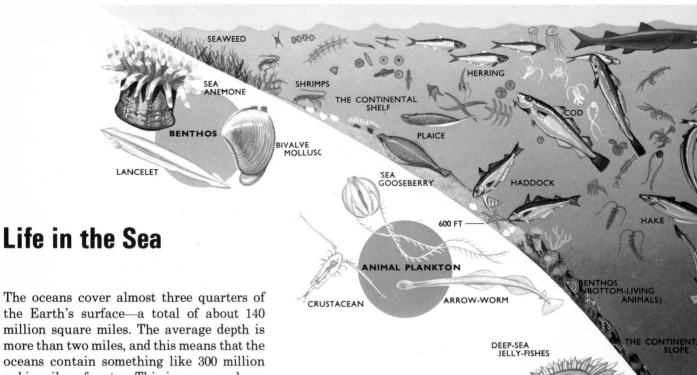

SEAWEED

SEA
ANEMONE

SHRIMPS

THE CONTINENTAL
SHELF

HERRING

BENTHOS

BIVALVE
MOLLUSC

PLAICE

COD

LANCELET

SEA
GOOSEBERRY

HADDOCK

HAKE

600 FT

ANIMAL PLANKTON

CRUSTACEAN

ARROW-WORM

BENTHOS
(BOTTOM-LIVING
ANIMALS)

DEEP-SEA
JELLY-FISHES

THE CONTINENTAL
SLOPE

Life in the Sea

The oceans cover almost three quarters of the Earth's surface—a total of about 140 million square miles. The average depth is more than two miles, and this means that the oceans contain something like 300 million cubic miles of water. This immense volume of water supports a fantastic array of animal life, from the surface right down to the deepest trenches. All the major animal groups are represented in the oceans, and some of them, such as the starfish group, are found nowhere else. Plant life in the ocean is less varied and consists only of algae and bacteria but, as on land, the plant life is all-important.

The Pastures of the Sea

Except in the coastal regions, where seaweeds can grow on the bottom, all life in the ocean depends upon the tiny plants that float in the surface layers. These plants, together with the little animals that float with them, make up the *plankton*. Most of the planktonic plants are confined to the uppermost hundred feet of the ocean. Some reach down perhaps to 600 feet, but below that there is not enough light for them to make food. For the same reason, seaweeds do not cover the whole of the sea bed. They grow only around the coasts where the water is shallow and light can reach right to the bottom.

The planktonic plants—collectively called *phytoplankton*—carry out photosynthesis just as land plants do, and they provide food for all the animals in the sea. Because of this, the surface layers are often called the pastures of the sea. These pastures are incredibly rich, and a given area can produce far more plant life than a similar area on land. To realise just how much plant material is produced, we have only to look at the amount of fish caught every year. The world's fisheries have recently been pulling 40 million tons of fish from the sea each year. This is only a fraction of the animal life in

Below: The Portuguese man o' war, a distant relative of the jelly-fishes which has numerous long stinging tentacles hanging down from a gas-filled float. Its poison is very powerful.

the sea, and all of it depends on the plants. It has been estimated that for every ton of fish produced the ocean must produce 1000 tons of planktonic plants.

The phytoplankton consists entirely of minute plants called algae. There are several different kinds, but all of them are under one tenth of a millimetre across. The most common kinds are called *diatoms*. Each diatom is a single-celled plant and it lives inside a glassy box which is often beautifully sculptured. Diatoms are generally brownish green or yellowish.

The animals of the plankton include a great many different types. Most of them are crustaceans, but there are also arrow worms, jellyfishes, and sea gooseberries. Many of these live permanently in the plankton, but others spend only part of their lives drifting at the surface. These 'part-time drifters' include the young stages of many crabs, barnacles, snails, bivalves, and starfishes. Planktonic animals often extend into deeper waters than the plants and many of them make daily journeys from one layer to another (see page 140). In common with the phytoplankton, many of the animals possess strange spines or feathery outgrowths. These help to keep the creatures afloat by increasing the resistance to the water.

FLYING FISH

OCEANIC SQUIDS

TUNNY

GIANT
SQUID

PRAWNS

LANTERN
FISH

2,000 ft.—

PLANKTONIC
PLANTS

Above : The various
regions of the sea, with
their different conditions
support very different
populations of animals.

SOME DEEP-SEA FISHES

few other creatures. Some of them live at
the surface and feed directly on the plankton,
while others live lower down and get their
food at second or third hand. In general, it is
the smaller nekton that live near the surface
and feed on the plankton. But this is not
always so. The whalebone whales, which are
the largest creatures in the sea, have cut out
the middle stages of the food chain and they
feed directly on the small planktonic animals. They filter them from the water with
great comb-like plates of baleen or 'whalebone' which hang in their mouths.

The Ocean Depths

The deeper one goes into the water, the
darker it becomes. Below about 2,000 feet it
is completely black and these deep waters
contain a number of very strange fishes,
squids, and prawns. Many of them have
light-producing organs (see page 143) which
help them to find their way about. Some of
the smaller animals feed on the debris falling
from above, but most of the deep-sea creatures are carnivorous. Many of the fishes
have huge mouths and formidable teeth and
they have been described as 'floating fish
traps'.

The Ocean Floor

The worms, starfishes, sea anemones, and
other creatures that live on the sea bed are
collectively called the *benthos*. The bottom-

The planktonic animals feed upon the
phytoplankton and upon each other. They
in turn are eaten by larger creatures, such
as herrings and other fishes.

The Swimmers

In contrast to the plankton, whose members merely drift with the currents, there are
many active swimmers in the sea. They are
collectively called the *nekton* and they include fishes, squids, whales, prawns, and a

dwelling creatures extend from the shore
right down to the ocean depths, but their
numbers get fewer as they get deeper. This
is because food gets scarce. Most of the
bottom-dwelling animals feed on the decaying matter that falls like rain from above.
In the deep sea regions very little material
falls to the sea bed: it is either eaten or completely decayed before it reaches the bottom.

207

Life on the Seashore

Conditions on the sea-shore are constantly changing. Many of the plants and animals on the shore cannot move about and they have to put up with the changing conditions. For part of the time they are covered with salt water. At other times they are exposed to the air and to fresh rain water. They also have to put up with the battering they receive from sand and pebbles hurled at them by the waves. Despite all these problems, many kinds of plants and animals do make their homes on the sea-shore. Nearly all the major groups of plants and animals are found there.

The type of shore varies enormously and depends very much on the rocks and the geography of the area. Exposed coasts with hard rocks will form craggy cliffs and rocky shores. Softer rocks, such as sandstones, will yield sandy beaches. Sandy beaches also develop in sheltered areas because the sea deposits the material that it has removed from somewhere else. Very sheltered inlets have muddy shores. This is especially true of river mouths. The rocky shores support the most varied life because only on the rocky shores can the seaweeds get a hold. The seaweeds provide food and shelter for the animals.

The Tides

The true sea-shore is the region between the highest and lowest tide levels. Most parts are covered and uncovered by the tide twice every day. The tides are caused by the pull of the Moon and, to a lesser extent, the Sun. When the Sun and Moon are in line their combined effect is stronger, and the resulting *spring tides* are of greater extent than normal tides. Spring tides occur every fort-

night. *Neap tides* also occur every fortnight, between the spring tides. They are rather weak tides, produced when the Sun and Moon are pulling at right angles to each other. The range of tidal movement varies from place to place. There are only a few inches between high and low tide levels in parts of the Mediterranean, but in some parts of the world the tide may rise and fall as much as 40 feet.

The Zones of the Shore

Many animals visit the sea-shore without actually living there. Fishes come into the shallow water when the tide is in, and birds flock down to the shore to feed when the tide is out. The tides do not worry these animals because they can move about easily. The

The brown pelican (top) lives in North America. It is 50 inches long. The common cormorant (below) is 36 inches long.

seaweeds, barnacles, molluscs, and other creatures that live permanently on the shore are greatly affected by the tides. They have to be able to live in the water and out of it. They are basically sea creatures and their main problem is to survive exposure to the air when the tide goes out.

Some plants and animals can withstand exposure better than others, and these creatures are usually found on the upper parts of the shore which are exposed for relatively long periods each day. Creatures which are less able to stand exposure are found lower down on the shore. The sea-shore therefore has a number of zones, distinguished by the length of time that they are exposed to the air and each supporting its own collection of plants and animals. These

Above: The common puffin, a member of the auk family, is 8 inches long. It can hold several fish at once in its broad beak.

Above: The king shag, a New Zealand member of the cormorant family, is 20 ins long.

Right: The bar-tailed godwit (top) and the spotted sandpiper (bottom) are wading birds which frequent marshes and sea-shores. In summer the male godwit is rich chestnut in color, but in winter it becomes much paler, with a greyish back and whitish underparts.

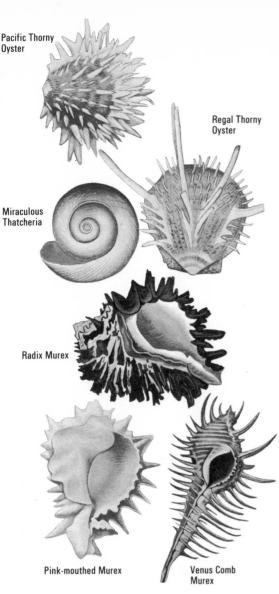

Pacific Thorny Oyster

Regal Thorny Oyster

Miraculous Thatcheria

Radix Murex

Pink-mouthed Murex

Venus Comb Murex

The Upper Shore

The upper shore is the region between the high water marks of the neap tides and the spring tides. There are days when the region is never submerged. The commonest seaweeds are the green ones that clothe the rocks and make them very slippery. The animals include the sea slater that we saw in the splash zone and several kinds of periwinkles. They hide among the seaweeds when the tide goes out.

The Middle Shore

This is the region between the low and high water marks of the neap tides. It is covered and uncovered every day and it supports much more life than the upper shore. The main seaweeds are the brown ones, particularly the bladder wrack and its relatives. They clothe the rocks and protect them from a good deal of the sand and gravel thrown about by the waves. The seaweeds also protect the many animals that live among them. These include more periwinkles, top shells, and whelks. Limpets and barnacles are also common animals of the middle shore. The limpets are snails with tent-shaped shells. They move about and scrape food from the seaweeds when the tide is in but they go back to their own resting sites when the tide goes out again. The shell is pulled down tightly and the animal remains quite safe until the tide returns. The barnacles have little white shells made of several plates. The plates close up when the tide is out, but they open when they are submerged and the animal 'combs' food from the water with its limbs. Barnacles are actually related to the crabs and, like the

Below: The razor shell has a very muscular foot with which it burrows through the sand. The whole shell can be buried in a few seconds.

Below: Wave-battered rocks do not shelter very many animals, but they make good vantage points for pelicans and other sea birds.

zones occur on all shores, but they are most obvious on rocky shores where there are plenty of seaweeds.

The Splash Zone

The splash zone is not strictly a part of the shore because it is never really submerged, except perhaps during storms. It is above the high water mark, but it is continually splashed by the spray, especially on rocky shores. There are no seaweeds in the splash zone, but many lichens and several flowering plants grow there. Among the flowering plants, we can mention the thrift (sea pink), which usually grows on rocks, and the sea campion, which usually grows on shingle. Animals of the splash zone include the sea slater, a little woodlouse-like creature, and some of the periwinkles. Both are found lower down the shore as well, but they can survive for long periods without water. They are well on their way to becoming land animals. The wood-lice and many of the land snails have already completed the change-over from shore-dwelling to land-dwelling, although most land-living animals probably arrived from the sea by way of fresh water.

Below: Many tropical shores are fringed with coral reefs, although few corals can withstand exposure to the air and they are therefore found mainly below low water mark. Coral reefs do not occur where cold currents wash the shore or where muddy water is washed out from estuaries. In the right of the photograph is a giant clam.

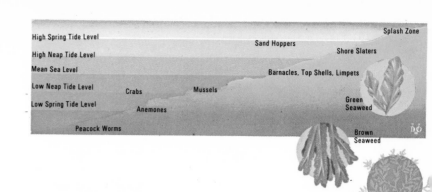

Above: The major zones of the shore. The lowest zones are uncovered for only a short time each day and there are some days when they are never exposed. Likewise, the uppermost regions of the shore are submerged for only a short period each day. Green seaweeds are most common high on the shore. Red seaweeds rarely grow above low tide level.

crabs, they have free-floating young. These young barnacles settle down eventually, but only those that settle in a suitable place will survive. If they settle too high on the shore they will perish because they cannot stand too much exposure. Those that settle too low down will also perish. Only those that land in the barnacle zone will survive.

The Lower Shore

This is the region below the low water mark of the neap tides. It is under water for most of the time and many kinds of animals make their homes here. The main seaweeds are the brown ones again, especially the long oarweeds and bootlace weeds. There may be some red seaweeds at the lowest levels, but most of these grow below the shore where they are always covered. Sea anemones cling to the rocks, shrinking to little blobs of jelly when uncovered. There are also mussels, whelks, and many tube-dwelling worms. Starfishes and sea urchins may crawl about on the sea bed, especially if there is some sand.

Below: Fishes generally move in and out with the tide but where rocky pools exist one can often find some of the shallow water fishes after the tide has gone out. Common fishes include the pogge or armed bullhead, the butterfish, and the three-bearded rockling.

Sandy Shores

Sandy shores look very empty when the tide goes out because there are no rocks or seaweeds. The animals that live there have to bury themselves in the sand when the tide goes out. There are not so many different kinds of animals on a sandy shore, but there may be huge numbers of individual animals. For example, there may be more than 1,500,000 clams under an acre of sand. Worms and bivalves make up most of the fauna of the sandy shore and they feed by drawing in currents of water and filtering out the organic particles. Sea urchins and starfishes are also common.

Armed Bullhead

Butterfish

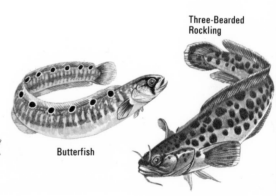

Three-Bearded Rockling

210

Life in Fresh Water

Rain falling on the land runs over and through the soil and finds its way into ponds and streams. On its journey it picks up a great deal of mineral matter and it also dissolves a certain amount of oxygen and carbon dioxide. These materials are just what plants need, and so it is not surprising that plants have colonized fresh water, closely followed by a rich variety of animal life. Fresh water has been colonized or invaded from two directions: some animals entered the rivers from the sea, and others entered the water from the land. The major problem of living in the water is to obtain oxygen. Animals coming from the sea had already solved this problem, but those coming from the land had to find ways of breathing under the water.

Pond Plants

A pond is a relatively small body of still water. There is no clear distinction between a pond and a lake, although we usually use the word lake to describe a larger stretch of water. Lakes and ponds present animals with the same sorts of problems and they contain the same sorts of living organisms. The pond community depends in the first place on the surrounding rocks. Ponds surrounded by acidic sandstones or granites are much poorer than ponds in chalk or limestone regions. Rocky shores provide little foothold for plants, but a pond resting on clay or river deposits will be completely surrounded by lush vegetation. This may consist of cattails, various rushes and reeds, and the splendid reed mace which is often wrongly called the bulrush. Some of these plants extend a fair way into the water, although they are not truly aquatic plants. The real aquatic species live further out in

Above: The kingfisher is a very striking bird of pond and stream. It sits on a branch overhanging the water and dives in after small fishes.

the water. They may be rooted, with floating leaves (e.g. water lilies) or with submerged leaves (e.g. Canadian pondweed) or with both kinds of leaves. Duckweed, frogbit, and some other plants float freely at the surface. And then there are the most important of the plants—the tiny floating organisms that form the plankton. These feed nearly all the small animals in the pond.

Pond Animals

The plants provide food and shelter for the animals, and they also provide oxygen. This is most important, both for the living animals and for the proper decomposition of dead material. Woodland ponds which are completely shaded by trees support very little plant life. There is little oxygen and leaves that fall into the water do not decay properly. They form a foul-smelling layer on the bottom and the only animals that can survive are various fly larvae. The rat-tailed maggot (a hoverfly larva) is one of these. It gets its oxygen through a telescopic tube which it pushes up to the surface.

Nearly every group of animals can be found in fresh water, and every region of the pond has its characteristic inhabitants. The surface of the water, although no different in

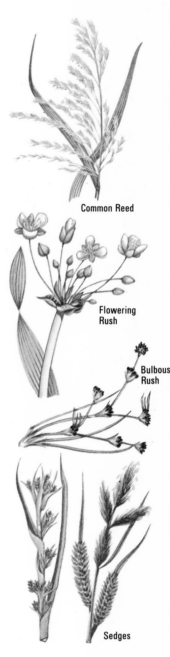

Below: The marshy ground around ponds and streams supports a number of characteristic plants.

Common Reed

Flowering Rush

Bulbous Rush

Sedges

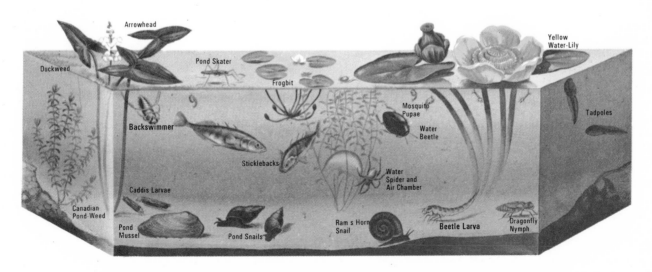

Arrowhead

Yellow Water-Lily

Duckweed

Pond Skater

Frogbit

Mosquito Pupae

Water Beetle

Tadpoles

Backswimmer

Sticklebacks

Water Spider and Air Chamber

Caddis Larvae

Canadian Pond-Weed

Pond Mussel

Pond Snails

Ram s Horn Snail

Beetle Larva

Dragonfly Nymph

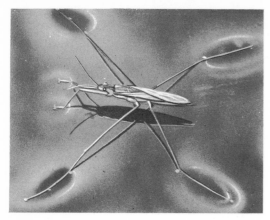

Above: The pond skater 'rows' itself over the water surface with its long legs. Its weight is supported by the surface film, but the legs make little depressions in the film.

composition from the rest, acts as a very thin skin and can support various small animals. Pond-skaters skim across the surface, often accompanied by shiny, black whirligig beetles. Both feed on flies and other small creatures that fall on to the surface. The whirligigs' eyes are each divided into two parts, so that they seem to have four eyes. The upper parts look across the surface, and the lower parts look down into the water. The underside of the water surface supports little black planarian worms and various small snails.

Among the free-swimming creatures there are fishes, water beetles, water bugs, such as the backswimmers and water boatmen, water spiders and mites, and a horde of little crustaceans. The latter include the water fleas and cyclops which live mainly in the surface waters and they feed by straining minute plants and other organisms from the water. They are especially common in ponds where cattle feed, because these ponds normally receive plenty of manure and they therefore support abundant plankton.

Living in or on the mud at the bottom of the pond, or crawling on the water plants, we find another group of animals. These include snails and mussels, together with the young stages of dragonflies, water beetles, and

caddis flies. There will also be various worms and fly larvae. One of the interesting worms is a little red creature called *Tubifex*. It is often sold in pet shops as fish food. The animal lives half buried in the mud and its tail sticks up into the water. By waving the hind end, the worm creates water currents which bring oxygen to it. Oxygen is never abundant near the bottom of the pond, and the worm contains the red pigment hemoglobin which helps it to absorb enough oxygen.

Breathing under Water

Small creatures, such as Hydra and the water fleas, get enough oxygen through their body surfaces. Larger creatures, however, need special breathing organs. The fishes, and the water snails which arrived directly from the sea, breathe by means of gills. These absorb oxygen directly from the water, and the animals therefore need well-aerated water. They cannot live in stagnant ponds. Other gill-breathing creatures include the young dragonflies and caddis flies.

The insects have invaded the water from the land and, although some young insects can get their oxygen direct from the water, the adults still have to breathe air. So do many water snails which have entered fresh water from the land. You will often see water boatmen and water beetles hanging upside down at the surface. They are renewing their air supply. They carry a bubble of air about with them, trapped under their wing cases or else trapped by a coat of fine hairs on the body. The air bubble is in contact with the openings of the breathing tubes. Water snails usually come to the surface every now and then to take a fresh air supply into their lungs. You can see them open the entrance to the lung as they hang upside down at the surface.

Some insects get their oxygen without actually coming to the surface. We have already mentioned the rat-tailed maggot. Another interesting example is the water scorpion—a flattened insect, unrelated to

Above: The nymph of the pond olive mayfly. The lobes on the sides of the body are the gills.

Bream

Many fishes live in fresh water. Most of them feed on insects and other small invertebrates. The pike, however, is a voracious fish with formidable teeth. No other fish is safe when a pike is about. The pike will also eat frogs and water birds.

Pike

Roach

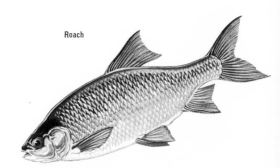

Many young insects live in water—dragonflies, mayflies, mosquitoes, and so on. Some adult bugs and beetles also live in the water. Most of them still breathe air, however, some have special breathing tubes which they push up to the surface. Others come to the surface every now and then to renew their air supplies.

real scorpions and quite harmless. It has a long breathing tube at the hind end and this carries air down to the breathing tubes.

Life in Streams

The moving waters of streams present quite different living conditions from those found in the still waters of ponds. What is more, the conditions vary as one moves up or down the stream. It is possible to divide a stream into several regions according to its animal and plant life.

The upper reaches of a river form the *headstream*. This is usually on high ground and it is a shallow, fast-flowing stretch. There is plenty of oxygen here and the main problem is to avoid being swept away. The only plants that manage to grow here are little algae that attach themselves to the rocks. A few small snails crawl on the bed of the stream, but the main animals are the young stages of various insects. Flattened nymphs of some mayflies and stoneflies cling to the stones and nibble away at the algae. There are also some caddis larvae, which make themselves little cases of sand grains.

The headstream gradually merges into the *troutbeck* region. The slope is not so steep here, but the bed is still rocky and the water still runs quickly. A few patches of water crowfoot may grow on the stream bed, but there is little vegetation apart from the

algae. The flattened insect nymphs from the headstream also live in this region, together with various snails and the river limpet. The latter has a conical shell, but it is not closely related to the limpets of the sea shore. Fishes make their appearance in the troutbeck. They are either strong swimmers, such as the trout, or they hide among the stones and boulders. The loach and the bullhead are among these bottom-dwelling species. They all feed on the insect nymphs or on adult insects that fall on to the water surface.

The next stretch of the river is usually called the *minnow reach*. The current is slower and a certain amount of sand and gravel covers the bottom. There are plenty of water plants, and animal life is also abundant. There are many different kinds of water snails, dragonfly nymphs, mayflies, and other insects. The minnow is the typical fish, but there are also sticklebacks, lampreys, dace, grayling, and salmon. The fishes of the troutbeck are found here as well. Another common creature is the freshwater shrimp, which darts around over the sand and among the plant stems.

Below the minnow reach, the river gets deeper and slower. This is the *cyprinoid reach*, characterised by fishes such as roach, rudd, perch, pike, chub, bream, and carp. The slow-moving water deposits plenty of mud and plants are able to grow along the edges. Water lilies may grow further out in the river. Worms, water snails, and mussels are common in the mud. There is a certain amount of floating plankton in this stretch of the river, especially in the weedy parts, but it is never as common as it is in the still water of a pond.

Above: The pond snail (top) and the ram's horn snail, two common pond dwelling species.

Above: The dipper, a bird of the troutbeck which actually walks on the stream bed in its search for insects.

Above: The beaver lives in woodland streams in North America and in parts of Europe. Its broad blade-like tail is used as a rudder when the animal is swimming.

Fossils

Fossils are the traces and remains of animals and plants which have been naturally preserved in the rocks, sometimes for many millions of years.

When an animal dies its body usually decomposes or else it is eaten by other animals. But the hard parts—the teeth and the bones or shell—are not so easy to destroy, and if they are buried quite quickly they stand a good chance of being preserved or fossilised. Most of the fossils that have been discovered are of marine creatures. This is because the best conditions for fossilisation occur in the sea, and also because most of our rocks were actually formed under the sea.

The story of a typical fossil can be followed by imagining some animals living on the sea-bed millions of years ago. Imagine also the sand and mud settling all around the animals. When the animals died their shells or skeletons became buried by this rain of sand and mud. In time the sediment became

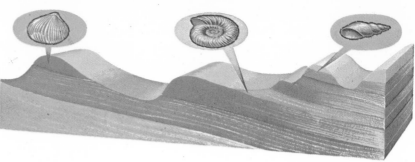

Rocks laid down at different times tend to contain different collections of fossils which enable the rocks to be recognised wherever they are found.

Fossil ammonites have been found over 8 feet in diameter.

Fossil remains of insects have been found in amber, the hardened resin of evergreen trees.

FOSSILS

Ammonites

The ammonites are among the best-known fossils, and many of the larger ones are sold as curios in seaside towns where they occur in the cliffs. Ammonites, which are all extinct, were marine creatures related to today's squids and cuttlefishes. They had coiled shells and it is the preserved remains of these that we find today as fossils. The original shell has almost always disappeared, and the fossils we find are either moulds or internal fillings as a rule. The internal fillings were formed when the shells became filled with sediment and then dissolved away, leaving the filling with a detailed pattern of the inside of the shell. Ammonites are very common in the Jurassic limestones and clays formed about 150 million years ago. Some of the fossils in the clays are composed of iron pyrites and they sometimes shine like gold.

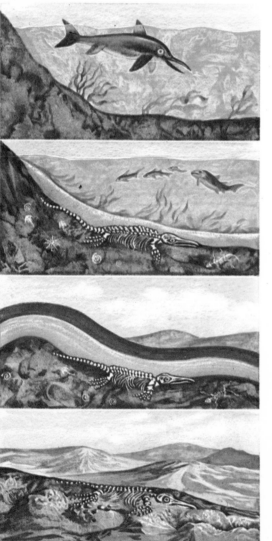

Diagram showing how an animal may be fossilised. A sea-dwelling ichthyosaur dies and sinks to the sea-bed. Soon its skeleton is covered with mud which gradually settles into solid rock. At a much later date earth movements buckle the sea-bed and cause it to rise above sea-level. Erosion then strips away the rock covering until the fossil is exposed.

compressed and hardened into rock and the animal remains became converted into fossils. At a much later date, movements of the Earth's crust lifted up these new rocks to form land. Wind and rain began to wear down the rocks and the fossils eventually came to light.

Fossils take on a number of different forms. Very occasionally the actual skeleton may be preserved. This has happened where animals have been trapped in bogs or tar pits and buried very rapidly. The Californian tar pits, for instance, have yielded a wealth of animal skeletons including those of several birds. Under certain very unusual conditions an entire animal may be preserved. Mammoths (relatives of today's elephants) have been preserved almost intact in the ice of Siberia and their flesh has been fit to eat after perhaps a million years in cold storage.

More often, however, the buried remains undergo a certain amount of change. The skeletons and shells are most commonly *petrified,* or converted into stone. Water percolating through the rocks gradually dissolves the original material and deposits mineral matter in its place. If the original material is dissolved away and not replaced it leaves a hollow in the rock. This hollow is called a *mold* and it is a very common form of fossil. If mineral matter is deposited in the mold it forms a natural *cast.* This type of fossil shows all the external features of the object, just as a jelly follows the shape of the mould it is made in, but it does not tell us anything about the internal structure of the object. It is possible to make artificial casts of fossils by filling the molds with modelling wax or some similar substance.

Many plant fossils are simply residues of carbon, which give the actual shapes of the leaves and stems. Some animals may be preserved in this way, too. Pieces of amber, which is a hardened or fossilised resin from coniferous trees, sometimes contain the remains of insects which had become trapped in the sticky resin. These are true fossils.

Ancient animals often left tracks and footprints in soft mud and the mud sometimes became baked hard before the print disappeared. The print then became filled with some other material and preserved as a fossil. Although a footprint is not actually

A petrified tree trunk. Each particle of wood has been replaced by mineral matter.

Unusual fossils are tracks left by prehistoric animals in mud which later hardened into solid rock.

Sometimes the fossilised skeleton of an animal may be surrounded by a film of carbon showing the actual outline of the animal's body.

Below: A fossil dragonfly

A mold of a trilobite, the impression that may remain after the animal itself has disappeared.

A cast of a trilobite formed by the later deposition of mineral matter in the original mold.

a part of an animal, it is a true fossil because it does tell us quite a lot about the animal that made it.

Looking for Fossils

Fossils are quite easy to find if you know the sort of places in which to look for them. You can often pick them up in fields and gardens, but these scattered finds are often incomplete and damaged because they have been rolling around for a long time. Better specimens are obtained from freshly exposed rocks. Granite and similar rocks contain no fossils because these rocks formed from molten material. If you want to find fossils you must look at the sedimentary rocks, which formed from sediments deposited in the seas and lakes. Sandy rocks are not very good for fossils and rarely contain complete specimens. Limestones, clays and shales are the best rocks to examine. The fossils are often harder than the surrounding rock and they stand out from the surface. They can then be chipped out with a hammer

Above: Archaeopteryx, the first known bird is known from fossils found in a Bavarian limestone quarry.

Right: Fossil crinoid or sea lily

Below: Fossil of Paradoxides, the 20-inch giant trilobite of the Cambrian seas.

and cold chisel. Always examine the heaps of clay and other rocks brought up during roadworks. Railway cuttings, quarries and seaside cliffs are also very good places to look.

The fossils are not usually scattered haphazardly through the rocks. More often they are concentrated in bands or layers. These layers of rock and fossils were laid down at times which were particularly favorable for the preservation of the remains. The best example of this concerns the formation of coal. Coal consists of the remains of giant fern-like plants that grew in coastal swamps some 250 million years ago. At that period the sea-level was unsteady and it kept flooding the forests. This killed the plants and buried them with sandy and muddy sediment. When the sea retreated again, a new forest sprang up on the deposits, and the remains of the earlier forests gradually turned into coal. This happened many times, producing numerous coal seams separated by beds of sandy rock.

Many limestones also consist almost entirely of fossils. Some are made up of old coral reefs, while others contain little but the remains of crinoids, or sea lilies. These were animals related to the starfishes, and they had skeletons composed of little limestone discs.

The rocks are like the pages of a history book to a geologist: they tell the story of the Earth. But they are far more difficult to read than the pages of an ordinary book because they are frequently torn, bent, or upside-down. Some of the pages are scattered over a wide area, and some are missing altogether. The key to the pages and their order lies in the fossils contained within the rocks.

If we take an undisturbed sequence of rocks it is fairly obvious that the oldest rocks will be at the bottom and the newest

216

rocks, which were laid down most recently, will be at the top. If we look at the fossils in such a sequence of rocks we will find that the fossils at one level are not exactly the same as those at another level. As one goes higher up the sequence, some of the fossils disappear, and new kinds come in to take their place. Each layer of rock thus has a characteristic collection of fossils.

This is of great importance to the geologist because if he finds the same assemblage of fossils in rocks from different regions he can safely say that the rocks are of about the same age. This helps him to piece together the jig-saw of the Earth's long history. The Earth's history is, in fact, divided up into a number of periods which are separated mainly on the basis of the fossils they contain.

But as well as showing us some of the Earth's history, the fossils in the rocks show us how the plant and animal life of our planet has changed. 500 million years ago there were plenty of shelled creatures related to today's crabs and cuttlefishes. There were also many other kinds of animals that are now completely extinct, but there were no animals with backbones as far as we know. Then, about 450 million years ago perhaps, the first fishes appeared. Only fragments of these early fishes have been found, but we can be fairly certain that the backboned creatures had arrived on the scene by this time. 100 million years later, the fossil record tells us that a great number of different kinds of fishes were in existence. Soon after this a new sort of creature began to live on the Earth. It was called an amphibian and it was one of the early forerunners of today's frogs and newts. Still later, about 250 million years ago, there appeared the animals that we now call reptiles. Birds and mammals did not appear until much later.

The fossils in the rocks therefore show us roughly when the various groups of animals (and plants) came into being. But they do more than this. There are some rare fossils that do not fit easily into any of the main groups. Some of them look partly like fishes and partly like amphibians: others look partly like amphibians and partly like reptiles. These fossils show us that the fishes —or, to be more accurate, some fishes— gradually changed into amphibians, and that some amphibians gradually changed into reptiles. In other words, the fossils show us that life *evolved* and that all of today's plants and animals have developed gradually from simpler forms of life.

THE AGES OF THE EARTH

Geological Periods	Plant and Animal Life Forms
Quaternary Period Began 2-3 Million Years Ago	Great Ice Age. Coming of Man.
Tertiary Period Began 70 Million Years Ago	Modern Mammals Appear.
Cretaceous Period Began 135 Million Years Ago	Dinosaurs Decline. Mammals Increase. First Flowering Plants.
Jurassic Period Began 180 Million Years Ago	Dinosaurs Abundant. Birds and Mammals Appear.
Triassic Period Began 225 Million Years Ago	Early Dinosaurs.
Permian Period Began 270 Million Years Ago	Reptiles Increase.
Carboniferous Period Began 350 Million Years Ago	First Reptiles.
Devonian Period Began 400 Million Years Ago	Age of Fishes. First Amphibians.
Silurian Period Began 440 Million Years Ago	Earliest Land Plants.
Ordovician Period Began 500 Million Years Ago	First Fishes.
Cambrian Period Began 600 Million Years Ago	First Abundant Fossils, All Invertebrates.

Formation of Earth About 4,500 Million Years Ago

Mammoths have been found preserved almost intact in the frozen wastes of Siberia and Alaska.

The First Animals

A somewhat magical line in the geological time scale separates the Cambrian Period from the vast expanse of Precambrian time. The first Cambrian rocks were laid down about 600 million years ago and they contain a wide variety of fossils. Nearly all the major animal groups were in existence at that time, but the Precambrian rocks contain very few fossils. It is almost as if life suddenly started up about 600 million years ago. But we know that this cannot be true because many of the Cambrian animals were quite advanced creatures. Life must have been in existence for a very long time before the opening of the Cambrian Period. Why then are there so few fossils? The most likely answer is that the earliest plants and animals had no hard parts which could be preserved. The early animals were probably all soft creatures like worms and jellyfishes.

Interest in the Precambrian rocks has increased in recent years, and intensive searching has brought a number of fossils to light. In 1947 rocks derived from a late Precambrian mudflat were discovered in the Ediacara Hills of South Australia. Pressed into these ancient deposits were the shapes of numerous soft-bodied animals, including worms, jellyfishes, sea pens, and many others. This remarkable find has shown that animals really were living in Precambrian times and that many of them were indeed soft-bodied creatures without shells.

The First Signs of Life

The earliest signs of life so far discovered are various lumps of limestone found in Rhodesia. There is little in the material to

Some Ordovician graptolites preserved in shale. The horny skeletons have been completely flattened.

Pre-Cambrian fossils from the Ediacara Hills in Australia: an annelid worm (bottom); a jellyfish (left); and an unidentified organism.

suggest that it was formed by living organisms, but its appearance is very similar to that of some of the limestones formed today by algae living on coral reefs. It is therefore reasonable to assume that these ancient limestones were formed by some primitive plant. They are believed to be about 3,000 million years old. Scattered deposits of carbon are probably of about the same age, indicating other sorts of plants. Sponge spicules and other very small fossils have been found in many Precambrian rocks, and all help to show what the earliest forms of life must have been like. Unfortunately, there is still a very big gap in our knowledge concerning the rather sudden appearance of the many shelled creatures in the Cambrian rocks. The development of shells must have

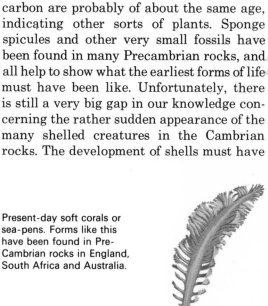

Present-day soft corals or sea-pens. Forms like this have been found in Pre-Cambrian rocks in England, South Africa and Australia.

taken millions of years, and shelled animals must have been living in Precambrian times, but the uppermost layers of the Precambrian rocks are missing in most places and we have yet to find fossils that show how the shelled animals evolved.

Trilobites

The earliest Cambrian rocks contain a wealth of fossils when compared with the Precambrian rocks, although they do not contain as many as some of the later rock formations. The most abundant of the Cambrian fossils are the *trilobites*. These creatures were distantly related to crabs and other arthropods and they looked rather like woodlice. They probably lived on the sea-bed and fed on the debris that accumulated there. Trilobites had several pairs of legs, but these are rarely preserved and the most common type of trilobite fossil is a mold of the body. This shows three distinct regions: a triangular or semi-circular head-shield, a thorax made up of several segments, and a tail section, or *pygidium*, which often looks like another head. The animals frequently broke up when they died, and the head and tail sections are often found by themselves. The trilobites

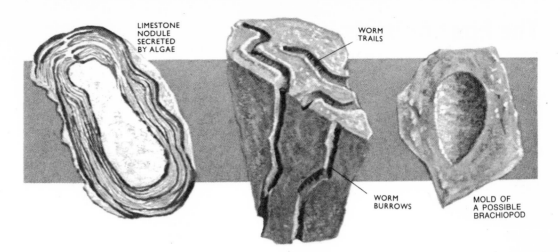

Some of the earliest traces of life: a limestone nodule perhaps secreted by algae; worm tracks and burrows; and a mold of a possible brachiopod.

LIMESTONE NODULE SECRETED BY ALGAE

WORM TRAILS

WORM BURROWS

MOLD OF A POSSIBLE BRACHIOPOD

were mostly small creatures, averaging less than an inch in length, but one species (*Paradoxides davidis*) reached nearly 20 ins. Various trilobites lived in the seas for about 400 million years, but they are now quite extinct.

Brachiopods

Also very common in the Cambrian seas were shelled creatures called *brachiopods*. Some species are found in the oceans today and they are called lamp shells. They look rather like cockle shells and other bivalves, but they belong to a very different group of animals. The shell is much more symmetrical than that of a bivalve mollusc and it conceals two 'arms' which are used to waft food particles into the mouth when the shell is open. The brachiopods lived on the sea-bed and were anchored to it by a muscular stalk.

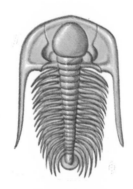

Above: Trilobites first appear in Cambrian rocks

Below: Artist's impression of life in the Cambrian seas.

Graptolites

Towards the end of the Cambrian Period there appeared a new kind of animal called a graptolite. Graptolites were very small animals and they lived in thread-like colonies. Each thread supported a number of tiny horny cups, and each cup contained a single graptolite animal. The early graptolite colonies were fan-like structures, but later ones became simpler and consisted of only one or two branches. We know very little about the soft parts of the graptolite animals because they are quite extinct and the fossils are usually only flattened impressions in shales. They probably floated freely in the sea and they rapidly spread throughout the world. They became abundant in the Ordovician Period but then they dwindled and disappeared completely about 400 million years ago.

The Age of Fishes

Below: A fossil fish from Devonian rocks. Fossils of marine creatures are more plentiful than those of land-dwelling animals since they stood a much better chance of being preserved (see page 214).

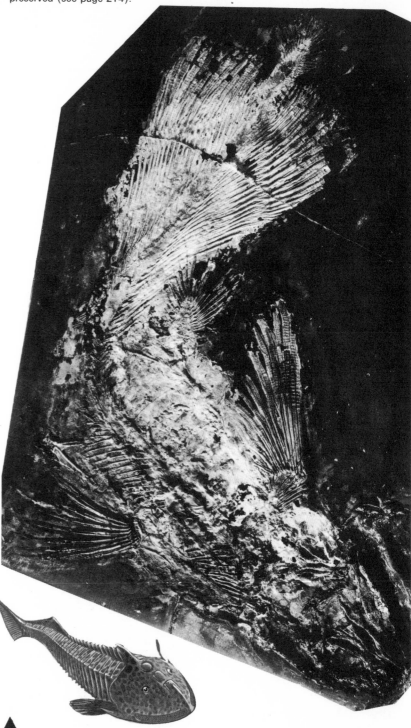

Although animal life was abundant during Cambrian times, and nearly all groups of animals had appeared by then, there were no backboned animals (vertebrates) as far as is known. The earliest traces of backboned animals have been found in some Ordovician rocks in America. They are about 450 million years old and they consist of various bone fragments and fish scales.

How the first fishes arose from earlier types of animal is not clear. Many groups of invertebrate animals have been suggested as ancestors of the vertebrates, but the most likely group is that containing the starfishes. This might seem surprising at first, but some members of the starfish group have larvae like those of the acorn worm, and the acorn worm is known to be related to the vertebrates.

Fishes without Jaws

Although the earliest traces of fishes are some 450 million years old, the oldest complete fossil fishes that we have are only about 400 million years old. These fishes lived in Silurian times and most of them were only a few inches long. Most of them were heavily armored with bony plates and scales, especially in the head region, but the most striking feature of all was the complete absence of jaws. The fishes had mouths, but they were quite unable to bite or chew. They probably fed by sucking up mud and other debris and filtering it through their gills

The first known jawed fishes were the placoderms. They were partly covered in armor and their teeth suggest that they fed on other fishes. Most were less than one foot long but *Dinichthys* (below) reached a length of thirty feet.

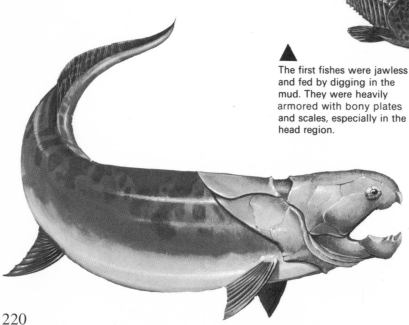

The first fishes were jawless and fed by digging in the mud. They were heavily armored with bony plates and scales, especially in the head region.

to extract food material. Many of them were flattened from top to bottom and they were adapted for life on the bottom of the sea or the rivers. The jawless fishes nearly all became extinct after the arrival of the jawed fishes, but a few rather special types survive to the present day. They include the lampreys and the hagfishes.

The First Jawed Fishes

Remains of the first known jawed fishes date from about 370 million years ago. We do not know the direct ancestors of the jawed fishes, but they almost certainly arose from some sort of jawless fish by the gradual transformation of some of the gill bones into the jaws. The early jawed fishes are known as placoderms and, like the primitive jawless fishes, they were partly covered with bony armor. They had teeth and they probably fed on other animals. As well as jaws, the placoderms had paired fins. These two features were important milestones in the evolution of the backboned animals. Most of the placoderms were small fishes, less than about a foot long, but there were a few giant species. *Dinichthys* reached a length of about 30 feet.

The Age of Fishes

The Devonian Period, which began nearly 400 million years ago, is often called The Age of Fishes. Many of the jawless fishes were still living, and the placoderms were plentiful, but evolution was moving rapidly and many new groups of fishes appeared at this time. In fact, all the major groups of fishes had appeared by the end of the Devonian Period. Two main lines emerged within the fishes: the shark-like fishes, with their soft cartilaginous skeletons, and the bony fishes. The early sharks were very much like their modern relatives, with powerful jaws and streamlined bodies. They lacked the heavy armor of their ancestral placoderms and they were much more mobile. Their hard teeth and spines are often preserved in the rocks.

Ray-finned Fishes

The bony fishes had skeletons of true bone and the early ones were quite well armored with bony plates and heavy scales. They also possessed air sacs which opened into the back of the throat. Two distinct lines soon emerged among the bony fishes. These were the ray-finned fishes, which include almost all the living fish species, and the lobe-finned fishes. The latter are represented today by just six species of lungfishes and the coelacanth. The fins of the ray-finned fishes are supported on narrow bony spines or rays, whereas those of the lobe-finned fishes are supported by a fleshy lobe in the center.

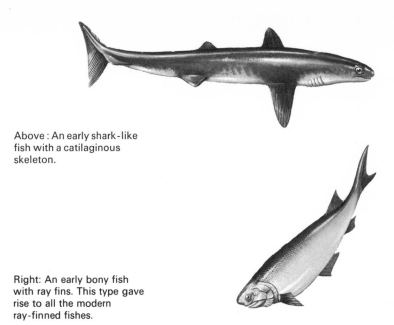

Above : An early shark-like fish with a catilaginous skeleton.

Right: An early bony fish with ray fins. This type gave rise to all the modern ray-finned fishes.

In most of the ray-finned fishes the scales became thinner and the air sacs gradually became modified to form the swim bladder. This is a gas-filled bag situated just above the food canal and its job is to regulate the buoyancy of the fish. A few ray-finned fishes, however, retained the air sacs and they are able to gulp air as well as breathe through their gills. The bichir is one of the fishes that retained the air sacs, and it also retained several other primitive features, such as the heavy scales. It lives in several African rivers.

Lobe-finned Fishes

Although we have very few lobe-finned fishes alive today, they were very common during the Devonian Period. Like the other bony fishes, they had air sacs but, whereas the nostrils of the ray-finned fishes were simply pits on the snout, those of the lobe-finned fishes opened into the throat. This meant that the fishes could breathe air, just as the modern lung fishes can. They could probably also move about on their lobed fins. Somewhere in this group of fishes there was the ancestor of the land-living vertebrates.

Below: The jaws of vertebrates are believed to have arisen from the bones that supported the third pair of gills in the earlier jawless fishes.

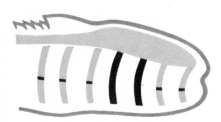 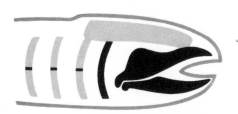 

Life Moves on Land

Although backboned animals came into existence something like 450 million years ago, it was a very long time before they managed to conquer the land. For about 100 million years the only backboned animals were fishes. Many kinds of fishes came and went before there was any sign of a move on to the land.

At the end of the Silurian Period, about 400 million years ago, there were extensive and violent movements of the Earth's crust. They resulted in the lifting up of large areas of the sea-bed, especially in the Northern Hemisphere. The new land was traversed by rivers and streams and it was dotted with lakes. As yet, the land surface was quite bare, with only a few plants here and there. Life was centered in the waters of the rivers and lakes, where numerous fishes swam. This was the Devonian Period, or the Age of Fishes.

The climate was mainly warm and dry: rivers and lakes periodically dried up. Many fishes were stranded on the mud, or they died through lack of oxygen in their shrinking pools. But some fishes had air sacs leading from their throats. They could obtain oxygen by gulping air into these sacs. These same fishes also had muscular lobes at the bases of their fins, and they could move about on their fins. Perhaps they found plenty of food in the form of

The skeleton of a ray fin (top) compared with the skeleton of a lobe fin (bottom) from which the typical five-fingered limb developed.

dead and dying fishes. Some of these lobe-finned fishes actually left their drying pools and wandered over the land in search of new stretches of water. Many of them died, but some of them managed to find a new home in which they could breed. The drought-resisting ability of these fishes and their descendants saved them many times, and gradually the ability to survive and move on land improved. Over a period of several million years the fins changed into legs and the air sacs became lungs. The new animals were the first amphibians.

Fossil Evidence

The story outlined above is a widely accepted theory of how land vertebrates evolved from fishes. But where is the evidence? There is actually quite a lot of evidence to support the theory. Many of the Devonian rocks are red and sandy. The hot, dry regions of the world today frequently contain red sands, and so it is fair to assume that the Devonian rocks were formed under mainly dry conditions. The presence of salt deposits also indicates that lakes were drying up.

The Devonian rocks contain fossils showing that the lobe-finned fishes were quite common, and careful examination of some of these fossils has shown that they did possess air sacs opening into the throat. In this respect they resemble the modern lungfishes, which gulp air and can also live out of water for a time. The Australian lungfish can use its fins to crawl about in the

When the waters became shallow and crowded, the lobe-finned fishes were at an advantage. They could breathe air and when the water dried up they could probably survive. This type of animal evolved into the primitive amphibians.

mud. It is reasonable to expect that the ancient lobe-finned fishes behaved in a similar way. This sort of evidence strongly supports the idea that the land-living amphibians evolved from fishes, but it does not prove it.

Much stronger evidence for the evolution of amphibians from fishes is provided by various fossils discovered in Greenland. These fossils, which are believed to be about 350 million years old, show striking resemblances to both lobe-finned fishes and amphibians. The bones of the skull and the spinal column are just like those of the lobe-finned fishes, but the animals definitely had five-fingered limbs. These limbs, and the way in which they were attached to the body, show that the animals definitely walked. Vertebrates had managed to invade the land.

Fossils in the Devonian and Carboniferous rocks indicate that the amphibians remained fish-like for millions of years, although the skeleton gradually became more suited for life on land. Many different types of amphibians appeared during the Carboniferous Period, but these animals never really managed to conquer the land. They could never completely escape from the water because, like most of the modern amphibians, they had to go back to the water to breed. The real conquest of the land was left to the next great group of vertebrates, the reptiles.

Some 300 million years ago, a group of amphibians developed a more waterproof covering and they began to lay eggs that could survive on land. Other changes took place gradually inside their bodies, and there came a time when the animals could no longer be called amphibians. They had evolved into reptiles. Freed from the need to return to the water to breed, the early reptiles began to spread out over the Earth. Many plants and insects had now appeared and there was plenty of food for the reptiles. They evolved quite rapidly and produced many different species.

The amphibians, which were rather slow and clumsy creatures, could not compete with their more active descendants, and most of the ancient amphibian groups disappeared. Today's frogs and newts are the descendants of a few specialised groups that managed to survive the competition with the reptiles.

Miobatrachus, an early frog-like amphibian.

Below: Fossil and reconstruction of Eryops, a prehistoric amphibian about the length of a crocodile.

Ichthyostega, an early amphibian, closely resembled the lobe-finned fishes.

The Age of Reptiles

Reptiles evolved from amphibians some 300 million years ago, at the time when the world's great coal deposits were being formed. They continued the invasion of the land that had been started by the amphibians millions of years earlier. Reptiles had two very big advantages over the amphibians as far as life on land was concerned: they had scaly, waterproof skins and they were able to lay eggs with tough shells. The eggs contained liquid and each young animal had what amounted to a private pond. The reptiles did not, therefore, have to go back to the water to breed.

The early reptiles had many of the characteristics of their amphibian ancestors, and they were generally only a foot or two long. They were flesh-eaters. During the Permian Period the reptiles improved their ability to live on land and they began to branch out into many different habitats. Many different species came into being. Some became plant-eaters. Some even returned to the sea and eventually became plesiosaurs and fish-like ichthyosaurs. One line evolved into the turtles, and another eventually produced the snakes and lizards. Mammals evolved from yet another group of reptiles, but it was a long time before the mammals became important. The Jurassic and Cretaceous Periods belonged to the reptiles, and these two periods are justly known as the Age of Reptiles.

The early reptiles walked on all four legs, but about 225 million years ago one group started to walk about on its hind legs. The hind legs then gradually became longer than the front ones. These animals were the *archosaurs*, the ancestors of the great dinosaurs which dominated the world for nearly 100 million years. The dinosaurs are now extinct, and have been for more than 70

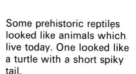
Some prehistoric reptiles looked like animals which live today. One looked like a turtle with a short spiky tail.

million years, but the archosaurs lived on in the form of the crocodiles and alligators. These animals have given up the habit of walking on their back legs and they have gone back to the water, but their back legs are still longer than the front ones. The crocodiles have changed very little during the millions of years that they have lived on the Earth, and they tell us a good deal about the structure of the early archosaurs.

Dinosaurs

The name 'dinosaur' is popularly applied to several groups of archosaur reptiles and it is not a particularly scientific name. It conjures up visions of huge, lumbering creatures, but not all the dinosaurs were large. Some were only two or three feet long.

The early dinosaurs were all flesh-eaters

Archaeopteryx, the first known bird, was about the size of a crow. It had wings with feathers just like those of modern birds. But it also had claws on its wings, teeth and a long reptilian tail. Archaeopteryx forms a link between the reptiles and modern birds.

Ichthyosaurs—the 'Fish Lizards'
The ichthyosaurs were reptiles that went back to the sea at an early stage. They then evolved a remarkably fish-like body, although they had flippers instead of fins and they still retained their air-breathing habits. In many ways, they were similar to the dolphins of modern seas. The skull was long and narrow and the snout carried numerous conical teeth. The animals must have fed on fishes, and perhaps on some of the numerous ammonites in the Jurassic seas. Many fossil ichthyosaurs have been found in the clays and limestones of the Jurassic Period. Some of the more complete remains have much smaller skeletons associated with them, showing that the ichthyosaurs brought forth living young instead of laying eggs.

The First Birds
Insects were flying in the air about 300 million years ago, and the pterosaurs followed them about 180 million years ago. But birds did not appear until perhaps 150 million years ago, making them the last of all the major animal groups to appear. The oldest fossil bird so far discovered comes from Germany and it is about 140 million years old. It is called *Archaeopteryx* and it forms a link between the reptiles and the modern birds. It had wings, with feathers just like those of modern birds. But it had claws on its wings. It also had teeth and a long reptilian tail. *Archaeopteryx* was clearly a bird because of its feathers, but it must have arisen from a group of reptiles. Its most likely ancestors were some early archosaurs that took to life in the trees.

and they walked on their back legs. Plant-eating types developed later, and some of them went back to walking on all four legs, although the back legs still remained much longer than the front ones. Some of these plant-eating dinosaurs reached enormous sizes. *Brontosaurus* and *Diplodocus* were about 80 feet long and must have weighed something like 30 tons. They lived in swamps, where the water supported the bulk of their great weight. Other plant-eating dinosaurs had bony armor or horns as a defence against the ferocious flesh-eaters such as *Tyrannosaurus*—a 50-foot monster with dagger-like teeth about 5 inches long.

Flying Reptiles

While the dinosaurs were busy colonising the land surface, another group of archosaur reptiles was beginning to conquer the air. They were small-bodied reptiles, but the fourth finger of the front limb was enormously long. It supported a thin membrane which grew out from the sides of the body and also enveloped the hind legs and tail. This membrane formed a wing, but the animals were not very efficient in the air. They probably did more floating and gliding than active flying. There were plenty of insects around by this time and the flying reptiles probably caught them in flight. They may also have swooped down to scoop fishes from the surface of seas and lakes. The flying reptiles are often called pterodactyls, but this name really belongs to only a few of them. The whole group are known as pterosaurs.

The Decline of the Reptiles

When the reptiles first appeared they had no competitors other than their own kind. They were able to spread rapidly and invade new areas. For about 100 million years they dominated the land, sea and air. Then, towards the end of the Cretaceous Period, they began to decline. By the end of that period, about 70 million years ago, the ruling reptiles had disappeared altogether. The only remaining reptiles were those relatively small groups that survive today. We do not know what caused the dinosaurs to disappear, but the most likely explanation is that they were unable to withstand the change in the climate that occurred at that time. There would have been a big change in the vegetation and this would have seriously affected the dinosaurs' food supplies. Whatever the reason, they did disappear and the field was clear for the mammals and birds to take over as the rulers of the land and air.

An early amphibian-like reptile.

Left: Tyrannosaurus, the great flesh-eating dinosaur.

Right: Stegosaurus, a plant-eating dinosaur, was protected by bony flaps on its back and large spikes on its tail.

Below: Triceratops, a horned dinosaur.

Left: Some of the plant-eating dinosaurs grew to huge proportions. Brontosaurus weighed about 30 tons. It probably spent much of its time in lakes where the water helped to support its vast bulk.

225

The Age of Mammals

Mammals are found almost everywhere on the earth—on the land, in the sea, and in the air. Men, whales and bats are all members of this large group. The mammals are the most intelligent of animals and they dominate the animal world nearly everywhere. But the mammals are comparative newcomers to the scene. One hundred million years ago the world was 'ruled' by the dinosaurs—huge cold-blooded reptiles with tiny brains. The dinosaurs then declined and the last of them disappeared about 70 million years ago. In their place came the warm-blooded and more intelligent mammals.

The mammals did not suddenly appear at this time, however: they had been in existence for nearly 100 million years, but they had been 'over-ruled'. The most obvious features of mammals are their hair, their warm blood, and their ability to feed their young on milk. These features are of little help to a palaeontologist dealing with fossilised bones, but there are other mammalian features which can be of help to him. Various bones in the skull and the lower jaw help to distinguish mammals from reptiles.

The Early Mammals

During the Triassic Period, some 200 million years ago, there were numerous reptiles that began to pull their sprawling limbs in under the body. They also evolved larger brains than the other reptiles. The fossil record shows how these mammal-like reptiles gradually changed into mammals. We do not know when hair and warm-bloodedness appeared, but true mammals were living in the Jurassic Period about

The early mammals were small shrew-like creatures. Their insignificance helped them to escape the huge flesh-eating reptiles alive at the time.

Below: During the last 60 million years the horse has evolved from a fox-sized creature. The foot bones show how all but the middle toe have diminished and disappeared.

160 million years ago. The earliest mammals were only the size of mice and rats and they probably ate fruit and insects. The early mammals retained the egg-laying habit, and at this stage one small group of mammals diverged from the rest. They survive today as the monotremes of the Australian region, and they still lay eggs. The rest of the mammals, however, soon began to produce living or active young.

We have very few fossils of the early mammals, probably because their small bones did not fossilise well, but it seems that they did not change much during the next 60 million years. Not until the dinosaurs became extinct did the mammals really begin to evolve into the many forms we know today. About 100 million years ago the mammals had reached the marsupial stage, bringing forth very tiny babies and keeping them in pouches until they were able to look after themselves. At this time the continents of the world were probably united as one large land mass, but this land mass was beginning to break up. The first continent to drift away from the others was Australia, and it took with it a selection of primitive marsupial mammals. Shortly after this, a new kind of mammal appeared on the main land mass. It did not have a pouch and the female kept her babies inside her body until they were well developed. She fed her babies through a special organ called a placenta, and these mammals were therefore called placental mammals.

The early placental mammals were small insect-eating creatures, not unlike their Jurassic ancestors in appearance, but they soon began to evolve into many forms as they filled the niches left by the dinosaurs. Plant-eating and meat-eating forms soon appeared, and some became very large and ungainly creatures. We might say that these

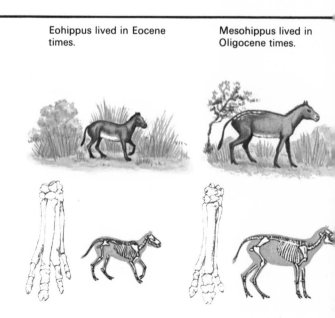

Eohippus lived in Eocene times.

Mesohippus lived in Oligocene times.

strange animals were experimental, for they soon gave way to more efficient types of mammals. By about 60 million years ago, early in the Tertiary Period, all the main groups of mammals were in existence. From then on they produced a vast range of species which gradually evolved into those we see around us today. We know precisely how some of today's mammals have evolved because they have left fossils of each stage of their evolution. We know that the horse, for instance, has evolved from a little fox-sized creature during the last 60 million years, and that the elephant has descended from a pig-sized animal.

The Mammals of Australia

The placental mammals replaced the marsupials (pouched mammals) in most parts of the world, but Australia had become separated from the rest of the continents before the placental mammals started their

The mammoth was an animal ideally suited for life on the margins of the ice sheets during the Great Ice Age of Pleistocene times.

Thylacosmilus, a sabre-toothed marsupial, one of the many strange animals which lived in South America while it was cut-off from the rest of the world. Most of the South American marsupials have now died out.

advance. A number of bats were able to fly to Australia, and various rats and mice reached it by drifting across from Asia on floating logs, but no other placentals arrived in Australia until men started to take them there. The marsupials therefore had the place almost to themselves for 100 million years. They produced a wide variety of

pouched mammals to fill all the available habitats. Most of the placental mammals have an equivalent marsupial living in a similar habitat in Australia. The two groups often look alike because they live under similar conditions.

There are also a few marsupials in South America, for this continent was also separated from the rest for a long period. The separation began perhaps 60 million years ago, when marsupials and some of the hoofed mammals had reached the area. Many strange animals evolved during the separation, but most of them died out when the continent was rejoined to North America about 10 million years ago. Carnivores from the north swept down into South America and killed off most of its more primitive inhabitants, but some marsupials survived together with the sloths and armadillos.

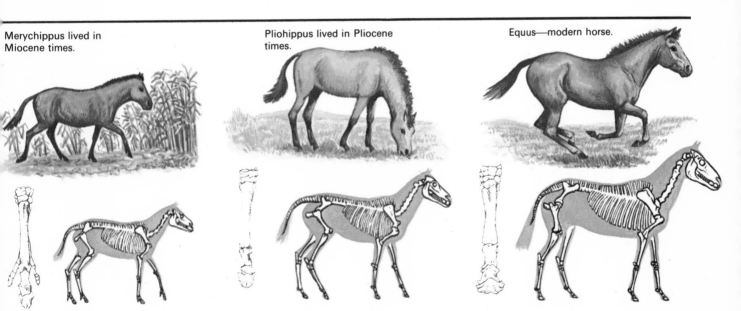

Merychippus lived in Miocene times.

Pliohippus lived in Pliocene times.

Equus—modern horse.

The Origins of Man

Man is an animal just as much as cats and dogs are animals. He belongs to a group of mammals called primates, and his nearest living relations are the apes. The main difference between man and the other animals is that man has a much better brain. It is this which enables him to talk, to think things out, and to make things with his hands.

The first primates appeared on the earth some 70 million years ago. They were small tree-living creatures, rather like the tree shrews that live today. In the next 10 million years the animals evolved in several directions and produced a number of creatures rather like today's lemurs and tarsiers. South America had become separated from the rest of the world by this time, and the tarsier-like creatures living there gradually gave rise to the New World monkeys. This happened perhaps 40 million years ago. Monkey-like creatures were also developing in the Old World—probably in Africa. By 30 million years ago these Old World primates had diverged into two groups—the monkeys and the apes. The early apes were not very specialized creatures and they lived mainly in the trees. Many kinds of small apes appeared during the next 20 million years and, by about 12 million years ago, the ancestors of today's apes had all branched off from the original group.

An ape called *Ramapithecus* was living in parts of Africa and Asia about 12 million years ago. We have fossils only of its teeth and jaws, but these show definite links with human jaws. Man was therefore beginning to separate from the apes at least 12 million years ago. We have not descended from chimpanzees or gorillas; we merely share a common ancestor with them. *Ramapithecus* probably looked little different from the other apes at first, although he may have been a more upright creature and he may have spent more time on the ground.

Ape Men

At the end of miocene times, about 12 million years ago, the apes began to dwindle and the fossil record shows that monkeys became more numerous. This is still so today. About two million years ago the climate began to cool down as the Great Ice Age approached. The forests shrank back towards the Equator and the monkeys and tree-living apes went with them. The ground-living descendants of *Ramapithecus*, however, were quite happy to stay in the open

Many of the Australopithecine fossils have been found in cave deposits. The presence of baboon skulls together with sticks and stones suggests that Australopithecus may have hunted the baboons with these primitive weapons.

Below: The family tree of man.

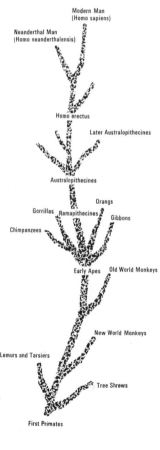

Modern Man
(Homo sapiens)

Neanderthal Man
(Homo neanderthalensis)

Homo erectus

Later Australopithecines

Australopithecines

Orangs

Gorillas Ramapithecines
Gibbons

Chimpanzees

Early Apes   Old World Monkeys

New World Monkeys

Lemurs and Tarsiers

Tree Shrews

First Primates

grassy country which replaced the forests. They became more common and many fossils of them have now been found in various parts of Africa. They are called australopithecines—a name which means 'southern apes'.

The first australopithecine fossils to be discovered were unearthed in South Africa in 1924. They were pieces of a skull and they showed a remarkable combination of human and ape-like features. More fossils have been found since and they show that the australopithecines were very much in between apes and men. They were real 'ape-men'. The skulls had the heavy brow ridges and the projecting jaws of apes, and the brain was just about the size of a gorilla brain. The teeth, however, were very like those of modern man, and the head was balanced on the top of the neck more or less as it is in modern man. The leg bones and the hip bones indicate that the creatures were about five feet tall and that they walked in an upright position. They were therefore creatures with the build and posture of men, but with the heads and intelligence of apes. They were undoubtedly hunting animals and they may well have used sticks and stones to kill their prey.

Several different kinds of australopithecines have been discovered, ranging between one million and two million years old. It is likely that there were several distinct lines and that one of them gradually gave rise to true men.

The First Real Men

It is difficult to say just when the ape-like creatures became men because the changes were so gradual. But it is commonly agreed that they became men when they began to make tools. On this basis, some of the later australopithecines were men, because simple stone tools go back at least one million years. The later australopithecines, however, were not direct ancestors of modern man: they were a sideline which became extinct. By 500,000 years ago real men were living all over Europe, Asia and Africa. They were much more advanced than the australopithecines, but they still had some way to go before producing modern man. One of the best known of these early men is known as Java Man, because his remains were first found in Java in 1891. He was about five feet tall and he walked in an upright position just as we do today. His skull was still apelike, however, with projecting jaws and heavy brow ridges. The brain had a volume of about 900 cc—much larger than that of the australopithecines, but still much smaller than that of modern man. Java Man was therefore probably not very intelligent, although he made simple stone tools. Java Man used to be called *Pithecanthropus*, which means 'ape man', but he is now called *Homo erectus*. This means 'upright man' and indicates that he was a true man quite closely related to ourselves. Men of this general

Some activities of Neanderthal Man—fashioning stone tools and hardening the points of wooden spears by fire.

A reconstruction of Java Man in his natural surroundings. He was a hunter and lived largely on various antelopes and other animals whose bones are common in the same rock deposits. He probably moved about in small bands in search of food.

type were widespread about 300,000 years ago. Modern man, called *Homo sapiens* or 'wise man', has descended directly from them, although there have been several other sidelines.

Cave Man

One of the best known of these sidelines lead to Neanderthal Man, who was living in many parts of the Old World until about 50,000 years ago. The ice age was still in progress and Neanderthal Man in Europe lived mainly in caves. His stone tools are quite common in cave deposits, but the famous cave paintings of Southern Europe were not done by him. These are the work of modern man who came on the scene just over 50,000 years ago. All of today's men belong to the one species *Homo sapiens*, although there are several different races.

Prehistoric Plants

Scattered traces in the rocks show us that plants were living in the world about 3,000 million years ago. These early plants were algae, related to the seaweeds that live today. Algae are soft-bodied plants and they have left few records in the rocks. The fossils that have been found are generally of algae that secreted layers of limestone around themselves. Algae of this sort help to build today's coral reefs.

The First Land Plants

Traces of mosses and ferns appear in the Cambrian rocks, but these groups must have been in existence long before the opening of the Cambrian Period about 600 million years ago. They presumably arose from the algae, but we have no fossils to prove it. It is likely that the algae started to live on the land and evolved first into liverworts and mosses, and then some of these changed gradually into fern-like plants. But again we have no fossils to prove this. The mosses and the ferns could well have descended from separate types of algae. The poor fossil record is due to two main factors: the earliest land plants would have had no hard parts and, being land plants, they would not have been buried very often.

After the Cambrian Period there is a big gap in the fossil record and we have to move forward about 100 million years to find some more plant fossils. These are of strange plants called psilophytes. They had no roots or leaves and the plants merely consisted of

Right: Fungi preserved in the wood of a Carboniferous tree (left) compared with hyphae of present-day fungi (right). There seems to have been hardly any change in their structures since at least Carboniferous times.

 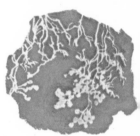

Below: A Carboniferous swamp-forest 300 million years ago. There were no flowering plants but the other main branches of the plant kingdom had emerged. The inset pictures show the modern relatives of these prehistoric plants. In most cases they are much smaller. The counterparts of our spindly horsetails for instance were woody and perhaps 100 feet tall.

branching stems up to a few feet high. Plants of this kind have been found perfectly preserved in some Scottish rocks about 400 million years old. Study of these fossils has shown that the plants had water-conducting tubes and that they carried spore capsules. These ancient plants are thus related to the ferns. Until the recent discovery of Cambrian land plants, the psilophytes were thought to be the ancestors of all land plants. It now seems more likely that they are 'cousins' of the ferns and not their ancestors. Rocks of about the same age as those containing the psilophytes have also yielded pieces of tree trunk, showing that much larger plants were already living.

The Coal Forests

Land plants probably evolved quite rapidly during the Devonian Period, when the vertebrate animals were also beginning to conquer the land. Mid-way through the Carboniferous Period, about 300 million years ago, there were extensive forests containing a varied assortment of woody plants.

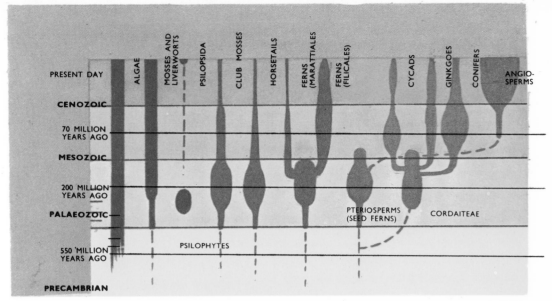

Right: A chart showing the sequence of the main plant groups in geological time. Because of an incomplete fossil record, the evolution from one group to another is uncertain. The origin of the angiosperms is particularly open to question.

ALGAL ANCESTORS?

We know a great deal about these forests because their remains gradually became converted into coal. Coal is really just a mass of fossilized plants, although the individual plants are not always recognizable. The coal forests grew in vast coastal swamps at a time when there were periodic changes in sea level. The sea gradually swept in through the forests and killed the plants. The plants fell and were quickly covered by mud and sand. Later on the sea retreated and the swamp forests spread out again over the fresh mud and sand. This happened several times, so that today's coal fields consist of alternating layers (seams) of coal and shale or sandstone. Some of the best plant fossils are found in the shale. They include stems and leaves washed out to sea from the swamps further inland.

Seed Ferns

The coal forests contained many ferns—some of them quite big—and they also contained plants called seed ferns. These were like ferns to look at, but they produced seeds instead of scattering tiny spores. There were some cone-bearing trees called Cordaitales. These were rather like today's monkey puzzle trees, but the main plants were huge horsetails and club mosses. Today's horsetails and club mosses are all small plants (see page 34), but the coal forest ones were woody trees reaching 100 feet in height. There were no flowering plants, although the seed ferns were possibly flowering plants in the making.

The Rise of the Conifers

Towards the end of the Carboniferous Period, in which the coal forests were living, the cone-bearing plants began to increase at the expense of the horsetails and club mosses. Conifers not unlike today's pine, larch, and cypress arose—probably from the Cordaitales. Cycads (page 37) and ginkgos arose at about the same time as well. The ginkgo or maidenhair tree has survived almost unchanged for millions of years and can certainly qualify for the title of 'living fossil'. It is very rare in the wild in its native China, but it is widely cultivated in gardens.

The First Flowering Plants

The ancient Cordaitales became extinct about 250 million years ago, but the seed ferns lived on until well into the dinosaur era about 125 million years ago. Flowering plants first became common at about this time but, although we have no fossil evidence, they must have been in existence long before that. They most probably descended from seed ferns, but we have no fossils to show us whether this is so or not. The lack of fossils is probably due to the fact that the early flowering plants appeared on high and dry ground where they would have little chance of being preserved. Whatever their origin, the flowering plants rapidly became the dominant plants in the world as a result of their more efficient reproductive methods (see page 38). The earliest flowers were pollinated by the wind, but many of them later formed associations with insects. Bees, butterflies, and other flower-feeding insects began to appear at this time. Their development went on side by side with the development of the flowers, for neither could survive without the other. The associations between some flowers and insects became so close that only one kind of insect can visit each kind of flower and collect nectar.

Fossil leaf of a seed fern and a reconstruction of the original plant. Seed ferns were common in Carboniferous times but died out during the Cretaceous period.

231

The Theory of Evolution

In the previous few pages we have seen how each group of backboned animals gradually arose from an earlier group. In other words, we have traced the evolution of the backboned animals. Evolution means the gradual change of one type of plant or animal into another. It is generally believed that a primitive form of life began on the Earth hundreds of millions of years ago and that all of today's plants and animals have evolved from this simple ancestor. The evidence to support this idea of the evolution of living things comes from several directions.

Evidence from Fossils

Fossils provide some of the strongest support for the idea of evolution, as we have already seen in the last few pages. *Archaeopteryx* was a primitive bird, but it had many reptilian features. This strongly suggests that birds evolved from reptiles, although we can only guess at the earlier stages in the transformation. There is even stronger evidence concerning the development of the horse. There are fossils showing almost every stage between a little fox-like animal and a modern horse. The only logical explanation is that the animals have gradually changed during the last 60 million years. Similar series of fossils taken from chalk deposits show how sea urchins gradually changed from one species into another. The changes from one generation to the next would have been imperceptible, but over millions of years they added up to large changes.

Evidence from Anatomy

A whale's flipper, a bird's wing and a man's arm all look different, and they are all used for different jobs, but the bones inside them

Convergent Evolution

The process of natural selection ensures that a plant or an animal develops the most efficient shape and structure for its particular way of life. Animals leading similar lives may thus come to resemble each other very closely, although they may be quite unrelated. This is known as convergent evolution. One of the best examples concerns the European mole and the marsupial mole of Australia. Both spend their lives tunnelling in the soil and they are remarkably similar, yet one is a pouched mammal and the other is a placental belonging to a very different group of mammals. The similarities are due solely to their similar modes of life. Convergent evolution also works with plants. Some of the cacti in the American deserts are almost indistinguishable from spurges growing in the deserts of South Africa.

Genetics and Evolution

Every cell in the body of a plant or an animal contains a number of minute thread-like structures called *chromosomes*. Each one carries a number of *genes*. The genes are the instructions' which ensure that the cells (and therefore the whole body) develop in the right way. When plants or animals reproduce they normally combine instructions from both parents. Slightly different combinations of genes are produced in the offspring, and these give rise to slight variations in the offspring. They might be variations in size, color, or resistance to disease. Natural selection then gets to work on these variations and favors those which are useful. The organisms thus become more efficient and better adapted to their surroundings. Sometimes there is a 'mistake' when the instructions are handed on to the offspring. Wrong instructions are usually harmful and the organism usually dies, but sometimes they result in major improvements. Organisms with a useful mistake, or *mutation* as it is usually called, will survive and breed. The useful feature will thus be handed on. Mistakes or mutations of this kind must have happened many many times during the course of evolution, resulting in the appearance of new kinds of animals.

Darwin did not know anything about genes and chromosomes when he put forward his theory of natural selection, but he did not need to. He could see that variations were being produced and that was enough for him to explain evolution. The discovery of the genes and chromosomes merely enabled us to explain *how* the variations were produced.

A fish, an ichthyosaur (reptile) and a whale (mammal)—three types of swimming vertebrate. The fore-limbs are homologous (similar in relative position and structure) showing that the three groups have an ancestry, however remote, in common.

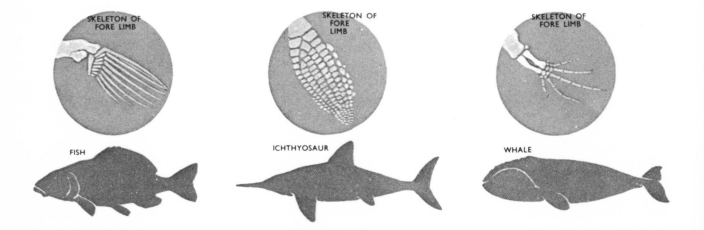

SKELETON OF FORE LIMB

SKELETON OF FORE LIMB

SKELETON OF FORE LIMB

FISH

ICHTHYOSAUR

WHALE

are remarkably similar. This suggests very strongly that all the animals have evolved from the same stock and that they have merely adapted the original skeleton for different jobs.

Evidence from Embryos

Lizards, birds, rabbits and men all look very different when they are fully grown. In their early stages, however, they all look much alike. All of them pass through a fish-like stage during their development. The only logical explanation for this is that the animals have all descended from some kind of ancestral fish. The fossil record points to the same thing and we can be quite certain that this is the correct explanation.

Young animals also provide evidence for evolution among the invertebrates. Some marine worms and molluscs have very similar young stages. This suggests that the two groups, very different in adult form, have descended or evolved from a single ancestral group.

An Old Idea

The idea that living things have evolved is not new. The Greeks suggested it about 2,500 years ago, but they had little evidence to back up the suggestion at the time. The idea cropped up again from time to time during the last thousand years, but most people preferred to believe in special creation. They believed that each kind of plant and animal was created by some supernatural power and that it did not change throughout its existence. The different kinds of animals at each period of the Earth's history were explained by suggesting that worldwide floods periodically wiped out all the animals and that new and slightly better forms were created in their place.

One of the main reasons why the idea of evolution was not readily accepted was that no-one could explain how or why the evolutionary changes might have taken place. In particular, no-one could explain the progressive improvement in the plants and animals. Why should living things have become more and more complex? Why did they never start going back the other way? Charles Darwin provided the answers to these questions in the middle of the 19th Century and the idea of evolution gradually came to be accepted.

Darwin's Voyage on the Beagle

In 1831, when he was only 22 years old, Charles Darwin began a round-the-world voyage in the survey ship *H.M.S. Beagle*. When the ship left England Darwin believed in special creation, but as the voyage went on he began to have doubts. He saw that animals and plants were beautifully adapted to their surroundings, but he also noticed that the animals and plants in one part of the

Darwin discovered a dozen or so species of finch inhabiting the Galapagos Islands of the Pacific. The ancestral species of finch is believed to have come from South America 600 miles to the east. In different parts of the archipelago new distinct species evolved from this original type, each adapted to a particular form of feeding.

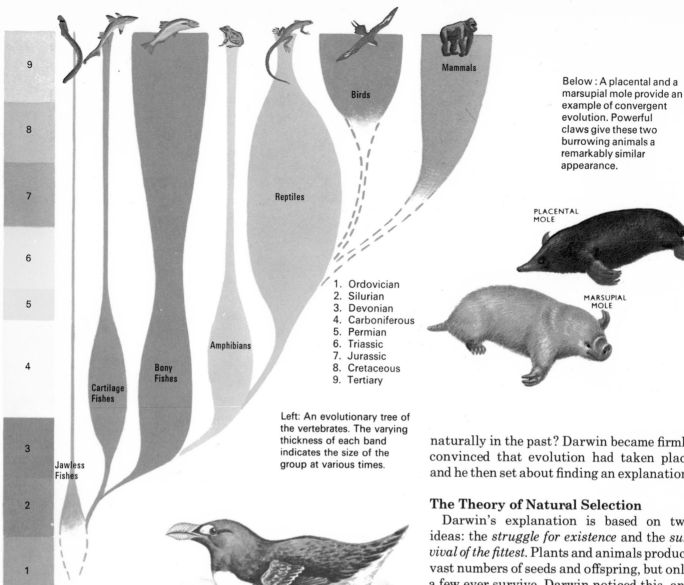

1. Ordovician
2. Silurian
3. Devonian
4. Carboniferous
5. Permian
6. Triassic
7. Jurassic
8. Cretaceous
9. Tertiary

Left: An evolutionary tree of the vertebrates. The varying thickness of each band indicates the size of the group at various times.

Below: A placental and a marsupial mole provide an example of convergent evolution. Powerful claws give these two burrowing animals a remarkably similar appearance.

PLACENTAL MOLE

MARSUPIAL MOLE

world were not the same as those in similar habitats elsewhere. Why should so many different types have been created to fill one type of habitat? In the Galapagos Islands of the Pacific he found numerous kinds of finches. They all had a general resemblance to each other and to the finches of the South American mainland, yet they were clearly different species. Why should so many species have been created in this small area, and why should they have so many similarities? Darwin came to the conclusion that the various species had not been created individually, but that they had all descended from a common ancestor. In other words, he came to the conclusion that evolution had occurred.

During and after the voyage Darwin collected much more evidence to support the idea—evidence from fossils, evidence from anatomy and embryology, and evidence from cultivated plants. He knew that crop plants had been developed from wild ones, so why should not similar changes have occurred

The flightless great auk became extinct in the last century.

Right: Evidence that molluscs and annelid worms had a common ancestor is provided by the existence of similar larval forms in some species.

naturally in the past? Darwin became firmly convinced that evolution had taken place and he then set about finding an explanation.

The Theory of Natural Selection

Darwin's explanation is based on two ideas: the *struggle for existence* and the *survival of the fittest*. Plants and animals produce vast numbers of seeds and offspring, but only a few ever survive. Darwin noticed this, and he also noticed that the members of a species vary among themselves. He reasoned that the individuals with the most useful variations would survive best. These would be the ones that would breed and produce the next generation. Offspring tend to resemble their parents, and so this next generation would possess the useful variations. There would still be a struggle for existence, and again the fittest would survive. This process would go on generation after generation and the plants or animals would gradually get better and better suited to their surroundings. They

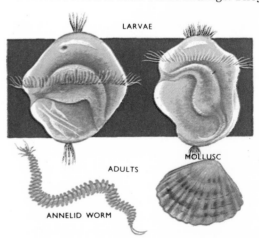

LARVAE

ADULTS

ANNELID WORM

MOLLUSC

would get more and more efficient at catching food and escaping from their enemies. Conditions are not the same everywhere, and so a variation that was useful in one place might not be useful in another. A given species might therefore change in two or more directions and give rise to two or more different species, each well suited to its own habitat. Progressive improvement is thus produced entirely by natural means. Darwin called his idea the Theory of Natural Selection, and we now believe that all plants and animals have evolved and become adapted to their surroundings by this method.

Evolution Today

It must not be thought that evolution is something that took place in the past and then stopped. It is taking place today just as much as ever, and it will continue for as long as conditions continue to change. One of the best examples of evolution that we can see taking place today concerns the spread of black (melanic) varieties of certain moths. The peppered moth is one of the best known. The normal form of this insect is black and white, but completely black individuals used to crop up every now and then – just as white (albino) varieties occur in other animals.

Until about 1850 the black moths were rare and they did not get much chance to breed. The moths rest on tree-trunks by day, and the normal forms were protected by camouflage among the lichens. The black ones, however, were easily spotted by birds and eaten. When the industrial revolution came and smoke began to blacken the tree-trunks, the black moths had more of a chance. They began to breed and get commoner. The black ones were now better camouflaged on the trees than the normal speckled forms. The original black-and-white form of the peppered moth is now quite rare in industrial regions. Man was responsible for the change in the surroundings, but the change in the moths came about entirely by natural selection.

Hedgehogs

Another interesting example of evolution in action concerns the behavior of hedgehogs. The normal reaction of a frightened hedgehog is to roll into a ball. This protects it from many of its natural enemies, but not from an approaching motor vehicle. There have probably always been some hedgehogs that have run away from danger, but there is evidence that this type of behavior is becoming more common now in built-up areas. 'Running' hedgehogs survive better under today's conditions and they are replacing 'rolling' hedgehogs, just as the melanic peppered moths replaced the speckled ones.

Evolution at work. The black (melanic) form of the peppered moth has increased in Britain during the last hundred years or so in industrial areas at the expense of the normal peppered moth because smoke pollution has blackened buildings and trees, thus providing it with better camouflage. The photograph below shows both forms on normal unblackened bark. In this case the normal peppered moth is far better camouflaged than the melanic form.

Living Fossils

In 1938 a strange fish was brought up by a trawler working off the coast of South Africa. The fish was heavily built and nearly five feet long. Its fins were most unusual because, instead of arising directly from the body, they were carried on muscular lobes. They were something of a cross between normal fish fins and the limbs of land animals. None of the fishermen had ever seen a fish like it before and they sent it to a museum. Scientists looked at it and realised that it was a coelacanth.

Scientists knew about coelacanths, but they knew them only as fossils. They were rather primitive fishes and they were common 350 million years ago, but they were thought to have become extinct more than 70 million years ago. Yet here they were alive in AD 1938 and very little different from their ancient ancestors. It is not surprising that the coelacanths have been called 'living fossils'.

The tortoise belongs to a group of reptiles which have changed very little since they first appeared some 200 million years ago.

The coelacanth was rediscovered in 1938 after it was thought to have become extinct some 70 million years ago.

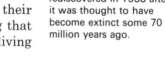

The tuatara is the only living representative of a group of reptiles more ancient than the dinosaurs.

Remains in the Earth's rocks show us that there has always been a gradual replacement of primitive plants and animals by more advanced and efficient types. Why then should the coelacanth have survived? The answer must be that the coelacanth, although a primitive type of fish, had become very well adapted to the conditions under which it lived, and that the later fishes have failed to produce a more efficient species for those particular conditions. Conditions in the sea have not changed a great deal during the Earth's history, and the features evolved by the ancient animals are still useful today.

There are many other marine animals that qualify for the title 'living fossils'. Examples include the nautilus and the king crab. The lung fishes are also living fossils, very similar to creatures that became extinct long ago.

It is not only in the sea that we find living fossils. The tortoises and crocodiles belong to very ancient groups of reptiles and they have changed very little since they first appeared some 200 million years ago. The tortoises have almost certainly survived until now only because of the efficient protection given to them by the shell.

Ancient animal types can also survive to become living fossils if they become isolated from the main centers of evolution. New Zealand, for example, is the home of the lizard-like tuatara. This reptile is about two feet long and it is only distantly related to the lizards. It is the only living representative of a group more ancient than the dinosaurs. New Zealand was separated from the rest of the world's land masses before the dinosaurs evolved, but some of these more ancient reptiles had already arrived there. Isolated from competition, they flourished for a long time. The tuatara was common in New Zealand until the middle of the 19th Century, but Europeans then started to bring in a variety of mammals which killed off the tuatara. It now lives only on a few off-shore islands, but it is quite plentiful there.

The Balance of Nature

No animal or plant lives alone: each one is intricately bound up with dozens of other organisms in what we can call the *web of life*.

Plants and animals have been on the Earth for a very long time and, in most places, nature's web has been perfectly developed. The members of each community have become perfectly adapted to each other. There are just the right number of plant-eating animals in each community to keep the plant life in check without any risk of over-grazing. Similarly, there are just enough carnivores to keep the plant-eaters in check. Any increase in the carnivores would push the herbivore population down. Many carnivores would then perish through lack of food, and the populations would return to the normal levels. The habitat as a whole would remain unchanged over a long period of time.

Large-scale changes, however, would have a serious effect on the community. Removal of one species would destroy much of nature's web, because all the species in a community are linked up. Removal of a major species could even destroy the habitat itself. Man is doing just this in many places today, often helped by the animals he introduces. Little harm is normally done by the occasional natural introduction of a species into a bal-

Artificial communities of plants do not support a large number of different species of animals. The result is that one species often increases to enormous numbers. In other words it becomes a pest. The Colorado beetle has ravaged potato crops in North America and Europe.

Removal of the natural vegetation has resulted in soil erosion in many parts of the world. The Tennessee Valley in North America was a particularly badly affected area. Now efforts are being made to repair the damage by replanting the natural grass and tree cover.

anced community. There is no room for a newcomer in a balanced community and it fails to survive as a rule. But man usually throws the community out of balance when he arrives, and the animals he introduces may become established.

Upsetting the Balance

The earliest men were hunters and gatherers. They roamed around in small bands, killing animals and collecting fruit for food. They were just like any other animal, fitting in with their surroundings and doing no harm to their environment. Each band probably had its own territory, but they did not stay in any one place for very long.

When men learned how to grow crops and to domesticate animals they started to settle down in permanent homes. From then on men have increasingly upset the balance of nature. Clearing the land to make way for crops and domestic animals led to the loss of many wild species, both plant and animal. Land clearance is still going on today, and it is going on more and more rapidly to

237

accommodate our increasing population. Land is needed for homes, schools, factories, and many other buildings as well as for agriculture. Animals are in danger all over the world because their habitats are being destroyed. Among these threatened animals are the orang utan, confined to Borneo and neighboring islands, and the unique lemurs of Madagascar.

Removal of the natural vegetation has resulted in soil erosion in many parts of the world. The land is bare and useless where this has happened.

Natural communities of plants and animals contain large numbers of different species. They are all kept in check by one another and their numbers remain steady. Artificial communities, however, do not have such stability. There are few species in a field of crops and one species often increases to enormous numbers. In other words, it becomes a pest. One way to keep down pests is to leave as much as possible of the natural community around cultivated land.

Strangers from Abroad

When men started to travel from place to place they opened another chapter in the destruction of nature's balance. Plants and animals went with them, sometimes taken deliberately and sometimes accidentally. Many of the species failed to survive in new areas, but some survived only too well. They had no natural enemies in the new lands and they multiplied rapidly, often with disastrous effects on the native plants and animals.

Mongoose

Above: A typical food chain beginning with a plant and ending with a flesh-eating mammal.

Food Chains

Lions eat zebras: zebras eat grass. This sequence of events is called a *food chain*. Every plant and animal is involved in one or more food chains. Some are longer than the one above, some are shorter, but every one starts with a plant. No matter what animal you start with, if you follow its food chain back, you will eventually arrive at a plant—usually in less than five steps. This is because the green plants are the only organisms able to incorporate the sun's energy into food. Nature's web is based very largely on these food chains.

The introduction of the rabbit to Australia has been called the most expensive mistake ever made. Twenty-four rabbits were released by an English settler about 100 years ago. They spread like wildfire and there were millions of them within a few years. They ate the grass down to its roots and the sheep farmers lost millions of pounds. Foxes, stoats, and other natural enemies were imported to try to control the rabbits, but they failed. They preferred to go for the native birds and mammals. The disease called myxomatosis has now brought the rabbits under some sort of control, but Australia will never be free of them.

A similar sort of story can be told about the mongoose which was introduced to the West Indies to control the rats that were destroying the sugar cane fields. The mongoose certainly destroyed the rats, but it also destroyed much of the native wildlife.

The rabbit and the mongoose, together with the grey squirrel and the muskrat, were deliberately taken to different areas. Many smaller animals have been carried around accidentally. Hundreds of insect species have become spread around the world through being carried unnoticed on plants and other materials. Rats have also been carried all over the world in ships. Both the black rat and the brown rat came from Asia, but they are now found almost everywhere and they do a great deal of damage. The black rat that was brought to Europe in the Middle Ages carried the germs of the Black Death, the plague that killed a large proportion of the population in the 14th century.

The introduction of animals, such as the mongoose, to new areas, whether by design or accident, frequently results in disaster to themselves or some other form of life. Having no natural enemies the animals sometimes multiply rapidly and become pests.

Conservation

Conservation means maintaining the balance of nature that has been built up over millions of years. Man has been destroying this balance over the last few centuries and he now has to make good the damage before it is too late.

The natural vegetation is the one that is most suited to a given area—otherwise it would not have developed there. Obviously we have to grow crops, but we must try to keep the natural vegetation as much as possible. Some of the lands around the Mediterranean can show us the results of over-grazing. The land has been reduced to unproductive desert in many places. Too many trees were removed to make way for crops and animals and too many goats were reared in a small area.

The animals are just as important as the plants. They can provide us with abundant food if they are managed properly. Many of the African antelopes could be herded like cattle and they would give better results than introduced cattle in many places. This is because they are better suited to the conditions and better able to stand up to the various pests and diseases.

The conservation movement has been building up rapidly in recent years and it is

The American bison was narrowly saved from extinction. Now there are thriving herds protected in National parks in North America.

Below: Wapiti were once common in both eastern and western parts of North America. The only herds that remain are in western reservations.

showing results in some places. Forests are being replanted and animals are being re-introduced to places from where they had disappeared. Nature reserves are being set up all over the world in an effort to save wildlife. The African game reserves and parks are famous tourist attractions as well as sanctuaries for the animals. Nature reserves cannot just be left alone. The right conditions must be maintained. Excess animals often have to be killed so that they do not destroy their food resources. Thinning the populations in this way is for the good

fences, but many have nothing between the animals and visitors except a deep ditch or a moat. Such an arrangement is much more pleasant for the zoo visitor, and it must be more pleasant for the animals themselves than being shut up in small cages.

Wildlife parks are becoming very popular today. These go even further than the zoological park, for the animals are completely free within the boundaries of the park. It is the visitors who are shut up: they drive around in cars or special 'safari trucks' just as they would in the wilds of Africa.

of the species, and it is also useful for man because he can use the excess animals for food.

Zoos

Zoos, now usually called zoological parks or gardens, are places where collections of wild animals are kept in captivity. Some zoos exist simply to provide entertainment for people, but a zoo can do much more than this. Many scientists work in zoos. They study the animals there and they learn a great deal about them, although the behavior of an animal in a zoo is not always the same as its behavior in the wild.

Zoos vary a great deal in size. A big zoo can obviously cater for more animals than a small zoo, but it is not necessarily a better zoo. Whether one is dealing with zebras or mice, the aim should be to provide the animals with conditions as near as possible to those they find in their native homes. Animals used to be cooped up in small cages with iron bars all around them, but most zoos are gradually doing away with cages of this kind. They now try to house their animals in paddocks. Some of the enclosures have

Releasing a rhino in an East African game reserve. Many areas are being restocked with native fauna by breeding centers.

Zoos and Conservation

A zoo which has interesting exhibits and which attracts a lot of visitors might be considered a successful zoo. But a zoo is not really a success unless it has a good breeding record. There is not much to be proud of in keeping an animal until it dies and then replacing it. It is much better if a zoo can breed its own replacements. It is quite an achievement to breed wild animals because many of them refuse to breed unless they are given just the right conditions.

Zoos which can breed wild animals can play an important part in saving rare species. They can exchange young animals, and this means that several zoos can display rare species without actually taking any from the wild. Many animals are threatened with extinction because they have been hunted too much or because their homes have been destroyed. If their numbers get too low they might be too scattered to meet and breed. The only way to save such species is to gather the survivors into a zoo or a wildlife park and breed them. When the numbers have risen, some of the animals can be

Père David's deer. Though extinct in the wild, a herd thrives at Woburn in England.

Elephants thrive in the East African reserves. Virtually their only enemy has been man.

returned to their homes or released into new homes.

The Hawaiian goose, or ne-ne, is one of several animals that have already been saved by breeding in zoos. In 1950 there were less than 20 of these birds alive in the wild and about the same number in captivity. Today, thanks largely to the efforts of the Severn Wildfowl Trust, there are about 800 birds. Many have already returned to their native Hawaii.

Another success story concerns Père David's deer. This fine deer is a native of China and it was unknown in the West until about 100 years ago. It was already extinct in the wild but a few animals survived in parks. Some were brought to Europe and the

Duke of Bedford formed a herd at Woburn. The original Chinese herds disappeared, but the deer thrived well at Woburn. Specimens have been sent to many other zoos and they have also been sent back to China, although they have not yet been released in the wild.

One of the latest attempts to save a vanishing species concerns the beautiful Arabian oryx. This animal was hunted almost to extinction in the sandy wastes of Arabia, but a few specimens were captured and taken to Phoenix Zoo in Arizona. The climate there is similar to that in Arabia and the oryx took quite happily to its new home. It began to breed and a small herd now exists. One day it might be possible to send specimens back to Arabia.

Domestic Animals

So-called 'wild' animals are those creatures which live outside Man's control. Most of them are frightened or suspicious of human beings. They cannot be approached, let alone touched. At a certain distance, which varies from animal to animal and even individual to individual, they retreat. This distance is called the *flight distance*. If cornered, the mildest of them will become frantic and fight.

Domestic animals in contrast will allow their owners or regular handlers not only to approach but to stroke them. The flight distance has been completely eliminated.

The process of domestication takes place by a very simple learning process called *habituation*. Once animals have overcome their initial fear they may soon come to tolerate, even enjoy human company. The original stimulus to run away disappears.

Domestication is further encouraged by another simple learning process called *association*. The wild creature perhaps comes to associate human beings with some form of advantage such as food delicacies.

When Man first started to domesticate wild creatures long ago in pre-historic times, almost certainly he did so by capturing young animals. Young wild animals are far more easily domesticated than their parents. Having no other experience of Man, they very readily accept him without fear. They are also more easily trained, for they have developed no patterns of behavior of their own. Thus it is far quicker and easier to train a puppy than a full-grown dog.

Breeding Domestic Animals

The emergence of so many varieties of domestic animals has been due solely to the influence of Man. By crossing carefully chosen animals he could gradually exaggerate certain qualities of appearance or behavior.

Animals particularly associated with selective breeding before the 18th and 19th centuries were horses, dogs and 'sporting' creatures such as fighting cocks and falcons. Before the age of mechanisation horses were almost indispensable. They were the chief beasts of burden and much in demand for conducting the numerous wars of the time. Great care was therefore taken in their

Many animals have been enlisted into the service of man which has aided their preservation. The Indian elephant is a valuable beast of burden in South-East Asia.

The horse has been one of the most valuable domestic animals. Numerous breeds have been developed to suit different purposes.

American Standard breed

Palamino

Arabian

Mongolian

Morgan

Pinto or Mustang

Selective breeding has brought many changes in domestic animals. The domestic sheep looks very different from its wild counterparts.

breeding. Even King John (King of England, 1199–1216) played his part in improving the stock. He passed laws prohibiting the excessive use of inferior and small horses in breeding.

Dogs were probably the first creatures domesticated and bred by Man. Their close relationship with Man dates from prehistoric times. Some of our present breeds are very ancient indeed. Bronze Age Man had Alsatian dogs 6,000 years ago while 5,000 years ago the Egyptian Pharaohs used greyhounds for coursing.

With the steady growth in human population during the 18th and 19th centuries more and more food was needed. Improved agricultural methods were employed. In Britain, about this time, fields first became enclosed for keeping cattle, pigs, sheep and goats. Man was then able to exercise far more control over the breeding of his livestock.

Perhaps the most remarkable of all changes have taken place in the pig. Three centuries ago, pigs in England were fierce creatures with most of their weight concentrated in their muscular shoulders. As a result of breeding, today's 'porkers' have their weight more in the middle of their bodies and towards their hindquarters.

Chickens, which probably descended from the Red Jungle Fowl of India, were probably introduced into Britain about 2,000 years ago. Breeding designed to increase output of eggs is quite a recent development. Formerly they were kept almost exclusively for their meat or cocks were bred for the 'sport' of cock-fighting.

All sorts of animals have been domesticated by man in various parts of the world—camels, yaks, elephants, reindeer, lamas. More are likely to be enlisted for Man's benefit.

Index

Aardvark 125, 145
Aardwolf 145
Abalone 145
Acacia 50
Accentor 145
Aceraceae 63
Acorn worm 145, 220
Addax 145
Adder 119, 145
Aestivation 141
African elephant 162
Agaric, fly 311, 134
Agave 57
Age, Great Ice 228
Age of fishes 220, 221, 222
Age of mammals 226, 227
Age of reptiles 224, 225
Agouti 145
Albatross 123, 145
    wandering 145
Alder fly 145
Algae 10, 12, 25, 26, 28, 29, 86, 134,
    230
Alismataceae 52
Alligator 119–121, 145
Almond 42
Alpaca 173
Alpine chough 155
Alpine plants 202, 203
Amaryllidaceae 57
Amber 214, 215
American bison 149, 239
American mink 176
American nightjar 141
Amino-acids 11
Ammonite 111, 145, 214
Amoeba 9, 74, 82, 86, 145
Amphibians 117, 118, 217, 222, 223
Anaconda 121, 145
Anchovy 145
Andean condor 157, 203
Anemone 55
Angiosperms 25, 231
Angler fish 143, 145
Ani 145
Animal cells 10
Animal classification 84
Animal electricity 144
Animal lights 143
Animal parasites 131, 132
Animal partnerships 133, 134
Annelids 90
Annual rings 16
Annuals 41
Anole 145
Ant 82, 95, 104, 105, 145, 146, 168,
    169, 189
Ant lion 146
Antarctica 205
Antbird 145
Anteater 145
Antelope 128, 145, 146, 194, 195
    Indian, see Blackbuck
    sable 194
Antlers 39
Apara, see Three banded armadillo
Ape 228
    Barbary 147
Aphid 83, 95–97, 101, 104, 146
Apocynaceae 65
Apodan 118
Apollo butterfly 203
Apple 40, 42

Arabian camel 153
Arabian oryx 178, 241
Araceae 52
Arachnids 106, 107
Araliaceae 61
Arapaima 146
Archaeopteryx 216, 224, 232
Archerfish 146
Archosaurs 224, 225
Arctic fox 146
Arctic, the 204, 205
Areoles 50
Argali 203
Argonaut 146
Aristotle's lantern 113
Armadillo 146, 227
Armed bullhead 210
Army ant 104, 146
Arrowpoison frog 146
Arrow worm 146, 206
Arrowhead 52, 211
Arthropods 92–94
Artichoke, globe 58
Arum 52
Asexual reproduction, see
    Vegetative reproduction
Ash 42, 61, 200
Asp 146
Ass 13, 147
Assassin bug 147
Association, symbiotic 32
Aster 58
Atlantic salmon 140
Atlas moth 147
Atoll 89
Aubretia 54
Auk, great 234
Australian lungfish 222
Australopithecus 228, 229
Auxin 23
Avens, mountain 202
Avocet 147
Axis deer 147
Axolotl 118, 147
Aye-aye 147

Babbler 147
Baboon 147, 228
Bacilli 27
Backboned animals 222
Backswimmer 147, 212
Bacteria 25, 26, 27
Bactrian camel 153
Badger 142, 147, 201
Bagworm 147
Balance of nature 237, 238
Bald eagle 147
Ball and socket joint 72
Balsa 52
Balsam 52
Balsaminaceae 52
Banana 52
Bandicoot 147
Baobab 52, 53
Barbary ape 147
Barbel 115
Barber fishes 147
Barberry 50
Barbet 147
Bark 15
Bark beetle 147, 200
Barn owl 148
Barn swallow, see Common swallow

Barnacle 92, 93, 147, 208, 209
Barnacle goose 148
Barracuda 148
Barracudina 148
Barrier reef 89
Basilisk 148, 170
Basket star 148
Basking shark 148
Bass 148
Bat 141, 142, 163, 189
    fish-eating 163
    fruit 164
    vampire 189
Batesian mimicry 138
Beaks 123, 124
Bean 40
    runner 49
Bear 126, 127, 150, 151, 154, 155,
    180, 185, 186, 201
Bear-cat, see Binturong
Bearded lizard 148
Beaver 129, 148, 201, 213
Beck 213
Bedbug 96, 148
Bedstraw 52
Bee 95, 102, 103, 132, 137
    dance of 102, 103
    honey 102, 103
Bee-eater 148
Bee hummingbird 123
Bee orchid 64
Beech 42, 52, 200
Beet 52
    sugar 20, 21
Beetle 94, 95, 96, 97, 100, 101, 129,
    143, 147, 150, 152, 156, 157,
    182, 191, 200
Begonia 52
Begoniaceae 52
Beisa oryx 178
Belemite 111
Bell bird 148
Bellflower 52, 54
Beluga 148
Benthos 207
Betulaceae 54
Bewick's swan 186
Bichir 116, 148, 221
Biennials 20, 41
Bighorn 148, 149
Bilberry 60
Bindweed 54
Binturong 149
Biological control 101
Birch 54
Bird 122–124
    bell 148
    bower 150, 151
    butcher 153
    elephant 122, 123
    feathers 124
    humming 169
    lyre 130
    man o' war, see Frigatebird
    mocking 176
    of paradise 149
    scrub- 183
    secretary 184
    tailor 186
    tropic 189
    weaver 191, 192
    widow, see Whydah
Bird-eating spider 149

Birds 122–124
first 224
Birdwing butterfly 197
Bison 149, 194, 239
American 149, 239
European 149
Bitterling 149, 150
Bivalves 77, 108, 109, 150, 210
Black bear 127, 150
Black bryony 54
Black kite 171
Black mamba 175
Black molly 150
Black moths 235
Black rat 181, 238
Black salamander 118
Black tetra 114
Black widow 107, 150
Black-footed ferret 162
Black-necked swan 186, 187
Blackberry 49, 50
Blackbird 124
Blackbuck 150
Blackfly 146
Blackgrouse 167
Blackthorn 50, 66
Bladder wrack 29, 209
Blesbok 150
Blight, potato 30
Blind snake 150
Blister beetle 150
Blood, mammalian 80, 81
Blood system, the 80, 81
Blue whale 150
Blue-tit 129
Bluebottle 97
Bluefin tuna 189
Bluefish 150
Boa 119, 121, 150, 197
water, see Anaconda
Bobcat 150, 191
Body cells, human 10
Boll weevil 191
Bollworm 150
Bombacaceae 52
Bonito 150
Bontebok 150
Bony fishes 114, 116, 221
Booby 150
Boomslang 150
Bootlace weed 210
Boraginaceae 59
Bower bird 150, 151
Bowfin 151
Box 54
Boxfish, see Trunkfish
Brachiopods 76, 219
Bracken 24, 34
Bracket fungus 31
Brain 74, 228
Bream 151, 213
Breathing 77, 78
Brine shrimp 151
Bristlemouth 151
Bristletail 95, 151
Bristleworm 91, 151
Brittle star 112, 113, 151
Bromeliads 197
Bromelidaceae 65
Brontosaurus 225
Broom 42, 65
Butcher's 62
Broomrape 54
Brown bear 151
Brown hyena 169
Brown rat 181, 238
Brush-tail opossum 151, 152
Bryony, black 54
Bryophyllum 24
Bryophytes 25
Buckbean 54, 55
Buckeye, see Horse chestnut
Buckthorn 55

Bud, subsidiary 22
terminal 16, 22
Budding 82
Buddleia 55
Budgerigar 152
Buffalo 133
Cape 153
Bufftip moth 99, 135
Bug, assassin 147
bed 148
Bulb 20, 21
Bullfinch 123
Bullfrog 152
Bullhead 116, 152, 213
armed 210
Bullrush, see Reed mace
Bumble bee 102, 132
Burchell's zebra 192
Burrowing owl 152
Burying beetle 152
Bush 43
Bush cricket 96, 158, 166
Bush dog 152
Bushbaby 128, 152
Bushbuck 152
Bushpig 152
Bustard 152, 153
Butcher Bird 153
Butcher's broom 62
Butomaceae 59
Buttercup 12, 27, 38, 55
Butterfish 153, 210
Butterfly 95, 96, 98, 99, 100, 139,
140, 141, 203
Butterwort 55
Buxaceae 54
Buzzard 123, 153

Cabbage 20, 21
Cabbage family 55, 56
Cabbage white 99
Cacomistle 153
Cactus 38, 50, 54, 55, 56, 198, 199,
232
Cactus, Christmas 56
prickly pear 199
Cactus coren 198, 199
Caddis fly 95, 96, 153, 211, 212, 213
Caecilian 117, 153
Caiman 120, 153
Cake urchin, see Sand dollar
Calanus 92
Calcium 19
Californian condor 157
Cambrian Period 218, 219, 230
Camel 153, 199
Camel, Arabian 153
bactrian 153
Camellia 70
Camouflage 135, 136
Campanulaceae 52
Campion, sea 209
Canada goose 139
Canadian lynx 174
Canadian pondweed 59, 211
Canary 153
Cane, sugar 44
Cannabinaceae 60
Cap, death 31
Cap root 22
Cape buffalo 153
Cape hunting dog 153, 195
Capercaillie 153, 167
Caprifoliaceae 60
Capuchin 153
Capybara 153
Caracal 153
Caracara 154
Carapace 92
Carbon 11, 19
Carbon dioxide 17, 18
Carboniferous Period 223, 230, 231
Cardinal 154

Care, parental 83
Caribou 139, 140, 154
Carnation, wild 65
Carnivores 76
Carrot 20
Carrot family 56
Carp 154, 213
Carpels 39, 40
Carpet beetle 100
Carpet shark, see Wobbegong
Cartilaginous fishes 114, 116
Caryophyllaceae 65
Cassowary 154
Cat 142
Cat, bear-, see Binturong
serval 9
wild 191
Cat-bear 154, 155
Cat-squirrel, see Cacomistle
Caterpillar 95, 96, 98, 99, 110
Catfish 155
Catkin 39
Catshark 155
Cattails 211
Cattle egret 133
Cauliflower 21
Cave fish 155
Cary 167
Cary, water, see Capybara
Cedar 36
red 37
Spanish, see Mahogany
Celandine, lesser 54, 55
Celastraceae 69
Cells 10, 11, 13
Centaury 59
Centipede 155, 200
Cephalopods 110, 111
Cereals 44, 45
Chameleon 120, 121, 136, 155
Chamois 139, 155, 203
Cheetah 73, 155, 195
Chemical energy 17
Chemical senses 75
Chenopodiaceae 52
Chestnut, horse 42, 60, 61
sweet 52
Chevrotain 155
Chickaree 155
Chicory 58
Chimaera 155
Chimpanzee 155
Chinchilla 155
Chipmunk 155
Chital 147
Chitin 72
Chlamydomonas 28, 87
Chlorine 19
Chlorophyll 17, 18, 25, 28, 29, 32,
38, 46, 47, 203
Chloroplasts 10, 17, 28
Chough, alpine 155
common 155
Christmas cactus 56
Chromosomes 10, 13, 232
Chub 213
Chuckwalla 155, 156
Cicada 156
Cichlid 156
Cinnabar moth 138
Cistaceae 67
Civet 156
Clam, giant 109
Classification of animals 84
Classification of plants 25
Clawed frog 156
Cleavers, see Goosegrass
Cleg-fly 169
Clematis 49, 65
Click beetle 143, 156
Climate and soil 14
Climbing perch 156
Climbing plants 49

246

Clothes moth 100
Clover 42
Club mosses 25, 26, 35, 231
Coal 34, 216, 231
Coalfish 156
Coati 156
Cobra 119, 120, 156
    spitting 156
Cocci 27
Cock-of-the-rock 156
Cockatoo 156
Cockle 108, 109, 210
Cockroach 96, 100, 156
Cock's foot grass 44
Cocoa family 56
Cod 83, 116, 156
Codling moth 99
Coelacanth 157, 221, 236
Coelenterates 88, 89
Coeloptera 95
Collared dove 157
Collared peccary 179
Collembola 95
Colobus monkey 157
Colorado beetle 100, 156, 157, 237
Colors, warning 137
Coltsfoot 58
Columbine 55
Comb jelly 157
Common bentgrass 44
Common chough 155
Common eland 161
Common octopus 110
Common swallow 139
Compositae 27, 57
Compound eyes 74, 95
Condor 157
    Andean 157, 203
    Californian 157
Cones 37
Conger 157
Coniferous forest 36, 200, 201
Conifers 25, 26, 36, 37, 200, 201,
    231
Conservation 239–241
Control, biological 101
Convergent evolution 232
Convolvulaceae 54
Convolvulus 49
Coot 157
Copepods 92
Copperhead 157
Coral 82, 88, 89, 134, 157, 210
Coral reef 88, 89, 210
Coral snake 157
Cordaitales 231
Cork oak 15
Corm 20
Cormorant 157, 208
Corn 41
Cornaceae 58
Cornflower 57, 58
Corylaceae 60
Cotton 63
Cottonmouth water moccasin 176
Cottontail 157
Coucal 158
Couch grass 44
Cougar, see Puma
Countershading 135
Courser 158
Courtship 129, 130
Cow, sea 161
Cowbird 158
Cowrie 158
Cowslip 67
Coyote 157, 158, 195
Coypu 158
Crab 72, 77, 92, 93, 133, 134, 162,
    168, 171, 176, 236
Crab spider 158
Crane 158
    whooping 129

Crane-fly 159
Cranesbill 59
Crassulaceae 69
Crayfish 158
Creeping jenny 67
Cretaceous 224, 225
Cricket 75, 158
    bush- 96, 158, 166
Crinoids 216
Crocodile 119, 120, 121, 133, 134,
    158, 224, 236
    Nile 133
Cross-pollination 39
Crossbill 158, 201
Crow 158
Crown of thorns 69, 89
Crustaceans 92, 93, 140, 212
Crutose lichens 32
Cuckoo 132, 139, 158
Cuckoo flower 55
Cucumber family 56
Cucumber, sea 113
Cucurbitaceae 56
Curassow 158
Curlew 158
Currant family 56, 57
Cuscus 158, 159
Cuttlefish 110, 111, 136, 159
Cycads 25, 26, 36, 37, 231
Cyclamen 67
Cyclops 92, 212
Cyperaceae 68

Dab 159
Dace 115, 213
Daddy-long-legs 106, 159
Daffodil 57
Dahlia 58
Daisy 57, 58
Damselfly 159
Dandelion 27, 38, 41, 57, 58
Darwin, Charles 232–235
Darwin's finches 159, 160
Dasyure 160
Date palm 199
Deadly nightshade 67
Deadnettle 57, 58
Death cap 31
Death-watch beetle 100, 101
Deciduous forest 200
Deciduous trees 16, 200
Deep sea angler fish with parasite
    male 115
Deer 135, 147, 162, 176, 190, 191,
    201, 241
Deer-mouse 160
Delphinium 55
Derby eland 161
Desert life 198, 199
Desert lynx, see Caracal
Desert plants 198, 199
Devil fish 160
Devil, Tasmanian 187
    thorny 119
Devonian Period 221, 222, 230
Dhole 160
Diatom 206
Dicotyledons 25
Digestion, feeding and 76
Dik-dik 160
Dingo 160
Dinichthys 221
Dinosaur 119, 120, 224, 225
Dioscoreaceae 54
Dioxide, carbon 17, 18
Diplodocus 225
Diplura 95
Dipper 160, 213
Dipsaceae 70
Diptera 95
Disease, foot and mouth 11
Diver 160
Diver, great northern 160

Dock 58
Dodder, 47, 48
Dog 159, 160
    bush 152
    Cape hunting 153, 195
    prairie 126, 180
Dogfish 114, 160
    lesser spotted 160
Dog's mercury 68, 69
Dogwood 58
Dollar, sand 182
Domestic animals 243
Donkey 13
Dormouse 126, 141, 160
    edible 160
    hazel 160
Douroucouli 160
Dove, collared 157
Dragon, komodo 120
Dragonfly 95, 96, 97, 211, 212, 213
Dragonfly, fossil 215
Driver ant, see Army ant
Dromedary 153
Drone 103
Drongo 160
Droseraceae 70
Duck 123, 139, 205
    shoveller 123
Duck-billed platypus 125
Duckweed 25, 58, 211
Dugong 160, 161
Duiker 161
Dunnock 145
Dyer's greenweed 64

Eagle, bald 147
    golden 203
    harpy 197
Ear 74, 75
Earshell, see Abalone
Earth, life on 11
Earthworm 82, 90, 142
Earwig 95
Eaved umber moth 137
Ebenaceae 58
Ebony 58
Echidna 125, 126
Echinoderms 112, 113
Echolocation 142
Edible dormouse 160
Edible turtle, see Green turtle
Eel 140, 144, 161, 175, 176
Eelgrass 58
Eelworm 90
Egg-eating snake 120
Egret 160, 161
Egret, cattle 133
Egyptian plover 133
Eider 161
Eland 146, 161, 162
    common 161
    Derby 161
Electric eel 144
Electric fishes 144
Electricity, animal 144
Elephant 125, 129, 161, 162, 194,
    227, 241, 242
Elephant bird 122, 123
Elephant seal 162
Elephant shrew 162
Elm 42, 58, 59
Emperor penguin 83, 103, 205
Emu 122, 161, 162
Endive 58
Endo-skeleton 72
Energy 17, 20
    chemical 17
    sun's 17
Eohippus 226
Epiphytes 64, 65, 196, 197
Equus 227
Ericaceae 60
Erosion, soil 237, 238

Eryops 223
Euglena 87
Euphorbiaceae 69
European bison 149
European eel 140
European lynx 174
European mink 176
European mole 232, 234
European spotted salamander 118
Evergreens 196
Evolution 13, 232–235
    convergent 232
    theory of 232–235
Excretion 9
Exoskeletons 72
Eye, 74, 75
    compound 74
Eye spot 138
Eyebright 47
Eyed hawkmoth 138

Fagaceae 52
Fairy fly 162
Fairy shrimp 92, 162
Falcon 123
    peregrine 179
Fallow deer 162
False acacia 50
False fruits 40
False oat grass 44
False scorpion 162
Families, flower 52–71
Fan worm 91, 151, 162
Fangs 119
Feathers 124
Featherstar 113, 162
Feeders, filter 109
Feeding 9
Feeding and digestion 76
Fence swift 185
Fennec fox 162
Fer de lance 162
Fern, bracken 24
Ferns 25, 26, 34, 35, 37, 197, 230,
    231
    seed 231
Ferret 162
    black-footed 162
Fertilization 82
Fertilizer 19
Fever nettle 51
Fibrous roots 15, 16
Fiddler crab 162
Fighting fish 162
Filter feeders 76, 109
Finch 200, 233
    Darwin's 159, 160
    weaver 124
    woodpecker 130
Fir 201
Firebelly 163
Firefly 143, 163
Fish, cave 155
    devil 160
    fighting 162
    flying 163
    hatchet 168
    Nile 144
    paddle 179
    parrot 179
    pilot 180
    porcupine 137, 180
    X-ray 192
    reproduction in 116
    respiration of 77
Fisheagle 163
Fish-eating bat 163
Fishes 114–116, 143, 207, 212, 213,
    222
    Age of 220, 221, 222
    bony 114, 116, 221
    cartilaginous 114, 116
    early 217

Fishes—Cont.
    electric 144
    jawed 220, 221
    jawless 220
    lobe-finned 221, 222, 223
    ray-finned 221, 222
Fission, binary 82
Flagella 87
Flamingo 163
Flatfish 116
Flatworm 83, 90
Flax 59
Flea 95, 96, 100, 132
    water 92
Flicker 163
Flies 75, 95, 96, 97
Flight 123, 124
Flounder 116, 163
Flower families 52–71
Flowering plants 36, 38–41, 231
Flowering rush 59
Flowering trees 42, 43
Flowers, food storage in 21
Flu virus 11
Fluke 90
    liver 131, 132
Fly agaric 31, 134
Fly, alder 145
    caddis 95, 96, 153
    cleg- 169
    crane- 159
    fairy 162
    horse 97
    ichneumon 96, 132
    lace-wing 95
    scorpion 95
    Spanish, see Blister beetle
    stable 97
    tsetse 189
Flycatcher 200
Flying fish 163
Flying fox 163
Flying reptiles 225
Flying squirrel 163
Foliose lichens 32
Food chains 238
Food storage in plants 20, 21
Foot and mouth disease 11
Forest, coniferous 200–201
    deciduous 200
    rain 196, 197
    northern 200, 201
Forget-me-not 59
Forsythia 61
Fossa 164
Fossils 214–217, 228, 230, 232
    living 236
Fowl, jungle 196
Fowl pest 11
Fox 142, 201
    Arctic 146
    Fennec 162
    flying 163
    raccoon, see Cacomistle
Foxglove 69
Freshwater life 211–213
Frigatebird 164
Frilled lizard 119, 164
Fringing reef 89
Fritillary 164
    snakeshead 62
Frog 73, 117, 118, 141, 223
    arrow poison 146
    clawed 156
    gliding 166
    hairy 167
    rain 181
    reed 181
Frogbit 59, 211
Fruit bat 164
Fruits 21, 38, 40, 41, 42, 43
    false 40
Fruticose lichens 32

Fuchsia 71
Fulmar 164
Fungus, bracket 31
Fungi 25, 26, 30, 31, 134, 201, 230
Fur seal 164
Furniture beetle 100, 101

Galagos, see Bushbaby
Galium 52
Gall wasp 164
Galls, plant 131
Game reserves, African 239
Gametes 82
Gannet 164
Gar 164
Garden ant 104
Garden snail 108
Garter snake 119, 121, 164
Gastropods 108
Gavial 120
Gazelle 146, 164
    Grant's 12, 164
    Mongolian 164
    Thomson's 164
    Tibetan 164
    Przewalski's 164
Gecko 164, 165
Gemsbok 178
Genes 10, 13, 232
Genet 165
Gentian 58, 59
Gentianaceae 59
Geotropisms 23
Geraniaceae 59
Geranium 59
Gerbil 165
Gerenuk 165
Germination 41
Gharial 165
Giant anteater 145
Giant clam 109
Giant owlet moth 98
Giant panda 178, 179
Giant snail 165
Giant squid 111
Gibbon 8, 165
Gila monster 120, 165
Gills 77, 93, 108, 109, 114, 116, 117,
    118
Ginkgo 25, 26, 231
Giraffe 165, 194
Glass snake 165, 166
Gliding frog 166
Glis glis, see Edible dormouse
Globe artichoke 58
Glow-worm 143, 166
Goat, Rocky Mountain 202
Goat 202, 203
Godwit 208
Goldcrest 201
Golden Eagle 203
Golden hamster 168
Goliath beetle 94
Goose, barnacle 148
    Canada 139
    greylag 167
    Hawaiian 241
Gooseberry 43, 50
    sea 206
Goosegrass 51
Gopher, pocket 180
Gorse, 43, 50
Goshawk 166
Grain weevil 100, 101
Gramineae 44, 60
Grant's gazelle 12, 164
Graptolites 218, 219
Grass, marram 44
    meadow 59
    rye 60
    sheep's fescue 203
Grass snake 119, 121, 166
Grasses 41, 44, 45, 59, 60, 202

Grasshopper 95, 106, 166
Grasslands 194, 195
Grayling 213
Great Auk 234
Great Barrier Reef 88
Great bustard 123
Great crested grebe 167
Great Ice Age 228
Great Northern diver 160
Greater kudu 171
Grebe 166, 167
   great crested 167
Green Mamba 175
Green turtle 167
Greenfly 96, 146
Greenweed, Dyer's 64
Grevy's zebra 192
Grey parrot 197
Grey squirrel 201, 238
Greylag goose 167
Grizzly bear 151, 167
Grossulariaceae 56
Ground squirrel 167, 203, 205
Grouper 167
   Queensland 167
Grouse 153, 167
   red 167
Growth 9
   plant 22, 23
Grubs 93
Guillemot 167
Guinea pig 167
Gull, herring 130
Gundi 167
Gunnel, see Butterfish
Guppy 167
Gurnard 167
Gymnosperms 25
Gymnure 167

Habitat 193
Haddock 115, 167
Hagfish 114, 167, 220
Hairs 75
   root 15
Hairy frog 167
Hairy hedgehog, see Gymnure
Hake 167
Halibut 168
Hammerhead shark 168
Hamster 141, 168
   golden 168
Hare, sea 183
Harpy eagle 197
Harrier 168
   hen 168
Hartebeest 167, 168
Harvest mouse 168
Harvesting ant 168
Harvestman 106, 168
Hatchet fish 168
Hawaiian goose 241
Hawk 124
   marsh 168
Hawkmoth 168
   eyed 138
Hawkweed 58
Hawthorn 50
Hazel 42, 60
Hazel dormouse 160
Heart 80, 81
Heartwood 15, 16
Heather 60
Hedge sparrow, see Dunnock
Hedgehog 137, 141, 142, 168, 200,
   235
   hairy, see Gymnure
Hellbender 168
Hen harrier 168
Henbane 67
Herb Robert 59
Herbaceous plants 16, 42, 44
Herbivores 76

Hermaphrodites 83, 108
Hermit crab 133, 134, 168
Heron 123, 213
Herring 116, 168
Herring gull 130
   king, see Dhimaera
Hibernation 141
Hibiscus 63
Hickory 71, 200
Hinge joint 72
Hip, rose 40
Hippocastanaceae 60
Hippopotamus 168, 169
   pigmy 169
Hoatzin 169
Hog deer 147
Hog louse 93
Hog, red river, see Bushpig
Hogweed 27
Holly 39, 42, 50, 51, 60
Hollyhock 63
Homo erectus 229
Homo sapiens 229
Honey 103
Honey ant 168, 169
Honey bear, see Kinkajou
Honey bee 102, 103, 140
Honey-parrot, see Lorikeet
Honeyguide 169
Honeysuckle 49, 60
Hop 60
Hormones 80, 130
Hornbill 169
Hornet 169
Horse 13, 128, 194, 226, 227, 243
Horse chestnut 42, 60, 61
Horse, evolution of 226, 227
   sea 183
Horsefly 97, 169
Horsetails 25, 26, 34, 35, 230, 231
Housefly 97, 100
Hoverfly 137, 138, 169
Howler monkey 168, 169
Human heart 80
Hummingbird 123, 169
Hummingbird bee 123
Humus 14
Hybrids 13
Hydra 82, 83, 89, 212
Hydrangea 61
Hydrangeaceae 61
Hydrocharitaceae 59
Hydrogen 11, 19
Hyena 169, 195
   brown 169
   laughing, see Spotted hyena
   spotted 169
   striped 169
Hymenoptera 95
Hypericaceae 68
Hyphae 30
Hyrax 169

Ibex 169, 203
Ibis 169, 170
Ice Age 203
   Great 228
Ichneumon fly 96, 132
Ichthyosaurs 214, 224
Ichthyostega 223
Iguana 170
Impala 75, 170
Indian antelope, see Blackbuck
Indian elephant 162
Indri 170
Infections, control of 80
Insect, breathing system of 78
Insect-eating plants 46
Insect pests 100, 101
Insect pollination 39, 40
Insect, scale 182
   stick 83
Insecticides 100, 101

Insects 72, 74, 75, 83, 94–105
   plant-feeding 96
   social 102
Instinct and hormone 130
Instinct and learning 129, 130
Internodes 22
Invertebrates 81, 84
Involuntary muscles 73
Iridaceae 61
Iris 20, 61
Iron 19
Isopods 93
Ivy 49, 61

Jacana 170
Jackal 170, 195
Jackass, laughing, see Kookaburra
Jackdaw 170
Jaguar 170
Jaguarundi 170
Jasmine, winter 61
Java Man 229
Jay 170
Jelly, comb 157
Jellyfish 77, 89, 143, 206, 218
Jerboa 198
Jet propulsion 110
Jird 165
Joint, ball and socket 72
   hinge 72
Joints 72
Juglandaceae 71
Juncaceae 68
Jungle fowl 196
Jurassic Period 224, 226
Jute 62

Kagu 170, 171
Kakapo 171
Kangaroo 126, 194
Kangaroo rat 171, 199
Kangaroo, red 126
Katydid 171
Kelp, see Oarweed
Kestrel 171
King crab 171, 236
King herring, see Chimaera
King shag 208
Kingfisher 123, 124, 171
Kinkajou 171
Kite 171
Kite, black 171
   red 121
Kiwi 122, 171
Knapweed 58
Knotgrass 58
Koala bear 126, 171
Kodiak bear 151
Kohl rabi 20
Kokoi 146
Komodo dragon 120
Kookaburra 171
Krill 92, 171
Kudu 171
   greater 171
   lesser 171

Labiatae 58
Laburnum 42
Lacewing fly 95, 171, 172
Lackey moth 99
Ladybug 137, 172
Ladybug beetle 96
Lady's smock 55
Lake 211
Lammergeier 172
Lamp shells 219
Lamprey 114, 172, 213, 220
Lancelet 172
Land migration 139
Land snail 77
Langur 172
Lantern, Aristotle's 113

Larch 201
Large white butterfly 100
Larvae 95, 96
Laughing hyena, *see* Spotted hyena
Laughing jackass, *see* Kookaburra
Lavender, sea 70
Learning, instinct and 129, 130
Leatherjacket 159
Leathery turtle 172
Leaves 17, 18, 20–22
  food storage in 20, 21
  scale 20, 21
Leech 90, 91, 172
Legionary ant, *see* Army ant
Leguminosae 27
Lemming 140, 172, 204
Lemnaceae 58
Lemur 172
  ring-tailed 172
Lentibulariaceae 55
Leopard 135, 172, 173, 194
Lepidoptera 95, 98, 99
Lesser celandine 54, 55
Lesser kudu 171
Lesser panda, *see* Cat-bear
Lesser spotted dogfish 160
Lettuce 58
  sea 29
Liana 197
Lice 95, 100
Lichens 32, 134, 201, 202, 204, 209
  crutose 32
  foliose 32
  fruticose 32
Life moves on land 222
  nocturnal 142, 143
  origins of 11
Ligaments 72
Light, ultra-violet 11
Lights, animal 143
Lilac 42, 61
Liliaceae 62
Lily 62
Lily of the valley 61
Lily, sea 113, 162, 183
Lily trotter, *see* Jacana
Lily, water 211, 213
Lime 22, 62
Limestone 218, 219
Limpet 108, 173, 209
  river 213
Linaceae 59
Linden 42, 43, 62
  European 22
Linnaeus 13
Linsang 173
Lion 127, 173, 194, 195
  mountain, *see* Puma
  sea 164, 183
Liver fluke 131, 132
Liverwort, flat 33
  leafy 33
Liverworts 24, 25, 26, 33, 230
Living fossils 236
Lizard, bearded 148
  frilled 119, 164
  monitor 176
  spiny 185
Lizards 83, 119, 120, 121, 141, 148, 169, 176, 185, 224
Llama 173, 174
Loach 213
Lobe-finned fishes 221, 222, 223
Lobelia 62
Lobeliacea 62
Lobster 77, 92, 93, 173
Locust 101, 140, 173, 174
Loganiaceae 55
London plane 43
Loon, *see* Diver
Loose strife 62
  yellow 67

Loranthaceae 63
Louse 131, 132
Louse, hog 93
  plant, *see* Aphid
Lorikeet 174
Loris 174
Lovebird 174
Lugworm 91, 151, 174
Lungfish 141, 173, 174, 221, 222, 236
  Australian 222
Lungs 77, 78
Lynx 174
  Canadian 174
  desert, *see* Caracal
  European 174
  Spanish 174
Lyrebird 130, 174
Lythraceae 62

Macaw 123, 174, 178
Mace, reed 211
Mackeral 115
Maggot, rat-tailed 212
Magnesium 19
Magnolia 42, 62, 63
Magnoliaceae 62
Mahogany 63
Maidenhair tree, *see* Ginkgo
Malaria 87, 100
Mallow 63
Malvaceae 63
Mamba 175
  black 175
  green 175
Mammalian blood 80–81
Mammals 125–128, 130, 226–227
  Age of 226–227
  early 226
  placental 125, 127–128, 226–227
Mammoth 214, 217, 227
Man, Java 229
  Neanderthal 229
Man o' war bird, *see* Frigatebird
  Portuguese 180
Man, origins of 228–229
Manatee 175
Mandrill 175
Mangrove 63
Mantis 175
Maple 42, 43, 63, 200
Marabou 175
Marigold 58
Marigold, marsh 55
Markhor 175
Marmoset 175
Marmot 175, 203
Marram grass 44
Marrow 39
Marsh hawk 168
Marsh marigold 55
Marsupial mole 232, 234
Marsupials 125, 126, 226, 227
Marten 175
Martin, purple 139
Martins 139
Matamata 188
Mayfly 175, 212, 213
Meadow foxtail 44
Meadow grass 59
Melanism 235
Meliaceae 63
Menyanthaceae 54
Merganser 175
  red-breasted 175
Merychippus 227
Mermaids' purse 114
Mesohippus 226
Mesozoic Age 145
Mesquite tree 198
Metabolism 9
Metamorphosis, complete 95, 98
  partial 95

Mice 200
Midwife toad 117, 118, 175
Migration 139–140
  land 139
  sea 140
Milkwort 63
Millipede 175–176
  pin 175–176
Mimicry 137–138
  Batesian 138
  Mullerian 138
Mineral nutrition 19
Mink 176
  American 176
  European 176
Minnow 213
  reach 213
Miobatrachus 223
Miraculous thatcheria 209
Mistletoe 47, 48, 63
Mites 131, 200
Mitten crab 176
Moa 123
Moccasin 176
Mocking bird 176
Mole 176, 232, 234
  European 232, 234
  marsupial 232, 234
Molluscs 108, 208, 234
Moloch 176
Monarch butterfly 139–140
Mongolian gazelle 164
Mongoose 175, 176, 238
Monitor lizard 176
Monkey, colobus 157
  howler 168, 169
  orchid 64
  owl, *see* Douroucouli
  proboscis 172
Monkey puzzle tree 36
Monkey, wooley 192
Monkeys 153, 157, 168, 169, 172, 192, 228
  New world 228
Monocotyledons 25, 44
Monotremes 125–126, 226
Monster gila 120, 165
Moon rat 167
Moorhen 213
Moose 176
Moray eel 175, 176
Morpho butterfly 197
Mosaic, tobacco 11
Mosquito 87, 96, 97, 100, 197, 211, 213
Moss, reindeer 204
Mosses 25, 26, 33, 201, 230
  club 25, 26
Moths 75, 95, 96, 98–99, 100, 130, 135, 137, 138, 141, 147, 235
  black 235
Motmot 176
Mouflon 176
Mould, pin 30
Moulds 25, 30–31
Moult 95
Mountains avens 202
Mountain life 202–203
Mountain lion, *see* puma
Mouse deer, *see* Chevrotain
Mouse, deer 160
  harvest 168
  sea 183
Movement 9
  plant 22–23
Mudpuppy 118, 176
Mule 13
Mullein 69
Mullerian mimicry 138
Müntjac deer 176
Murex 209
Musaceae 52
Muscles 72–73

250

Muscles—*Cont.*
  skeletal 73
  involuntary 73
  voluntary 73
Mushroom 30, 31
Musk ox 176
Musk rat 176–177, 238
Mussel 108, 109, 210, 213
  pond 211, 212
Mutation 232
Mute swan 123, 186, 187
Mycorrhiza 134
Mynah 176, 177
Myxomatosis 11, 238

Narwhal 177
Nasturtium 23, 63
Natterjack toad 177
Natural selection 232, 234
Nature, balance of 237, 238
Nature reserves 239
Nautilus 230
  paper *see* Argonaut
  pearly 110
Ne-ne, *see* Hawaiian goose
Neanderthal Man 229
Neap tides 208
Nectar 102
Nekton 207
Nematode 90
Nervous system 75
Nests 124, 129
Nettle, fever 51
  stinging 51, 69
Neuroptera 95
Newts 83, 117, 118, 223
Nightjar 177
  American 141
Nightshade, deadly 67
  woody 27, 67
Nile crocodile 133
Nile fish 144
Nine-banded armadillo 146
Nitrates 19, 134
Nitrogen 11, 19
Nocturnal 142, 143
Northern forests 200, 201
Nucleus 10
Numbat 177
Nuthatch 177, 200
Nutrition, mineral 19
Nymph 95
Nymphaeaceae 71

Oak 42, 200
Oak apple 131
Oak, cork 15, 42
Oarweed 29, 210
Ocelot 177, 178
Ocotillo 50, 199
Octopus 108, 109, 110, 111, 176, 178
  common 110
Okapi 178
Oleaceae 61
Olive 61
Olm 143, 178
Omnivores 76
Onagraceae 71
Opossum 126, 151, 152, 177, 178
  brush-tail 151, 152
  Virginia 126, 178
Orang utan 177, 178, 238
Orange family 63, 64
Orb-web 106
Orchid family 64
Orchid, bee 64
  monkey 64
Orchidaceae 64
Orchids 14, 64, 197
Ordovician Period 219, 220
Organs, sense 74
Origins of life 11
Oriole 177, 178

Ormer, *see* Abalone
Orobanchaceae 54
Orpine 69, 70
Oryx 146, 178, 241
  Arabian 178, 241
  beisa 178
  scimitar 178
Osprey 123, 178
Ostrich 122, 123, 178
Otter 178
Ovenbird 179
Ovule 40
Owl 142, 148, 152, 200, 201, 204
  barn 148
  burrowing 152
Owl monkey, *see* Douroucouli
Owl, snowy 204
Ox, musk 176
Oxalidaceae 71
Oxlip 67
Oxpecker 133
Oxygen 11, 17, 19
Oyster 108, 109, 209

Pack rat 179
Paddle fish 179
Palm, date 199
  family 64
Panda, giant 178, 179
  lesser, *see* Cat bear
  red, *see* Cat bear
Pandorina 28
Pangolin 179
Papaveraceae 67
Paper nautilus, *see* Argonaut
Paradoxides 216, 219
Parasites 30, 48, 131–132
  animal 131–132
  social 132
  plant 47–48
Parasol ant 105
Parental care 83
Parks, wildlife 240
Parrot 124
  fish 179
  grey 197
  honey-, *see* Lorikeet
Parthenogenesis 83
Partnerships 133–134
Pea 40, 42, 49
  family 42, 64, 134
  sweet 39
Peacock 129
  worm 91, 162
Pear 40
  prickly 56
Pearls 109
Pearly nautilus 110
Peccary 179
  collared 179
  white-lipped 179
Pelican 208
Pellia 33
Penguin, emperor 83
Penicillin 30
Pepper family 64
Peppered moth 235
Perch 115, 116, 213
  climbing 156
  sea, *see* Grouper
Pere David's deer 179, 241
Peregrine falcon 179
Perennials 41
Period, Cambrian 218, 219, 230
  Carboniferous 223, 230–231
  cretaceous 224, 225
  Devonian 221, 222, 230
  Jurassic 224, 226
  Ordovician 219, 220
  Permian 224
  Silurian 222
  Tertiary 227
  Triassic 226

Periwinkle 65, 209
Persimmon 58
Pest, fowl 11
Pests 238
  insect 100–101
Petals 39
Petioles 49
Petrel 145
Phalanger 179
Pheasant 123, 129
Phohippus 227
Phosphorus 19
Photosynthesis 17–18, 20, 32, 38
Phototropisms 23
Phytoplankton 206
Piddock 179
Pig 243
  guinea 167
  water, *see* Capybara
Pigeon, wood 200
Pigmy hippopotamus 169
Pika 179, 203
Pike 212, 213
Pilot fish 180
Pin millipede 176
Pin mould 30
Pine 36, 37, 201
Pineapple family 65
Pink family 65
Piperaceae 64
Pit-viper 180
Pitcher plant 46, 66
Pithecanthropus 229
Placental mammals 125, 127–128, 226–227
Placoderms 220, 221
Planarian worms 90, 212
Plane, London 43
Plankton 206, 211, 212
Plant cells 10, 22, 23
  galls 131
  louse, *see* Aphid
  pitcher 46
Plantaginaceae 66
Plantain 39, 66, 67
Plant growth 22–23
  movement 22–23
Plants, alpine 202–203
  classification of 25
  climbing 49
  desert 198–199
  evolution of 231
  flowering 36, 38–41
  food storage in 20–21
  herbaceous 16, 42, 44
  insect-eating 46
  parasitic 47–48
  prehistoric 230–231
  protection in 50–51
  sugars in 16, 17, 18, 20
  woody 16
Platypus, duck-billed 125
Plesiosaurs 224
Plover, Egyptian 133
Pneumatophores 197
Pocket gopher 180
Pogge 210
Poinsettia 69
Polar bear 180
Polar life 204–205
Polecat 180
Pollack, *see* Coalfish
Pollarding 22
Pollen 37, 38, 39, 40, 102
Pollination 28, 39, 40
  cross 39
  insect 39–40
  self- 39
  wind 39, 42
Pollution 240
Polyanthus 67
Polygalaceae 63
Polygonaceae 58

251

Polyp 88
Pond life 211–212
    mussel 211, 212
    skater 180, 211, 212
    snail 108, 211, 212, 213
Pondweed, Canadian 59, 211
Poorwill 177, 180
    see also Whip-poor-will
Polar 42, 43
Poppy 40, 67
Porcupine 136
    fish 137
Porpoise 180
Portuguese man o' war 89, 180, 206
Potamogetonaceae 66
Potassium 19
Potato 21
    blight 30
    family 67
Potto 180
Powder post beetle 97
Prairie dog 126, 180, 194
Prawn 93, 136, 143, 207
Prehistoric plants 230–231
Prickles 50–51
Prickly pear cactus 56, 199
Primates 228
Primeval soup 11
Primrose 27, 67
Primula 67
Primulaceae 27, 67
Privet 61
Proboscis 96
    monkey 172
Pronghorn 180–181
Protection in plants 50–51
Protective resemblance 137
Proteins 11
Prothallus 36
Protista, see Protozoa
Protorema 33
Protoplasm 10, 16, 18
Protozoa 10, 145
Protozoans 77, 86–87, 105, 134
Protura 95
Przewalski's gazelle 164
Psilophytes 230
Ptarmigan 167, 181, 205
Pteridophytes 25, 34
Pterodactyl 225
Pterosaurs 224
Puffball 31
Pufferfish 137
Puffin 181, 208
Puma 181
Pupa 95, 96, 99
Purple martin 139
Purple moor grass 44
Python 121, 181

Quagga 192
Quail 181
Quaking grass 44
Queensland grouper 167
Quelea 181

Rabbit 180, 181, 238
Raccoon 181
    fox, see Cacomistle
Rafflesia 48
Ragged robin 65, 66
Ragworm 90, 151, 181
Ragwort 58
Rail 181
Rain forest 196–197
    frog 181
Ramapithecus 228
Ram's horn snail 211, 213
Ranunculaceae 27, 55
Rat, black 181
    brown 181
    kangaroo 171, 199
    moon 167

Rat—Cont.
    musk 176–177
    pack 179
    rice 182
    sand, see Gerbil
Rat-tailed maggot 212
Rats 167, 171, 176–177, 179, 181,
    182, 199
Rattlesnake 181, 199
Raven 124, 181
Ray 114, 116, 144
Ray-finned fishes 221, 222
Razor shell 209
Razorbill 181
Red cedar 37
Red grouse 167
Red kangaroo 126
Red kite 171
Red panda, see Cat bear
Red river hog, see Bushpig
Red squirrel 200
Red underwing 136
Redstart 129
Redwoods 25, 196
Reed frog 181
Reedmace 67, 211
Reeds 211
Reef, barrier 89
    coral 88–89
    fringing 89
    Great Barrier 88
Regeneration 83
Reindeer 181–182, 205
    moss 204
Remora 182
Reproduction 9, 82–83
    asexual 82–83
    sexual 24, 31, 82–83
    vegetative 24, 31
Reptiles 119–121, 217, 223, 224–225
    Age of 224–225
    flying 225
Reserves, African game 239
    nature 239
Respiration 77–78
Rhamnaceae 55
Rhea 182
Rhinoceros 128, 182, 194, 240
Rhizome 20, 21, 24, 34, 35, 44
Rhizophoraceae 63
Rhododendron 43
Rhyssa 96
Ribes 57
Rice 45
    rat 182
Right whale 182
Ring-tailed lemur 172
Ringed snake, see Grass snake
Rings, annual 16
Ringtail, see Cacomistle
Ringworm 30
River limpet 213
Rivers 213
Roach 213
Roadrunner 182
Robin 130
Rock 14, 316–317
Rockling, three-bearded 210
Rockrose 67–68
Rocky Mountain goat 202
Rodents 125, 126, 199, 203
    desert 199
Roller 181, 182
Rook 123
Root cap 22
    hairs 15
    tubers 20, 21
Roots 15, 16, 20, 21, 22, 23, 24, 41
    fibrous 15, 16
    food storage in 20, 21
    tap 15, 16, 20
Rorqual 182
Rosaceae 27, 68

Rose 27, 49, 51, 68
    hip 40
    of Sharon 68
Rosebay willowherb 71
Roundworms 90, 131, 132
Rove beetle 182
Rubiaceae 52
Rudd 115, 213
Runner bean 49
Runners 24
Ruff 182
Rush family 68
    flowering 59
Rushes 211
Rust 30
Rutaceae 63
Rye grass 60

Sable antelope 194
Saiga 182
Sailfish 115
St. John's-wort 68
Salamander 117, 118, 182
Salamander, black 118
Salamander, European spotted 118
Salicaceae 71
Salmon 116, 140, 213
Salmon, Atlantic 140
Sand dollar 182
Sand rat, see Gerbil
Sandpiper, spotted 208
Saprophytes 30, 48, 64
Sapsucker 182
Sapsucker, Williamson's 182
Sapsucker, yellow-bellied 182
Sapwood 16
Sarraceniaceae 66
Sauro 198
Savanna 194
Sawfish 182
Sawfly 182
Saxifragaceae 68
Saxifrage 68
Scabious 70
Scale insect 182
Scale leaves 20, 21
Scallop 109, 182–183
Scarab 183
Scimitar oryx 178
Scorpion 106–107, 137, 183
Scorpion, false 162
Scorpion fly 95
Scorpion, water 191, 212
Scorpionfish 183
Scorpion fly 183
Scrophulariaceae 69
Scrub-bird 183
Sea anemone 12, 88, 133, 134, 145,
    210
Sea bass, see Grouper
Sea biscuit, see Sand dollar
Sea campion 209
Sea cow 161
Sea cucumber 113
Sea elephant, see Elephant seal
Sea gooseberry 206
Sea hare 183
Sea horse 116, 183
Sea lavender 70
Sea lettuce 29
Sea life 206–207
Sea lily 113, 162, 183
Sea lily, fossil 216
Sea lion 164, 183
Sea migration 140
Sea pen 218
Sea perch, see Grouper
Sea-slater 93, 209
Sea slug 183
Sea snake 183
Sea squirt 183
Sea urchin 112, 113, 210
Seal 205

Seal, elephant 162
Seal, fur 164
Seashore life 208–210
Seaweeds 25, 29, 208
Sedge 68–69
Seed ferns 231
Seeds 21, 36, 37, 40, 41, 42, 43
Segmented worms 90
Selection, natural 232, 234
Self-pollination 39
Sense organs 74
Senses 74
Senses, chemical 75
Sensitivity 9
Sepals 38
Sequoia 196
Serval cat 9, 184
Sexual reproduction 24, 31
Shag, king 208
Shark, basking 148
Shark, blue 115
    carpet, see Wobbegong
    hammerhead 168
    thresher 130
    whale 116, 190
Shearwater 183, 184
Sheep 203
Sheep's fescue grass 203
Shelduck 184
Shells 108–109, 219
Shoots 41
Shore, lower 210
    middle 209
    upper 209
Shoveller duck 123
Shrew, elephant 162
    tree 188
Shrews 162, 184, 188, 200, 201, 205
Shrike 130
Shrimp 92, 93
    brine 151
    fairy 162
Shrubs 43
Sidewinder 184
Silicon 19
Silk moth 130
Silky anteater 145
Silurian Period 222
Silverfish 95
Siphonaptera 95
Siren 184
Sisyrinchium 61
Skate 114, 116, 144, 184
Skeletal muscles 73
Skeleton 72–73, 116
    soil 14
Skimmer 124, 184
Skink 184
Skua 184, 185, 204
Skull 72
Skunk 137, 184
Slater, sea- 93
Sleeping sickness 100
Sloe, see blackthorn 50
Sloth 185, 227
Slow worm 120, 185
Slug 108, 109, 142, 200
    sea 183
Smallpox virus 11
Snail, garden 108
    giant 165
    pond 108, 211, 212, 213
    ram's horn 211, 213
Snails 108–109, 141, 142, 200
    water 77, 212, 213
Snake, blind 150
    copperhead 157
    coral 157
    egg-eating 120
    garter 164
    glass 165–166
    grass 119, 121, 166
    moccasin 176

Snake—Cont.
    pit-viper 180
    ringed, see Grass snake
    sea 183
    water 191
Snakes 73, 81, 119–121, 141, 224
Snakeshead fritillary 62
Snapdragon 40, 69
Snapping turtle 185
Sneezewort 57
Snipe 185
Snowdrop 57
Snowy owl 204
Social insects 102
Social parasites 132
Sodium 19
Softwood 37
Soil 14
    climate and 14
Soil erosion 237, 238
Soil skeleton 14
Soil, types of 14
Solanaceae 27, 67
Sole 185
Solomon's seal 62
Soup, primeval 11
Spadefoot toad 185, 199
Spanish cedar, see Mahogany
Spanish fly, see Blister beetle
Spanish lynx 174
Sparrow, hedge, see Dunnock
Sparrowhawk 185, 200
Species, plant and animal 13
Spectacled bear 185
Speedwell 68
Spermatophytes 25
Spermophile, see Ground squirrel
Sphinx moth, see Hawkmoth
Spider, bird-eating 149
    crab 158
    trapdoor 188
    water 106, 191
Spiders 106–107, 149, 158, 187, 188,
    192, 200
Spindle 69
Spines 50, 137
Spiny-anteater 185
Spiny-lizard 185
Spiny-skinned animals 112, 113
Spirilli 27
Spirogyra 28
Splash zone 209
Sponge 89
Spoonbill 185, 186
Spores 30, 31, 32, 33, 34, 35, 37
Spots, eye 138
Spotted deer, see Chitral
Spotted dogfish, lesser 160
Spotted hyena 169
Spotted salamander, European 118
Spotted sandpiper 208
Spring tides 208
Spruce 36, 201
Spurges 69, 232
Squid 110–111, 136, 143, 207
    giant 111
Squirrel 141, 200
    cat-, see Cacomistle
    flying 163
    grey 201
    ground 167
    tree 188
Squirt, sea 183
Stable fly 97
Stamens 39
Star, basket 148
    brittle 112, 113
    feather 113
Starch 20, 23
Starfish 83, 89, 112–113, 206, 210,
    220
Stegosaurus 225
Stems 15, 16, 20, 21, 22, 24

Sterculiaceae 56
Stick insect 83
Stickleback 116, 211, 213
Stigma 39, 40
Sting ray 115
Stinging cells 89
Stinging hairs 51
Stinging nettle 51, 69
Stings 137
Stock 40
Stomata 17
Stonecrop 69–70
Stonefly 213
Stork, marabou 175
    white 139
Storksbill 59
Strawberry 24, 40
Streams 213
Strepsiptera 95
Striped hyena 169
Subsidiary buds 22
Succulents 198
Sugar beet 20, 21
Sugar beet yellows 11
Sugar cane 44
Sugars 11, 16, 17, 18, 20
Sulphur 19
Sun bear 186
Sundew 12, 46, 70
Sunflower 57
Surgeonfish 186
Surinam toad 117, 186
Survival 135–138
Suslik, see Ground squirrel
Swallow 139, 186
    bar, see Common swallow
    common 139
Swan 123, 186, 187
    Bewicks 186
    black-necked 186, 187
    mute 123, 186, 187
    whooper 186
Swarming 103
Sweet chestnut 52
Sweet pea 39
Swift 139, 186
Swordfish 186
Swordtail 186
Sycamore 63
Symbiosis 133
Symbiotic association 32
Syrian bear 151

Tadpoles 117
Tailor-bird 186
Takahe 187
Takin 187
Tamandua, see Two-toed anteater
Tanager 187
Tansy 57
Tap roots 15, 16, 20
Tapeworm 90, 131, 132
Tapir 187, 188, 196
Tarantula 187
Tarpon 187
Tarsier 187
Tasmanian devil 187
Tea family 70
Teal 123
Teasel 70
Tendons 73
Tenrec 188
Terminal bud 16, 22
Termites 104, 105, 134, 195
Terrapin 119, 120, 188
Tertiary Period 227
Thallophytes 25
Thatcheria, miraculous 209
Theaceae 70
Thick-lipped mullet 115
Thistle 16, 38, 41, 50
Thomson's gazelle 164
Thorns 50, 51

253

Thorny devil 119
Three-banded armadillo 146
Three-bearded rockling 210
Thresher shark 130, 188
Thrift 70, 209
Thrips 188
Thunderbolt 111
Thylacine 188
Thylacosmilus 227
Thysanura 95
Tibetan gazelle 164
Ticks 132, 188
Tides 208
Tiger 135, 188, 194
Tiger moth 99
Tilapia 114, 188
Tiliaceae 62
Timothy 44
Tit, blue- 129
Titmice 200, 201
Toad, midwife 117, 118, 175
    natterjack 177
    spadefoot 199
    surinam 117, 186
Toads 76, 82, 117, 118, 175, 177, 186,
    199
Toadstools 25, 30, 31
Tobacco mosaic 11
Tomato 67
Tools 228, 229
Toothwort 48
Top shells 209
Tortoise 119, 120, 141, 236
Toucan 188
Touch, sense of 75
Trapdoor spider 188
Traveler's joy 55
Tree frog 117, 188, 189, 197
Tree, maidenhair, see Ginkgo
    monkey puzzle 36
Tree shrew 188
Tree squirrel 188
Tree, tulip 43
    deciduous 16
    flowering 42, 43
Triassic Period 226
Triceratops 225
Trichoptera 95, 153
Trilobites 215, 216, 218, 219
Tropaeolaceae 63
Tropic bird 189
Tropisms 23
Trotter, lily, see Jacana
Trout 189, 213
Troutbeck 213
Trunkfish 189
Trypanosome 87, 189
Tsetse fly 87, 97, 100, 189
Tuatara 236
Tubers, root 20, 21
Tubeworm 189
Tubifex 212
Tulip 38
Tulip bulb 20
Tulip tree 43
Tuna 189
Tundra 204
Turaco 189
Turkey vulture 153
Turnstone 189
Turtle, edible, see Green turtle
Turtle, green 167
    leathery 172
    snapping 185
Turtles 119, 120, 167, 172, 185, 224
Two-toed anteater 145
Typhaceae 67
Tyrannosaurus 225

Ulmaceae 58
Ultra-violet light 11
Umbelliferae 27, 56

Urchin, cake, see Sand dollar
    sea 112, 113
Urticaceae 69

Vacuole 10
Vallisneria 59
Vampire bat 189
Vegetative reproduction 24, 31
Velvet ant 189
Venus fly trap 70
Vertebrates 72, 81, 84, 222
    evolution of 234
Vetch 27, 65
Vine family 70
Violaceae 71
Violet 40, 42, 71
Violet family 42
Viper 119, 145
Viperfish 189
Virginia creeper 49, 70
Virginia opossum 126, 178
Virus, flu 11
    smallpox 12
Viruses 11, 12
Viscacha 189
Volcano 11
Voles 194, 200, 201, 203, 205
Voluntary muscles 73
Vulture 123, 189, 195

Waders 139
Wahoo 69
Wallaby 126, 189, 191
Wall barley 44
Walnut family 71
Walrus 127, 189
Wandering albatross 145
Wapiti 189, 191, 239
Warblers 139, 189, 200
Warthog 189
Wasp 95, 96, 137
Wasp beetle 129
Wasp, gall 164
Water 16, 17, 18
Water beetle 212
Water boatman 211, 212
Water boa, see Anaconda
Water cavy, see Capybara
Water flea 92, 212
Water hole 195
Water life, fresh 211–213
Water lily 71, 211, 213
Waterbuck 76
Waterpig, see Capybara
Water scorpion 191, 212
Water snails 77, 212, 213
Water snake 191
Water spider 106, 191, 211, 212
Water vascular system 112
Waterbuck 76, 189, 191
Wattlebird 191
Waxwing 191
Weasel 191, 205
Weaver ant 104
Weaver bird 191, 192
Weaver finch 124
Web, orb 106
Weed, bootlace 210
Weevil, boll 191
    grain 100, 101
Wels, see Catfish
Whale, blue 150
    right 182
    white, see Beluga
Whalebone whales 207
Whales 140, 150, 182, 207
Wheat 45, 59
Whelk 108, 191, 209, 210
Whip-poor-will 177, 180
Whirligig beetle 191, 212
Whiskers 75
White ant 105

White stork 139
White whale, see Beluga
White-lipped peccary 179
White-tailed deer 190, 191
Whiting 191
Whooper swan 186
Whooping crane 129
Whortleberry, see Bilberry
Whydah 191
Widow bird, see Whydah
Widow, black 107
Wild carnation 65
Wild cat 191
Wildebeest 195
Wildlife parks 240
Williamson's sapsucker 182
Willow 39, 42, 43, 71
Willowherb, Rosebay 71
Wind-pollination 39, 42
Winkle 108
Winter jasmine 61
Winter wren 123
Wireworm 156
Wisent, see European bison
Wobbegong 191
Wolf 191, 192, 205
Wolf spider 187, 192
Wolverine 192
Wombat 192
Wood ant 104, 105, 192
Wood mellick 44
Wood millet 44
Wood pigeon 200
Wood rush 68
Wood sorrel 71
Woodchuck 192
Woodcock 192
Woodlouse 92, 93, 129, 142, 192
Woodpecker 124, 192, 200, 201
Woodpecker finch 130
Woodworm 101
Woody nightshade 27, 67
Woody plants 16
Woolly monkey 192
Worm, acorn 145, 220
    arrow 146
    bristle 91
    fan 91, 151
    glow- 166
    peacock 91, 162
    planarian 90
    slow 120
    flat- 90
    round 90
    segmented 90
Worms 90, 91, 200, 210, 213, 218,
    219, 234
Wrack, bladder 29, 209
Wrassell 115
Wren, cactus 198, 199
Wryneck 192

X-ray fish 192

Yak 192, 203
Yeast 30
Yellow flag 211
Yellow loosestrife 67
Yellow-bellied sapsucker 182
Yellows, sugar beet 11
Yucca 199

Zebra 127, 192, 194, 195
    Burchell's 192
    Grevy's 192
Zinnia 58
Zone, splash 209
Zoos 240–241
Zorille 192
Zosteraceae 58